Surface Characteristics of Fibers and Textiles

(in two parts)

Part I

FIBER SCIENCE SERIES

Series Editor

L. REBENFELD

Textile Research Institute
Princeton, New Jersey

Volume 1 CLOTHING: COMFORT AND FUNCTION
by Lyman Fourt and Norman R. S. Hollies

Volume 2 ESSENTIAL FIBER CHEMISTRY
by M. Carter

Volume 3 INSTRUMENTAL ANALYSIS OF COTTON CELLULOSE AND MODIFIED COTTON CELLULOSE
Edited by Robert T. O'Connor

Volume 4 CHEMICAL MODIFICATION OF PAPERMAKING FIBERS
by Kyle Ward, Jr.

Volume 5 HIGH-MODULUS WHOLLY AROMATIC FIBERS
Edited by W. Bruce Black and J. Preston

Volume 6 THE FINE STRUCTURE OF COTTON: AN ATLAS OF COTTON MICROSCOPY
by I. V. DeGruy, J. H. Carra, and W. R. Goynes
Edited by Robert T. O'Connor

Volume 7 SURFACE CHARACTERISTICS OF FIBERS AND TEXTILES
Edited by M. J. Schick

Other Volumes in Preparation

Surface Characteristics of Fibers and Textiles

(in two parts)

Part I

edited by

M. J. Schick

Diamond Shamrock Chemical Company
Process Chemicals Division
Morristown, New Jersey

MARCEL DEKKER, INC. New York and Basel

MARCEL DEKKER, INC.

270 Madison Avenue, New York, New York 10016

LIBRARY OF CONGRESS CATALOG CARD NUMBER: 75-10346

ISBN: 0-8247-6316-5

Current Printing (last digit):
10 9 8 7 6 5 4 3 2 1

PRINTED IN THE UNITED STATES OF AMERICA

CONTENTS

LIST OF CONTRIBUTORS

H. R. BILLICA, Textile Fibers Department, Fiber Surface Research Section, E. I. du Pont de Nemours & Co., Inc., Kinston, North Carolina

KERMIT E. DUCKETT, Fiber Research Laboratory, University of Tennessee, Knoxville, Tennessee

RUTH L. GALBRAITH, School of Home Economics, Auburn University, Auburn, Alabama

GEORGE GOLDFINGER, Department of Textile Chemistry, School of Textiles, North Carolina State University at Raleigh, Raleigh, North Carolina

S. P. HERSH, School of Textiles, North Carolina State University at Raleigh, Raleigh, North Carolina

KWOK CHING LAU,* Department of Textile Chemistry, School of Textiles, North Carolina State University at Raleigh, Raleigh, North Carolina

K. RACHEL MAKINSON, Division of Textile Physics, Commonwealth Scientific and Industrial Research Organization, Sydney, Australia

RALPH McGREGOR, Department of Textile Chemistry, School of Textiles, North Carolina State University at Raleigh, Raleigh, North Carolina

FRANCIS L. SCARDINO, Textile Department, Philadelphia College of Textiles and Science, Philadelphia, Pennsylvania

M. J. SCHICK, Diamond Shamrock Chemical Company, Process Chemicals Division, Morristown, New Jersey

---

*Present address: Color Science Laboratory, Lehigh University, Bethlehem, Pennsylvania.

JOHN P. SIBILIA, Chemical Research Center, Allied Chemical Corporation, Morristown, New Jersey

R. D. VAN VELD, Textile Fibers Department, Fiber Surface Research Section, E. I. du Pont de Nemours & Co., Inc., Kinston, North Carolina

S. H. ZERONIAN, Division of Textiles and Clothing, University of California, Davis, California

CONTENTS OF PART II

PREFACE

Within the past two decades the phenomenal growth in the production of man-made fibers has focused interest on the surface properties of fibers and textiles. A perusal of the literature indicates that papers dealing with the surface properties of fibers were relatively sparse until the 1950's. However, more sophisticated probings into the fundamental concepts are now in progress. Coupled with this research program has been a continuing effort to apply the results of this research to practical problems. Although some of these developments have been covered in other books, the need for a comprehensive treatise on the surface characteristics of fibers and textiles has been felt for a long time. Our treatise endeavors to fill this need by presenting the substantial progress in science and technology related to the surface properties of fibers and textiles. The investigation of these surface characteristics requires an interdisciplinary approach which combines surface science with textile technology and eliminates the myopia of classical-discipline centered work so commonly seen in the past.

Since the subjects to be discussed represent such a broad spectrum, no single person could write a critical review on more than a very limited number of topics. Thus, the treatise is a collection of chapters written by outstanding specialists. The contributions, although self-contained, are interrelated. This treatise offers critical reviews, which describe experimental facts, theories, and processes and which handle these in a rigorous way. Symbols are clearly defined in each chapter. It was impossible to publish a work of this magnitude with all chapters in a logical order and in a single book. However, every effort was made to group related chapters in sequence. A certain amount of overlap is unavoidable, but was kept to a minimum. The indexes to the entire book will appear at the end of Part II.

The treatise encompasses the surface properties of both natural and synthetic fibers. Emphasis has been placed on the frictional, geometrical, electrical, wetting, adhesive, and optical properties of fibers and fabrics as well as on phenomena related to these properties. In addition, experimental procedures used to assess the surface properties of fibers and fabrics are described. It is the sincere hope of the contributors and editor that much exciting and stimulating research will result from this effort. Likewise, it is hoped that this presentation of the most up-to-date thinking in this area of surface characteristics of fibers and textiles will lead to further advances in textile technology.

It is with great pleasure that I express my sincere gratitude to the contributing authors in carefully preparing their chapters and their splendid cooperation with the editor. Likewise, the guidance and enthusiastic support of Dr. Ludwig Rebenfeld, consulting editor of the Fiber Science Series, is gratefully acknowledged. Thanks are also due to the academic, government, and industrial organizations with which the contributing authors and the editor are associated for the assistance given to us, and to the publishers who have given their permissions for the reproduction of illustrations. Finally, thanks are expressed to the following staff members of the Diamond Shamrock Chemical Company: Miss Marion Kearney and Mrs. Evelyn Fornale for secretarial assistance and Miss Sandy Haines for preparation of illustrations.

M. J. Schick
Morristown, New Jersey
October, 1975

Surface Characteristics of Fibers and Textiles

(in two parts)

Part I

Chapter 1

FRICTION AND LUBRICATION OF SYNTHETIC FIBERS

M. J. Schick

Diamond Shamrock Chemical Company
Process Chemicals Division
Morristown, New Jersey

I. INTRODUCTION

During the last decade advances have been made in understanding the basic mechanisms of friction, lubrication, and wear of solid surfaces. The adhesion theory of metallic friction has been extended to include high polymers and ceramic materials. The mode of deformation, real area of contact, and the role of junction growth between sliding surfaces have received more attention. New liquid lubricants with unusual physical properties have been synthesized and tested in various applications. Great emphasis has been placed on the preparation of heat-stable lubricants. For detailed discussions of the fundamentals of friction and lubrication of solids the reader is referred to standard texts by Bowden and Tabor [1,2] and Rabinowicz [3], and reviews by Bowers and Murphy [4] and Tabor [5,6].

It must be noted at this point that Howell, Mieszkis, and Tabor [7] published a treatise entitled Friction in Textiles in 1959. Those authors covered mainly the friction of natural and hydrophilic synthetic fibers. The treatise also included a review of friction in fabrics. In contrast, this chapter deals mainly with the progress made in the decade following 1959 in advancing the knowledge of friction and lubrication of hydrophobic synthetic fibers. This emphasis is in line with the phenomenal growth of hydrophobic fibers in the man-made fiber industry over the period covered. The friction of natural fibers and of fabrics is reviewed in subsequent chapters. In order to limit the scope of this chapter no attempt has been made to review again the material already covered by Howell et al. [7].

II. GENERAL MECHANISM OF FRICTION

When one body slides over another under a normal load W, a finite force is resisting motion; this force is termed the frictional force and is designated F_s for static and F_k for kinetic conditions. The magnitude of the static frictional force generally exceeds the magnitude of the kinetic frictional force. The latter is the force required to maintain sliding. Friction generally follows two basic laws, which are referred to as Amonton's classical laws [8]: The frictional force is (1) proportional to the load and (2) independent of the geometric area of the sliding bodies.

Let us first review some basic concepts of metallic friction. It is generally accepted that friction of unlubricated surfaces arises from two main factors. The first, which is usually the predominant factor, is the adhesion at the real areas of contact. Such areas are usually much smaller than and independent of the geometric area of the sliding body. Thus, the major component of friction between sliding solids is the force necessary to shear all these junctions. Consequently, if A is the true area of contact and s the shear strength of the weaker material, the adhesion part of friction may be written $F_{adhesion} = As$. The second factor arises from the plowing and grooving of the softer material by the harder one. It is denoted as the deformation term, and under conditions of negligible interaction between these two processes the total frictional force F may be expressed by the summation of the adhesion and deformation terms:

$$F = F_{adhesion} + F_{deformation} \tag{1}$$

Since most frictional behavior of metals can be approximated by the adhesion mechanism, its quantitative treatment is given below.

If the contacting asperities deform plastically, as is generally observed in metal friction, the real area of contact is equal to the ratio of the normal load W and the mean yield pressure p of the softer material:

$$A = \frac{W}{p} \tag{2}$$

Consequently, the coefficient of friction, f, will equal the ratio of the shear strength to the yield pressure of the softer material:

$$f = \frac{F}{W} = \frac{s}{p} \tag{3}$$

This relationship must be considered an oversimplification, since the shear strength s and the yield pressure p in the contact region may differ from those of the bulk material [9,10].

Let us now review some basic concepts of high-polymer friction. As stated in the introduction the adhesion theory of metallic friction has been extended to include high polymers [11]. For example, it has been confirmed experimentally that during sliding strong adhesion at the areas of contact occurs, and fragments sheared off from one polymer have been observed adhering to the other polymer [6,12]. Consequently, with most high polymers reasonable agreement has been found between s/p and the measured coefficient f in line with Eq. (3) [4,7]. Polytetrafluoroethylene (Teflon*) is a striking exception. The low surface energy of this material causes a low level of intrinsic adhesion [13,14]. Thus, with respect to the adhesion theory no discrepancy seems to exist between metallic and high-polymer friction. In contrast, a discrepancy between these two types of friction has been observed in the nature of the deformation. Whereas metallic friction exhibits only plastic deformation, in high-polymer friction the deformation is partially elastic over a wide range of loads [10]. Consequently, high-polymer friction is affected by hysteresis losses and the viscoelastic nature of high polymers. Similarly, as a result of the elastic deformation the plowing term in Eq. (1) and the surface roughness play a greater role in high-polymer than in metallic friction. Finally, it must be stated that the applicability of Amonton's classical laws [8] is restricted in high-polymer friction.

Before proceeding to the discussion of fiber friction, the main topic of this chapter, reference is made to two investigations dealing

---

*Registered trademark for DuPont's TFE-fluorocarbon resin.

with the lubrication of high-polymer films [11,15]. In both investigations friction was determined on polymer films that were made from materials generally used in fiber manufacture. Therefore, these findings must be considered also applicable to fiber friction and are given below. Bowers, Clinton, and Zisman [15] determined the friction of nylon and steel sliding on nylon films when lubricated by 16 liquids varying with respect to the types of polar end groups and the chain length. In support of the adhesion theory these authors demonstrated that the most effective lubricants formed the most strongly adherent film with the highest intermolecular cohesion. Similarly, Fort [11] has reported, as a result of an investigation of boundary lubrication of polyethylene terephthalate, that the effectiveness of liquids of the same chain length decreased in the order: acids, alcohols, and alkanes, which corresponds to their ability to adsorb onto the polymer surface. In line with the adhesion theory, in each homologous series the most effective lubricant formed the most strongly adherent film with the highest intermolecular cohesion.

III. FRICTION AND LUBRICATION OF SYNTHETIC FIBERS

A. Introduction

Textile yarns pass over guide surfaces in spinning and winding operations and rub against each other during drafting, weaving, and knitting operations. In these processes textile lubricants are used essentially to reduce friction, wear, and static electricity. Moreover, friction must not only be reduced, it has to be kept uniform to render an even product. In addition to the processes cited above the frictional properties appear to play an important part in the stability of packages.

The complexity of textile operations and the large number of variables that influence yarn friction have made it difficult to correlate many of the original data found in the literature. Pioneering work to develop quantitative relationships has been carried out by Hansen and Tabor [16], Schlatter et al. [17,18], Olsen et al. [19,20], Wegener and Schuler [21-23], and Roeder [24]. Hansen and Tabor [16] have shown that the frictional behavior of a fiber passing over a cylindrical guide can be considered analogous to that of the journal bearing, and these authors thereby established a link between fiber and metal lubrication.

In metal lubrication two distinct regions are recognized: boundary and hydrodynamic [1-6]. Under boundary conditions the sliding surfaces are separated by a lubricant film of molecular dimensions, and solid-solid contacts frequently penetrate the protective film. In contrast, under hydrodynamic conditions, the sliding surfaces are completely separated by a continuous fluid film. Semiboundary conditions exist in the intermediate region. The present review clearly documents that these general concepts of metal lubrication also apply to the frictional behavior of synthetic textile fibers.

B. Theoretical Background

Figure 1 presents the general frictional behavior of liquid-lubricated textile yarns, relating the frictional coefficient with yarn speed, lubricant viscosity, and pressure between yarn and guide [16,19,25]. The dashed lines represent the contribution of the boundary and hydrodynamic component, respectively. The frictional behavior that is observed in practice is the solid line, which is the result of the boundary and hydrodynamic contributions to the friction. It is apparent that boundary friction reaches a significant value at low speeds, whereas hydrodynamic friction reaches a significant value at high speeds. The semiboundary region represents the transition between boundary and hydrodynamic frictions. It comprises the region of minimum friction for the system, and its position is determined by the relative contributions of boundary and hydrodynamic friction.

Boundary lubricants function by interposing a layer of low shear strength material between the sliding surfaces and are more effective at the fiber/metal (F/M) than at the fiber/fiber (F/F) interface [20]. The existence of a continuous fluid film separating the surfaces sliding at appreciable speed was first demonstrated by Hansen and Tabor [16]. These authors discovered this phenomenon from analysis of

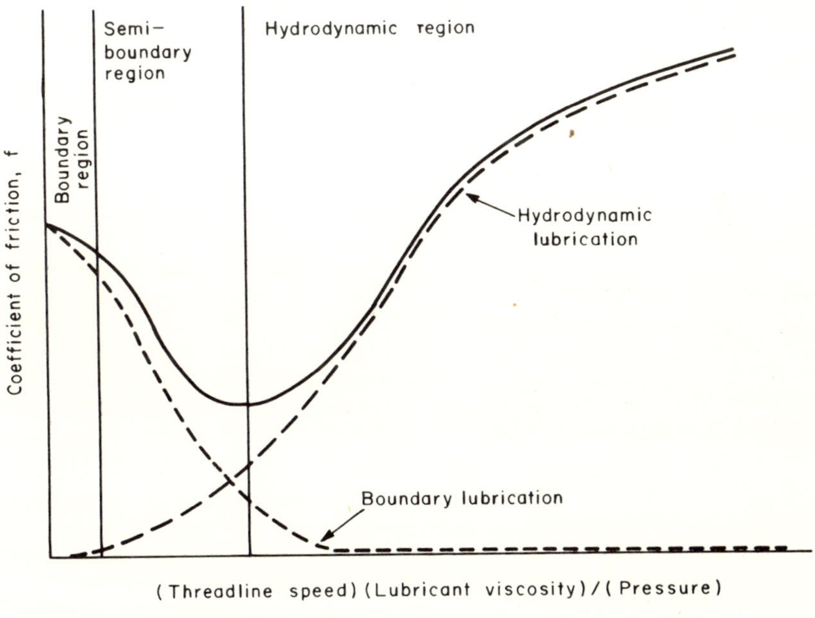

FIG. 1. General frictional behavior of liquid-lubricated textile yarns. Reprinted from ref. 25, p. 105, courtesy of Textile Research Journal.

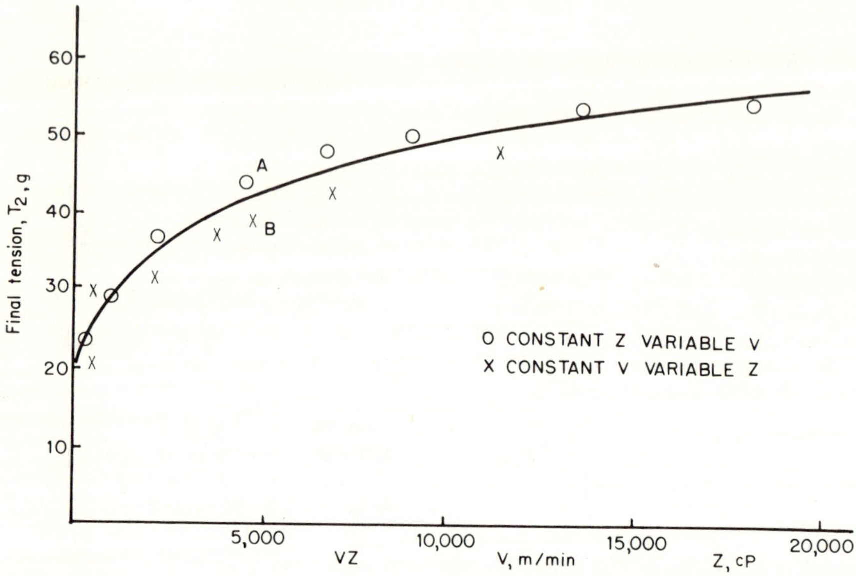

FIG. 2. Plot of Lyne's data for an acetate yarn sliding over a cylindrical chromium pin using white-oil lubricants of known viscosities. The initial tension T_1 is 10 g. The final tension T_2 is plotted as a function of the product VZ, where V is the velocity and Z the viscosity of the lubricant. The points A and B are nominally identical experiments. Reprinted from ref. 16, p. 301, courtesy of Textile Research Journal.

Lyne's experimental data as shown in Fig. 2. Lyne [26] determined the effect of speed, viscosity, and pre-tension on friction of acetate yarn sliding over a chromium pin of 1/4-in. diameter using mineral oils of known viscosity as lubricants. The interchangeability of speed and viscosity in the plot shown in Fig. 2 supports the view that hydrodynamic effects play an important part in the frictional behavior of yarns at appreciable speeds. Likewise, Hansen and Tabor [16] found that plots of the effective frictional coefficient vs viscosity × velocity/load follow a single curve when either speed or pre-tension is varied, which reinforces their original postulate.

According to Olsen [19] the important factors in boundary lubrication are the chemical nature of the lubricants and sliding surfaces, shear strength of the lubricants, rigidity of substrates, and pressures at areas of contact. In contrast the important factors in hydrodynamic friction are the yarn speed, lubricant viscosity, temperature, concentration of applied lubricant, denier per yarn, pre-tension, guide and yarn surface roughness, and guide diameter.

C. Test Methods

A great number of methods for measuring friction of yarns and fibers have been described in the literature. The original instruments were designed for research only [7,27], whereas presently automatic recording instruments predominate [17-19,25,28]. The latter are applicable both to research and to lubricant quality control for spinning operations and processing of yarn in the mill. For the measurement of drafting forces of staple the Shirley draftometer or a similar instrument is used [29]. Simplified instruments are also available for practical use with machine-knitting yarns [30,31].

In order to assess the friction of lubricants on multifilament yarns, finish-free yarns are required. These may be produced if the necessary spinning equipment is available. However, more frequently, commercial yarns are used in which the producer's finish is solvent-extracted until complete removal is indicated by the antistatic properties of the yarn. The spin finish may then be applied from aqueous or isopropyl alcohol solutions with a simple padder arrangement. A superior method requires an Atlab applicator with a padder attachment [32], which assures a 100% takeup of the finish by the yarn. For frictional measurements in the low- and medium-speed regions, our laboratory uses the Chemstrand friction meter which was originated by the Chemstrand Research Center [17,18,25,28]. This automatic recording instrument allows measurement of yarn friction both at the fiber/metal (F/M) and fiber/fiber (F/F) interfaces with the highest accuracy. Its principle of operation is illustrated in Figs. 3 and 4, and friction is expressed in terms of frictional force $F = (T_2 - T_1)g$, where T_1 is the input tension and T_2 is the output tension. For the F/F measurements $2\frac{1}{2}$ turns over a distance of 2.5 in. and five pulleys are generally used. Fort and Olsen [20] measured F/F friction simply on a 1/4-in. diameter steel cylinder wrapped with a complete cover of the test yarn with a helix angle of about 15°. For frictional measurements in the medium- and high-speed regions the Rothschild friction meter is used in our laboratory [25,31,33]. This instrument allows measurement of yarn friction mainly at the F/M interface with easily interchangeable chrome and ceramic guide pins. Its principle of operation is illustrated in Fig. 5. Results are generally expressed in terms of the kinetic coefficient of friction f, which is calculated from the capstan equation $T_2/T_1 = e^{\theta f}$, where T_1 (g) signifies the input tension, T_2 (g) the output tension, and θ (radians) the angle of wrap on a cylindrical pin [34,35]. For a derivation of the capstan equation the reader is referred to the treatise of Howell et al. [34]. The dimensionless frictional coefficient f is useful for comparison of results between laboratories. However, it must be emphasized that the capstan equation is applicable only for correlations involving a constant area of yarn contacting the cylindrical pin, which limits its overall usefulness. In our laboratory an electrostatic voltmeter has been attached to the Rothschild friction meter for the measurement of the

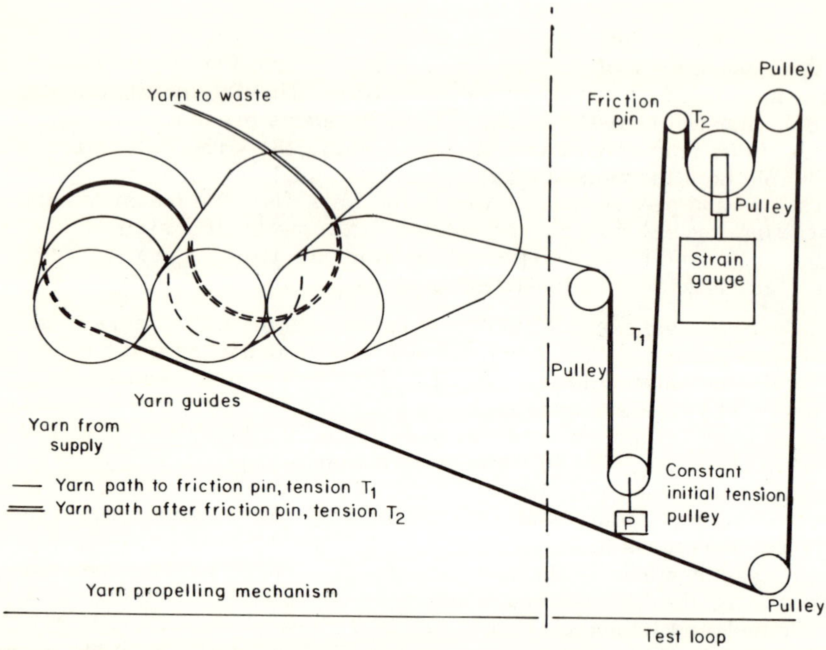

FIG. 3. Chemstrand yarn friction measuring instrument. Reprinted from ref. 25, p. 104, courtesy of Textile Research Journal.

charge generated by a moving yarn under controlled tension and at a specific speed. In general, the instrument is operated at 300 yd/min for 10 sec, the voltage buildup is recorded, and the meter is grounded without stopping the yarn. Results are reported for a specific yarn length in volts for three 10-sec and 40-sec intervals.

Since the measurement of antistatic properties of textiles is the subject of a subsequent chapter, only brief reference is made to antistatic test instruments for fibers at this point. In our laboratory the antistatic properties of yarns are also assessed by measurement of the antistatic half-life from the decay of an applied voltage on skeins of 10 filaments [25]. The method essentially consists of applying 160 V to the skeins and measuring the time required for this potential to drop to its half-value of 80 V. Condensers of 0.01 μF may be used to allow the drop to take place in a reasonable time limit. The electrodes are 2 in. apart, and the wiring circuit of the instrument is illustrated in Fig. 6. Other instruments have been described by Steiger, Zaukelies, Loebel, and others [36-38]. It is mandatory that frictional and antistatic measurements be carried out under conditions of controlled temperature and humidity.

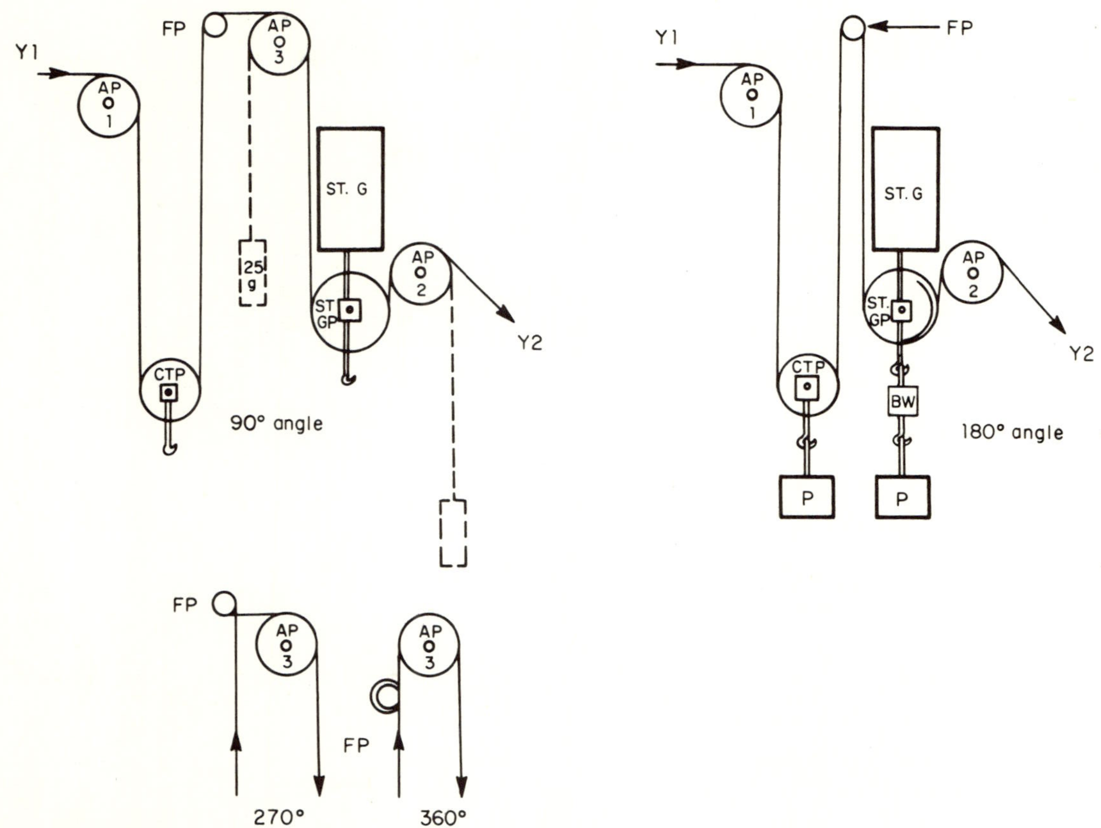

FIG. 4. Test loop arrangement to measure friction of yarns over pins as final tension minus initial tension (T_2 - T_1) for various angles of contact. ST. G-strain gauge, CTP-constant tension pulley (floating pulley), FP-friction pin, AP-auxiliary pulleys, P-additional tension weights, ST. GP-strain gauge pulley, BW-balance weight to compensate for constant tension pulley assembly, Y1-yarn from propelling mechanism, Y2-yarn back to propelling mechanism. Reprinted from ref. 25, p. 104, courtesy of Textile Research Journal.

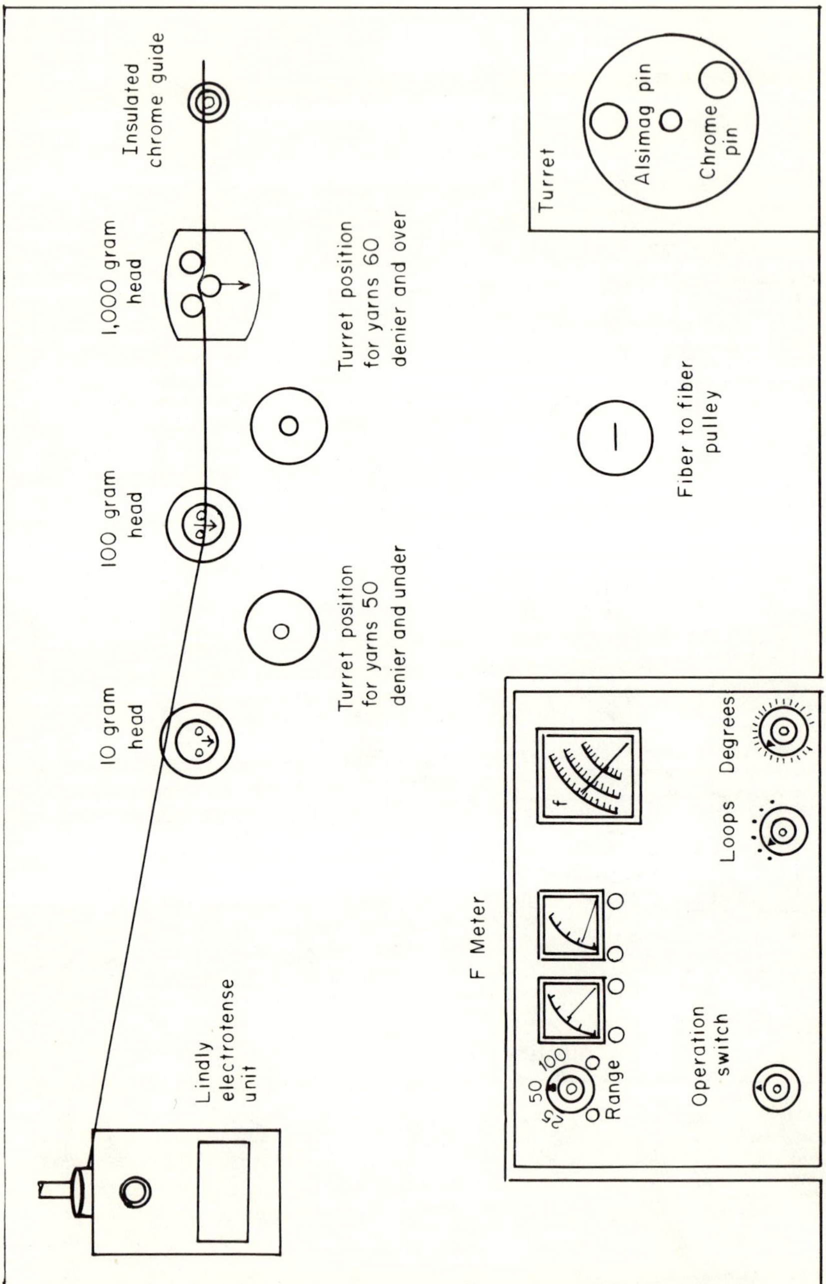

FIG. 5. Rothschild F meter. Reprinted from ref. 25, p. 104, courtesy of <u>Textile Research Journal</u>.

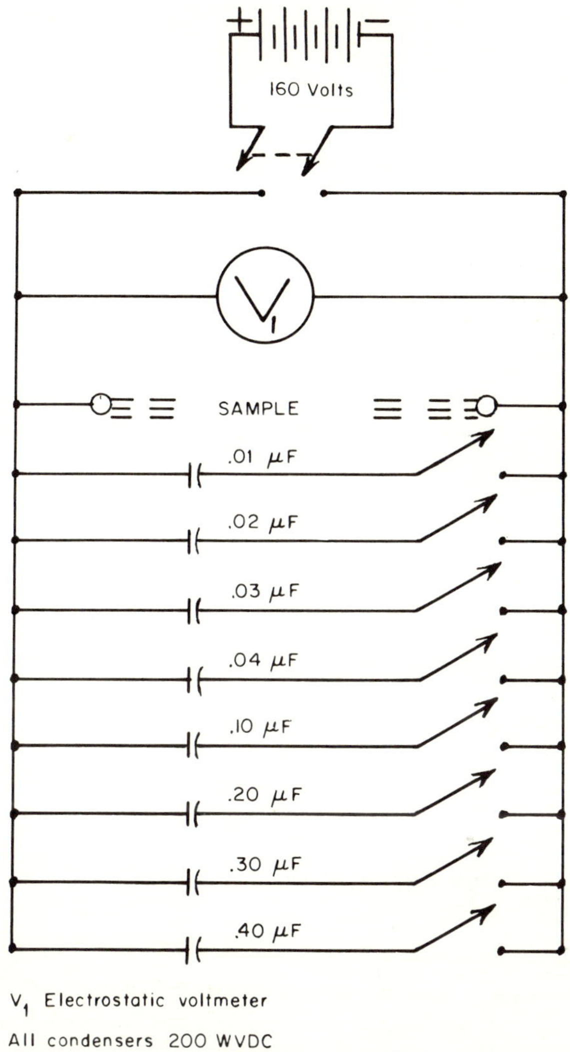

FIG. 6. Wiring for antistatic measurements. Reprinted from ref. 25, p. 104, courtesy of Textile Research Journal.

D. Experimental Results

1. Stick-Slip

An appropriate subject with which to begin the presentation of experimental results on friction and lubrication of synthetic fibers is the stick-slip phenomenon, which is commonly observed in lubrication at low sliding speeds. Stick-slip is characteristic of a boundary and semiboundary region wherein the frictional force decreases with increasing speed [19]. Figure 7 illustrates the frictional data of a polyester yarn at the F/F interface as a function of pre-tension, that is, the pressure existing at the F/F interface. It is evident that the magnitude of the stick-slip, i.e., the discrepancy between the two branches of the friction curves, decreases with increasing yarn speed and finally disappears abruptly. At this point, friction changes from semiboundary to hydrodynamic, reaches a minimum for that particular system, and is now positively dampened so that stick-slip disappears. In addition, the data indicate that the magnitude of the stick-slip increases with increasing frictional forces as well as with increasing pre-tension. The latter may be attributed to increased pressure at areas of contact in the F/F interface with subsequent increase of F/F contacts. It will be demonstrated subsequently in this chapter that by judicious choice of good boundary lubrication conditions, both at the F/F and F/M interfaces, stick-slip may be minimized or eliminated down to a speed as low as 10^{-4} m/min, and the slope of the friction vs. speed plots will be either 0 or slightly positive [25].

The data of Fig. 7 have been amplified with frictional data for stepwise increases in pre-tension from 10 to 75 g on lubricated nylon

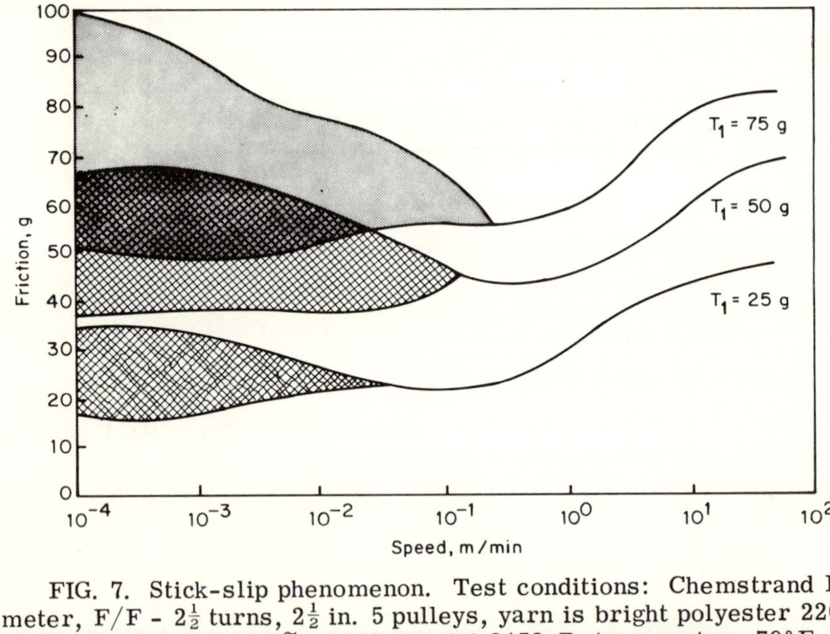

FIG. 7. Stick-slip phenomenon. Test conditions: Chemstrand F meter, F/F - $2\frac{1}{2}$ turns, $2\frac{1}{2}$ in. 5 pulleys, yarn is bright polyester 220/50 (duPont), finish is 1% wof Nopcostat 2152-P, temperature 70°F, RH 50%. Reprinted from ref. 25, p. 105, courtesy of Textile Research Journal.

yarn at the F/F interface and are presented in Fig. 8 [39]. No stick-slip was observed at the lowest pre-tension level of 10 g. Subsequent increases in pre-tension resulted in stick-slip formation with the magnitude of the stick-slip increasing with increasing pre-tension and a migration of stick-slip as a function of pre-tension. At pre-tension levels of 15 to 50 g, stick-slip is discontinuous, but becomes continuous at the highest pre-tension level of 75 g. As stated before, it is generally accepted that the abrupt disappearance of stick-slip at the higher speed level signifies the transition from semiboundary to hydrodynamic friction. Thus, stick-slip disappears with the onset of a continuous fluid film, which separates the sliding surfaces. In contrast, the abrupt disappearance of stick-slip at the lower speed level is attributed to changes in the elastic deformation of the fiber. Thus,

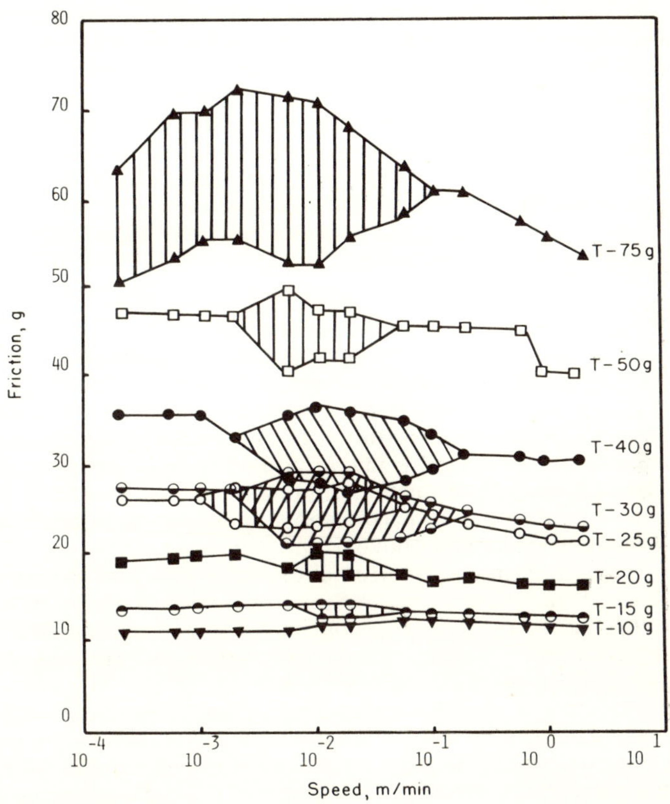

FIG. 8. Effect of pre-tension on stick-slip. Test conditions: Chemstrand F meter, F/F - $2\frac{1}{2}$ turns, $2\frac{1}{2}$ in. 5 pulleys, yarn is bright nylon 66 200/34 (duPont), finish is 1% wof n-butylstearate, temperature 70°F, RH 50%. Reprinted from ref. 39, p. 765, courtesy of Textile Research Journal.

at very low speed levels the elastic deformations are not significant enough to result in stick-slip formation. However, on further increase in speed the elastic deformations reach a magnitude that causes the formation of stick-slip. Similarly, the continuous stick-slip at the highest pre-tension level of 75 g may be explained on the basis of elastic deformations. At this pre-tension level the pressure in the F/F interface reaches such a magnitude that the elastic deformations become significant enough to cause stick-slip formation even down to the lowest speeds investigated.

2. Surface Roughness

It has been shown that variations in surface roughness of both guide and fiber surfaces have a marked effect on friction [19,25]. Olsen [19] postulated that an increase in guide surface roughness can be considered analogous to an increase in the pressure between yarn and guide as a result of the decrease in F/M surface contact area. This results in a shift toward or into the semiboundary region with consequent lowering of the friction from the high values observed on polished guide surfaces. This concept has been reinforced by the data of Figs. 9 to 11, which illustrate the effect of guide surface roughness on friction of nylon using the identical chrome pins of different surface roughness on the Chemstrand and Rothschild F meter [25]. Figure 9 covers the low-speed region, whereas Figs. 10 and 11 cover the high-speed region.

It is evident from the data of Fig. 9 that increasing the surface roughness of the chrome pins sequentially from 4 to 20 and to 60 μin. (RMS-4 to 60) at speeds below 10 m/min, that is in the boundary and semiboundary region, resulted in an increase in friction. Furthermore, the minimal stick-slip exhibited by the smooth chrome pin (RMS-4) rises to an appreciable magnitude with the rougher chrome pins (RMS-20 and 60). Both these phenomena are attributed to increased wear of the fiber by the rougher guide surface in the boundary region. In contrast at speeds above 10 m/min the exactly opposite phenomenon occurs, namely, friction decreases with increasing surface roughness of the chrome pins. Since this speed definitely falls into the hydrodynamic region, it is postulated that increased surface roughness disrupts the continuous fluid film between yarn and guide. Thus, the decrease in surface contact area results in a shift toward conditions simulating a semiboundary region with consequent lowering of friction. The data of Fig. 10, which cover the hydrodynamic region, corroborate these findings or in other words friction decreases with increasing roughness of the guide surfaces. Furthermore, it is evident that, with a smooth guide surface, the fiber melted at a speed above 55 yd/min, whereas on rougher surfaces a speed of 300 yd/min could easily be reached. Friction at this high speed was of a magnitude approaching semiboundary conditions, because the guide surface

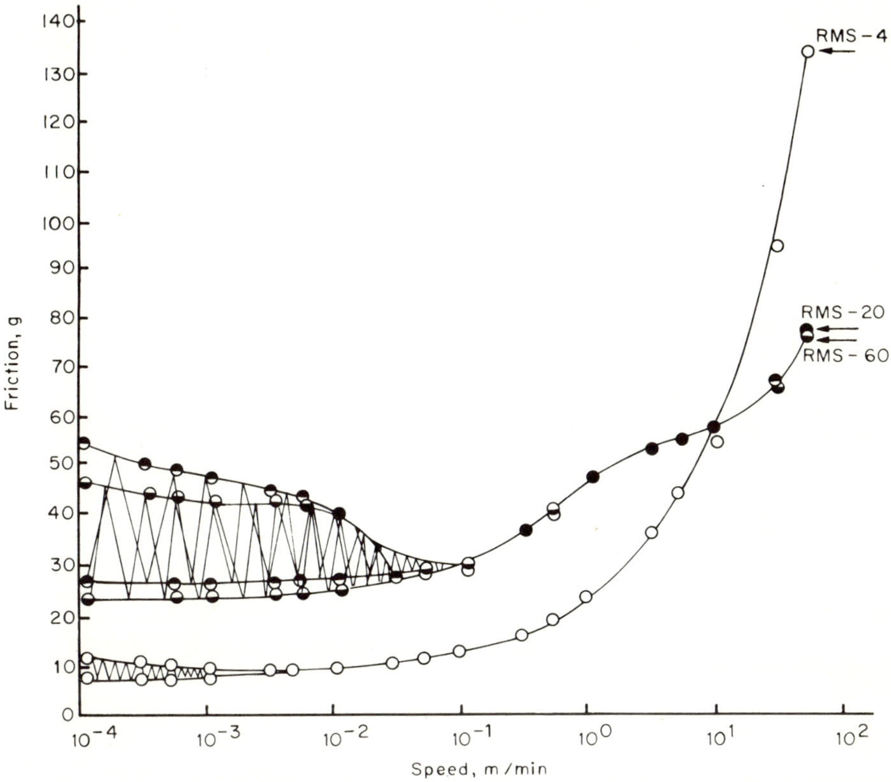

FIG. 9. Effect of guide surface roughness on friction. Test conditions: Chemstrand F meter, yarn is bright nylon 66 200/34 (duPont), finish is Merge 18698, chrome pin 0.5-in. diameter, contact angle 360°, T = 10 g, temperature 70°F, RH 50%. Reprinted from ref. 25, p. 106, courtesy of Textile Research Journal.

was too rough to permit the establishment of a continuous fluid film. The data of Fig. 10 have been replotted for two speed levels in terms of f vs. RMS and are given in Fig. 11. Friction decreases rapidly in the range of 5-20 μin. of surface roughness and then levels off to a constant value. These findings on guide surface roughness clearly demonstrate that fundamental differences exist in the frictional properties of textile yarns in the boundary, semiboundary, and hydrodynamic regions, respectively. It is concluded that intermittent solid-solid contacts exist in the boundary and semiboundary regions, whereas a continuous fluid film separates fibers from guide in the hydrodynamic region.

The effect of variations in surface roughness of fiber surfaces is illustrated in Figs. 12 and 13 [25]. A false-twist polyester feeder

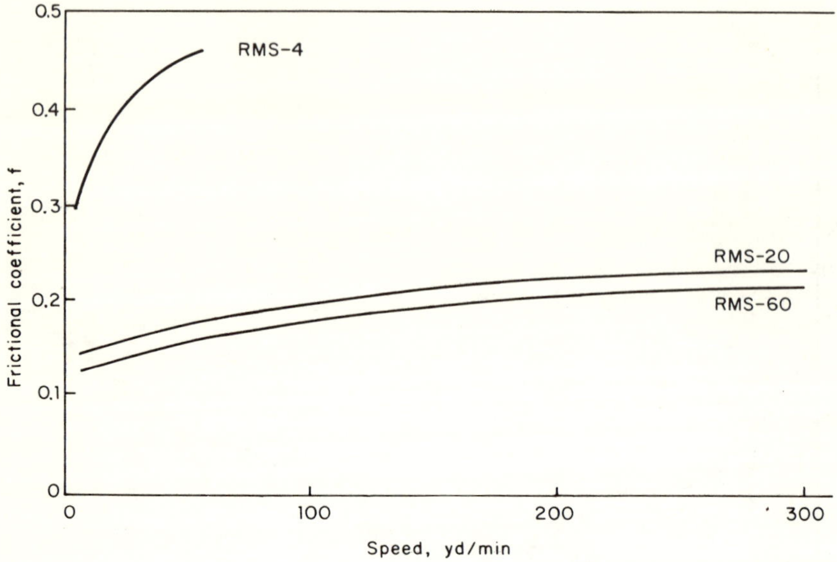

FIG. 10. Effect of guide surface roughness on friction. Test conditions: Rothschild F meter, yarn is bright nylon 66 200/34 (duPont), finish is Merge 18698, chrome pin 0.5-in. diameter, contact angle 360°, T = 10 g, temperature 70°F, RH 50%. Reprinted from ref. 25, p. 106, courtesy of Textile Research Journal.

yarn and the corresponding textured yarn have been selected for such a comparison. The electron micrographs shown in Figs. 14 and 15 clearly indicate that the surface of the textured yarn is much rougher than that of the original feeder yarn [25]. In order to offset loss of finish in the texturizing process, the textured yarn was extracted and the finish reapplied to the original level. Conforming with the results on guide surface roughness, the friction in the hydrodynamic region was higher for the smooth feeder yarn as compared to that of the rough textured yarn, while passing over a smooth chrome pin, as shown in Fig. 12. It is concluded that the effect of surface roughness on friction at the F/M interface is the same, whether the changes in surface roughness are ascribed to the guide or fiber surfaces. In practice proper selection of guide surface roughness is an important factor in reducing friction to acceptable levels in spinning operations.

Finally, the effect of surface roughness on friction encompassing both sliding surfaces is shown in Fig. 13 [25]. Friction of the smooth feeder yarn was again compared with that of the rough-textured yarn in the low-speed region at the F/F interface. Both yarns exhibited stick-slip under these test conditions, but the reverse behavior was observed in the boundary and semiboundary region in contrast to the

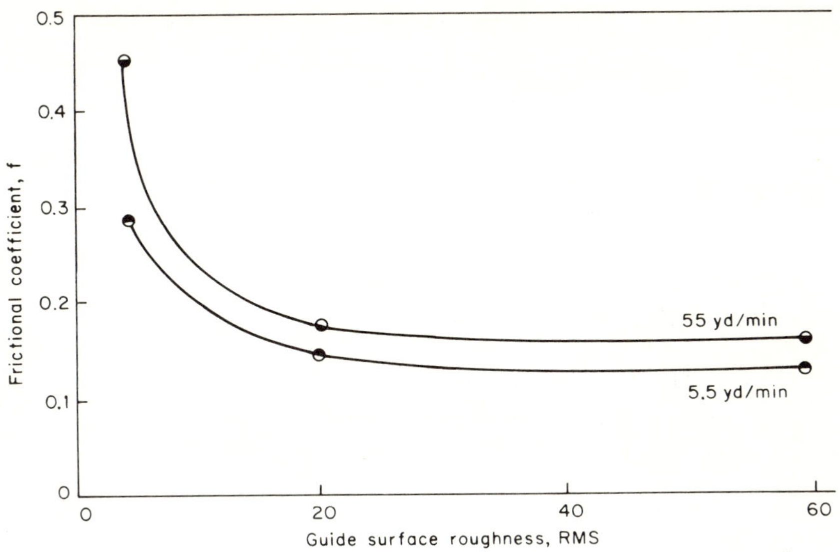

FIG. 11. Effect of guide surface roughness on friction. Test conditions given in Fig. 10. Reprinted from ref. 25, p. 106, courtesy of Textile Research Journal.

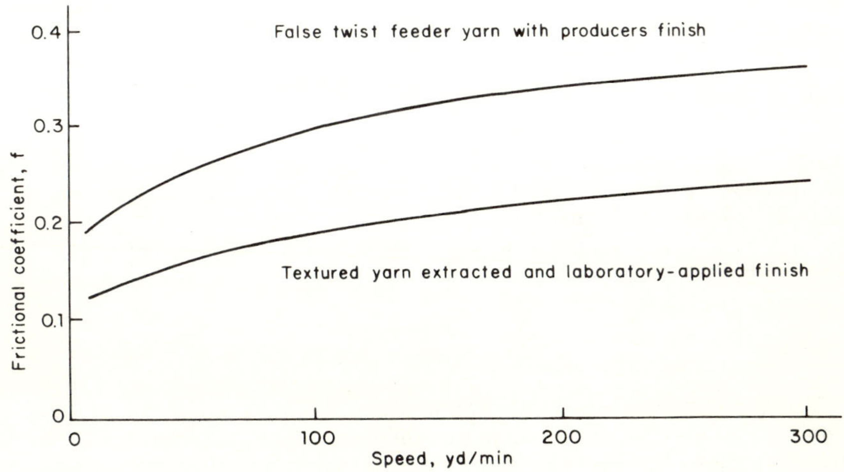

FIG. 12. Effect of yarn surface roughness on friction at the F/M interface. Test conditions: Rothschild F meter, yarn is polyester 150/34 (AVC), finish is Nopco false twist, chrome pin RMS-4, 0.5-in. diameter, contact angle 360°, T = 7.5 g, RH 50%. Reprinted from ref. 25, p. 107, courtesy of Textile Research Journal.

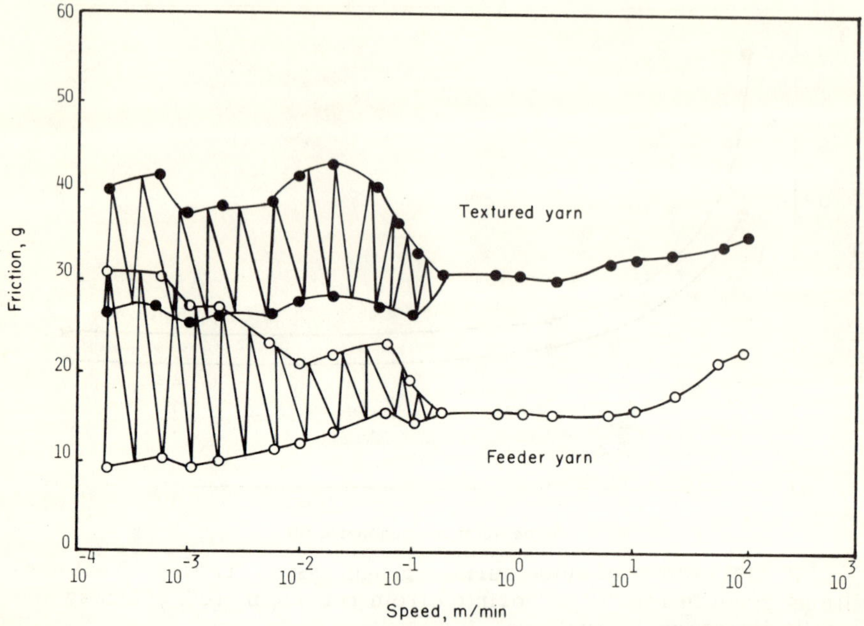

FIG. 13. Effect of yarn surface roughness on friction at the F/F interface. Test conditions: Chemstrand F meter, F/F $2\frac{1}{2}$ turns, 2.5 in., 5 pulleys, yarn is polyester 150/34 (AVC), finish is Nopco false twist, T = 7.5 g, temperature 70° F, RH 50%. Reprinted from ref. 25, p. 108, courtesy of Textile Research Journal.

hydrodynamic region as shown in Fig. 12. The increase in friction with increasing fiber surface roughness in the boundary and semi-boundary region conforms with the findings on guide surface roughness cited above and is attributed to increased wear of rough fiber surfaces in the boundary region.

3. Speed

The trend in modern textile technology is to increase the speed of spinning. Some operations have reached speeds as high as 4,000 yd/min. For these high-speed operations, thermally stable lubricants are required, which prevent melting of the fiber at the high temperatures developed at the F/M interface. Thus, both the effect of speed and temperature on the frictional behavior of multifilament yarns must be considered important factors in spinning operations and are discussed below. The friction of a nylon yarn over a wide speed range using three conventional finishes is illustrated in Fig. 16

FIG. 14. Electron micrograph of polyester false-twist feeder yarn. Reprinted from ref. 25, p. 107, courtesy of Textile Research Journal.

covering the low-speed region, and in Fig. 17 covering the high-speed region [25]. These tests were performed under comparable conditions of guide surface roughness, pre-tension, and angle of contact. No marked difference in friction was observed among these three lubricants (see Fig. 16) in the low-speed region despite the differences in viscosity. In contrast, in the high-speed region (see Fig. 17), friction increased with increasing viscosity. The data demonstrate that friction in the boundary and semiboundary region is independent of viscosity, whereas friction in the hydrodynamic region is highly dependent on viscosity. The former is attributed to intermittent fiber-to-metal contacts penetrating the interfacial film, whereas the latter is attributed to an increase in shear stress of the continuous liquid film separating the sliding surfaces. These conclusions will be reinforced by data presented in Section 10. It is worth noting that the friction vs. speed plots of Fig. 17 show an initial rise followed by a leveling off to a plateau at a speed above 300 yd/min. Similar results have been

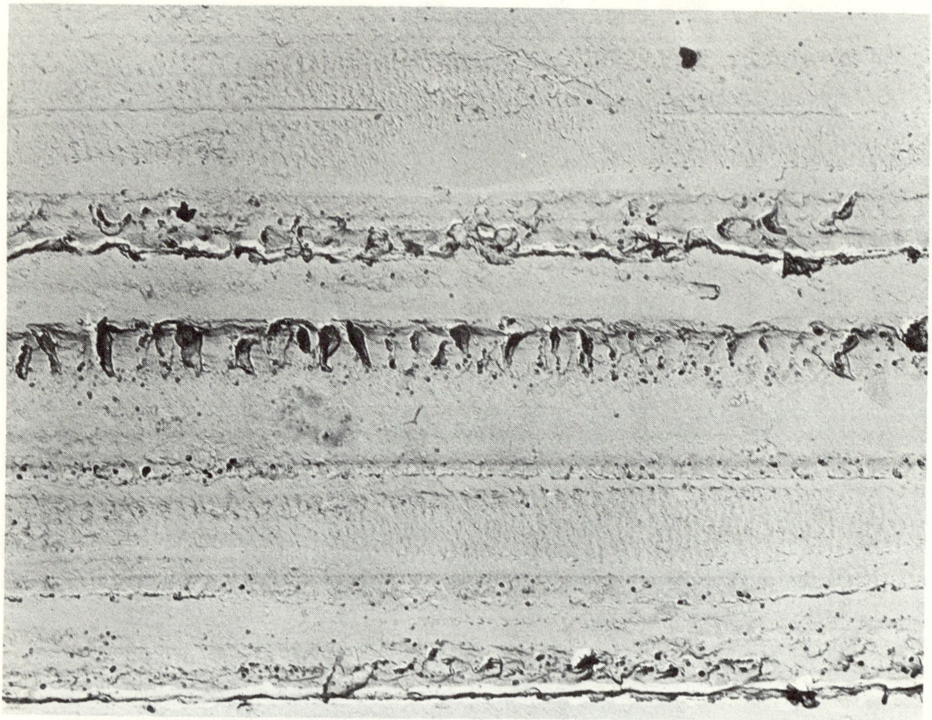

FIG. 15. Electron micrograph of polyester false-twist textured yarn. Reprinted from ref. 25, p. 107, courtesy of Textile Research Journal.

reported by Olsen [19] confirming a consistent pattern. These findings on the effect of speed on friction corroborate the postulate that fundamental differences exist in the frictional behavior of textile yarns in the boundary, semiboundary, and hydrodynamic regions.

4. Guide Temperature

As pointed out before, high-speed spinning and draw spinning require heat-stable lubricants, which prevent melting of the fiber at the high temperatures developed when passing over guides or heated plates. Therefore, studies of the temperature dependence of friction of these heat-stable lubricants are of vital importance. In our laboratory this is being studied over a wide temperature range using a heated chrome pin with a thermocouple attached to its surface, which records temperature on a digital indicator.

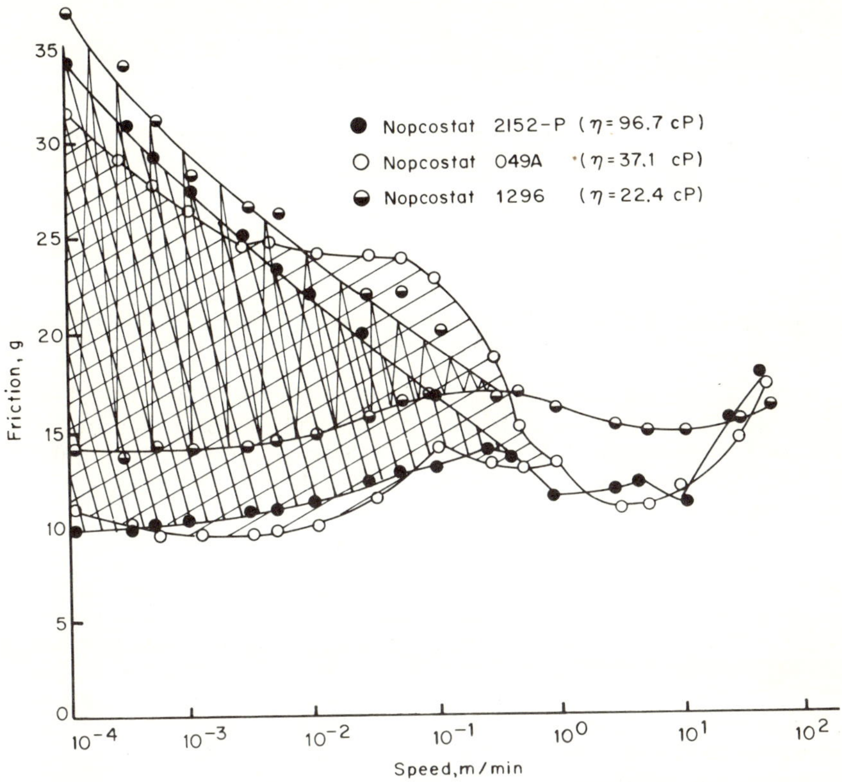

FIG. 16. Friction vs. speed, low-speed region. Test conditions: Chemstrand F meter, yarn is bright nylon 66 200/34 (duPont), finish is 1% wof, chrome pin RMS 60, 0.5-in. diameter, contact angle 360°, T = 10 g, temperature 70° F, RH 50%. Reprinted from ref. 25, p. 108, courtesy of Textile Research Journal.

The temperature dependence of friction of a polyester yarn over a wide temperature range using n-butylstearate and Nopcolube* HT as lubricants is illustrated in Fig. 18 [40]. n-Butylstearate is a highly volatile lubricant showing a 97% loss on exposure of a 2-g sample to a temperature of 200° C for 1 hr in a draft oven, whereas Nopcolube HT is a heat-stable lubricant showing a loss of only 8% under comparable conditions. It is evident that the friction of n-butylstearate initially decreases with increasing temperature, passes through a minimum,

*Registered trademark for textile lubricant of Diamond Shamrock Corporation.

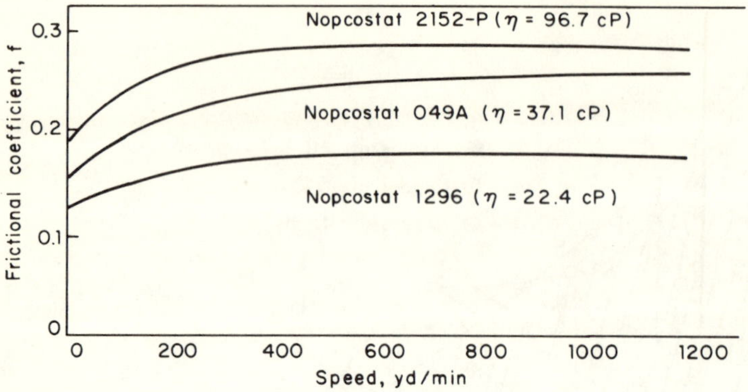

FIG. 17. Friction vs. speed, high-speed region. Test conditions: Rothschild F meter, yarn is bright nylon 66 200/34 (duPont), finish is 1% wof, chrome pin RMS-60, 0.5-in. diameter, contact angle 360°, T = 10 g, temperature 70° F, RH 50%. Reprinted from ref. 25, p. 108, courtesy of Textile Research Journal.

and rises again. Olsen [19] attributes an initial decrease in friction to a decrease in lubricant viscosity, and subsequent increase in friction to volatilization and/or thermal decomposition of the lubricant causing softening of the polymer surface. A more detailed examination of friction as a function of speed is given in Fig. 19 [19]. Friction, as anticipated, increases with increasing speed and the upswing of the curves, after passing through the minimum, increases with

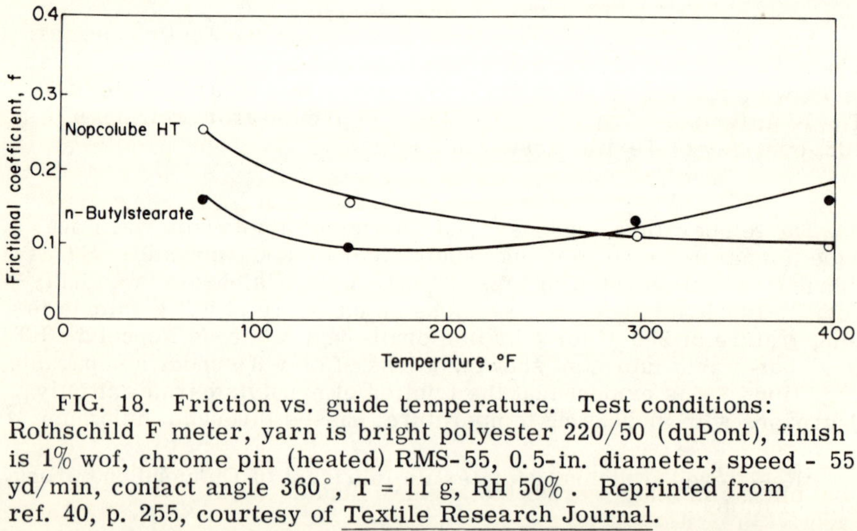

FIG. 18. Friction vs. guide temperature. Test conditions: Rothschild F meter, yarn is bright polyester 220/50 (duPont), finish is 1% wof, chrome pin (heated) RMS-55, 0.5-in. diameter, speed - 55 yd/min, contact angle 360°, T = 11 g, RH 50%. Reprinted from ref. 40, p. 255, courtesy of Textile Research Journal.

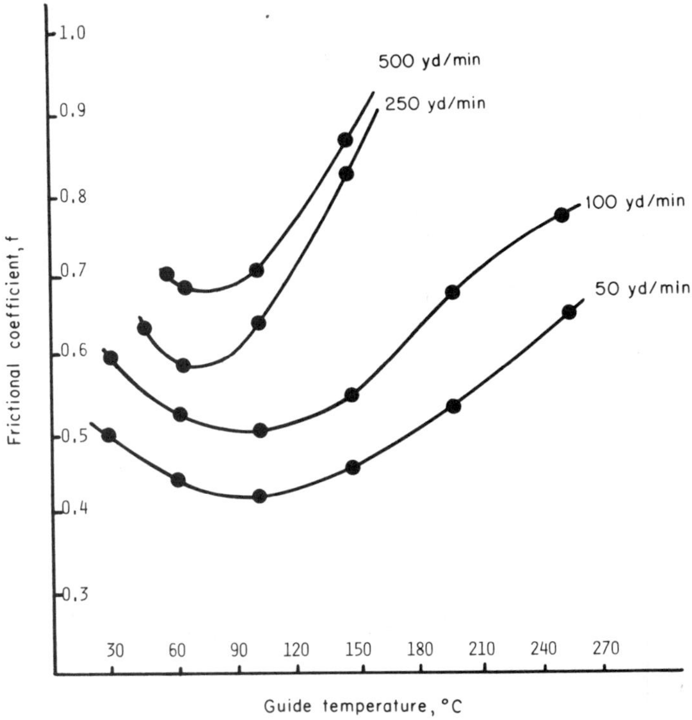

FIG. 19. Friction vs. guide temperature at various speeds. Test conditions: yarn is semidull nylon 70/34 (duPont), finish is 0.8% wof, chrome pin 5AA μin., 0.375-in. diameter, T = 10 g, RH 65%. Reprinted from ref. 19, p. 35, courtesy of Textile Research Journal.

increasing speed. If a matte chrome pin (RMS-120) is substituted for a polished chrome pin (RMS-15) at comparable speed, the overall frictional level is lowered and the temperature can be greatly increased before the point of rapid lubrication failure is reached as shown in Fig. 20 [19].

5. Area of Contact

This section covers the effect of loop size, twist, guide diameter, denier, and moisture regain on fiber friction. All these variables influence the area of contact at the F/M interface.

a. Loop Size. The effect of loop size of a yarn around a cylindrical pin, or in other words, the angle of contact, on friction at the F/M interface is illustrated in Fig. 21 for a rough chrome pin (RMS-60)

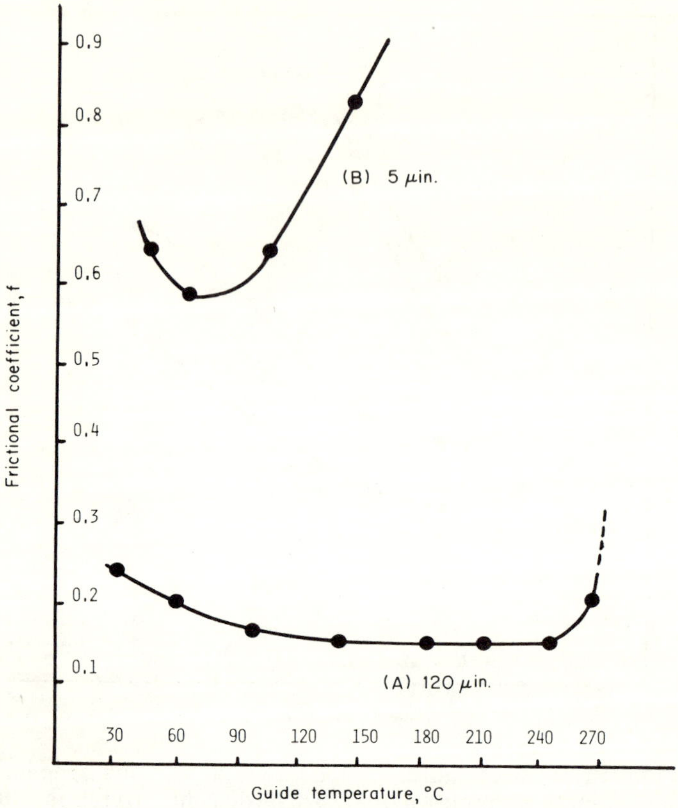

FIG. 20. Friction vs. guide temperature and surface roughness. Test conditions given in Fig. 19, speed 250 yd/min. Reprinted from ref. 19, p. 35, courtesy of Textile Research Journal.

and in Fig. 22 for a smooth chrome pin (RMS-4) [40]. The data observed with the rough chrome pin (see Fig. 21) indicate that in the boundary region the appreciable stick-slip observed at $\theta = 720°$ (two loops around pin), decreases on reducing θ to 360° (one loop), and completely disappeared at $\theta = 90°$ (one-quarter loop). In the hydrodynamic region, i.e., at speeds above the abrupt disappearance of stick-slip, friction is roughly proportional to the angle of contact. Since a change in the angle of contact causes a proportional change in area of contact, it is concluded that in the boundary region, an increase in the area of contact leads to an increase in fiber wear with consequent increase in stick-slip, whereas in the hydrodynamic region, in line with the prediction of the equation by Schlatter and Demas [18]

$$F = \eta(AV/d) \tag{4}$$

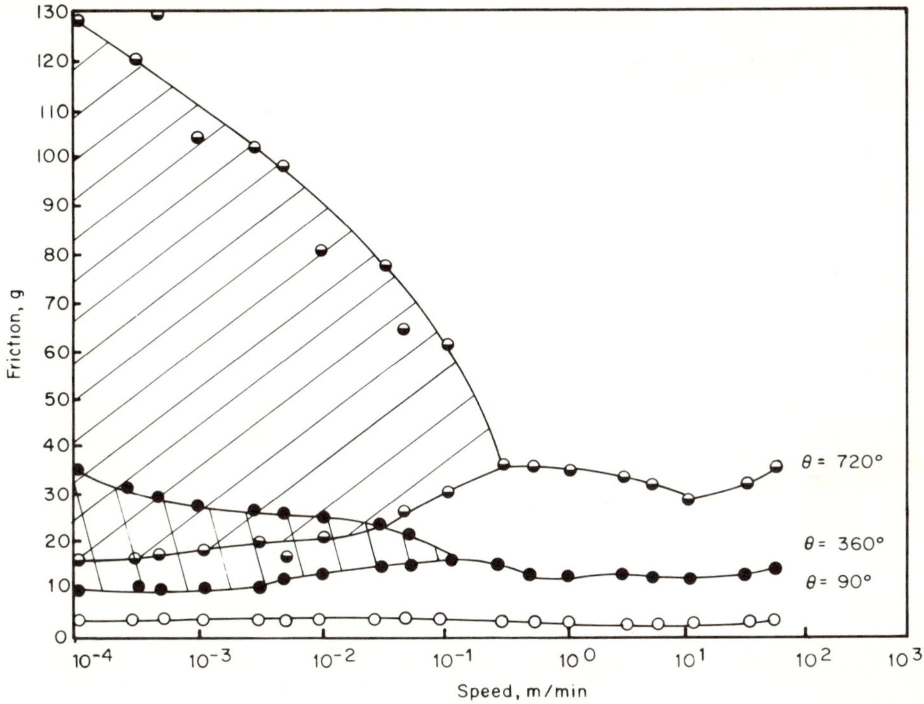

FIG. 21. Effect of contact angle (loop size) on friction with rough pin. Test conditions: Chemstrand F meter, yarn is bright nylon 66 200/34 (duPont), finish is 1% wof Nopcostat 049-A, chrome pin RMS-60, 0.5-in. diameter, T = 10 g, temperature 70° F, RH 50%. Reprinted from ref. 40, p. 255, courtesy of Textile Research Journal.

friction is proportional to the area of contact. In Eq. (4) F is the frictional force, A the area of contact between fiber and guide, V the yarn speed, d the thickness of the lubricant film, and η the viscosity of the lubricant. Frictional data observed with the smooth chrome pin, see Fig. 22, conform to those observed with the rough chrome pin. However, the magnitude of the stick-slip was substantially reduced and the curves tended to indicate higher friction at speeds above the abrupt disappearance of stick-slip. These findings on the effect of loop size of a yarn around a cylindrical pin on friction conform again with the contention that intermittent contacts between fiber and substrate exist in the boundary region, whereas a continuous fluid film separates fiber from substrate in the hydrodynamic region. It is worth noting that stick-slip at the F/M interface may be eliminated or at least minimized by the use of an angle of contact of 90°, which is of vital importance in the testing of lubricant components.

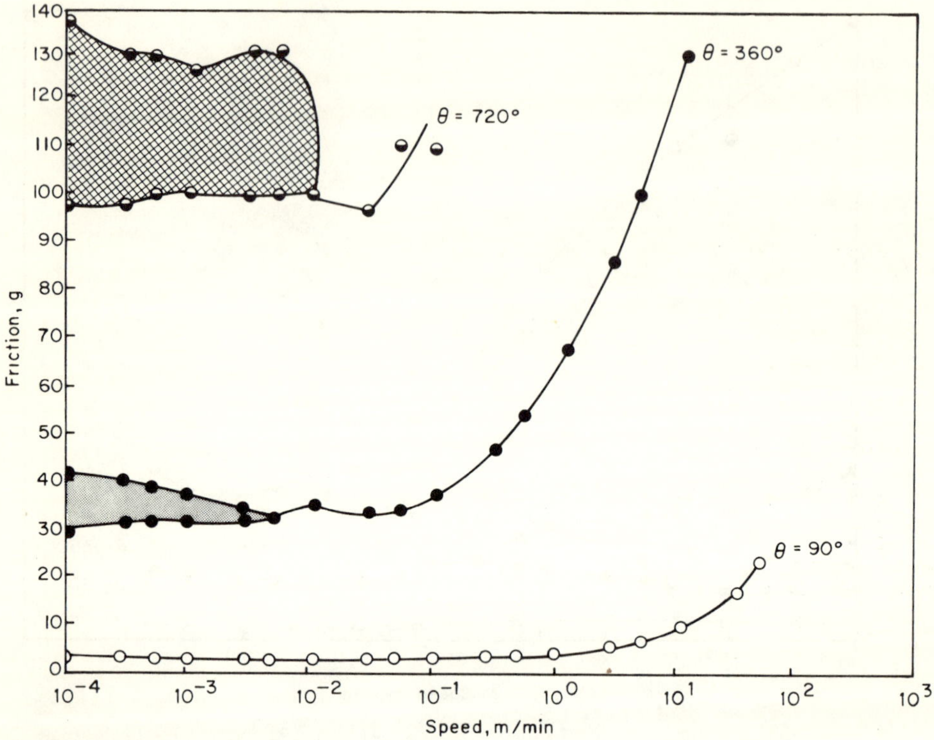

FIG. 22. Effect of contact angle (loop size) on friction with smooth pin. Test conditions: Chemstrand F meter, yarn is bright nylon 66 200/34 (duPont), finish is 1% wof Nopcostat 049-A, chrome pin RMS-4, 0.5-in. diameter, T = 10 g, temperature 70°F, RH 50%. Reprinted from ref. 40, p. 255, courtesy of Textile Research Journal.

b. Twist. The effect of yarn twist on friction is illustrated in Table 1 [41] for multifilament yarns with varying degrees of twist, passing over chrome pins of two different degrees of surface roughness. On both pins friction decreases with increasing twist. However, the effect is more significant on the smooth than on the rough pin. Likewise friction of nylon decreases with increasing twist when passing over alumina pins and remains unchanged when passing over rough alumina pins as shown in Fig. 23 [41]. The degree of smoothness of the alumina surfaces was measured with a Polysurf instrument [41a] and was expressed as H_{CLA} values in microinches. The filaments in high-twist multifilament yarns are bound together and are not moving side by side when rubbing against the guide. Consequently, the area of contact decreases with increasing twist, resulting

TABLE 1

Coefficient of Friction of Nylon Yarn on Metal Guides[a]

	Chromium-plated steel guides	
Type of yarn	Polished	Matte finish
30/10/3/4Z	0.47	0.31
30/10/7Z	0.35	0.29
30/10/30Z	0.28	0.28

[a]Speed 300 ft/min, initial tension 1 ± 1/2 g. Reprinted from ref. 41, p. 105, courtesy of Journal of Textile Institute.

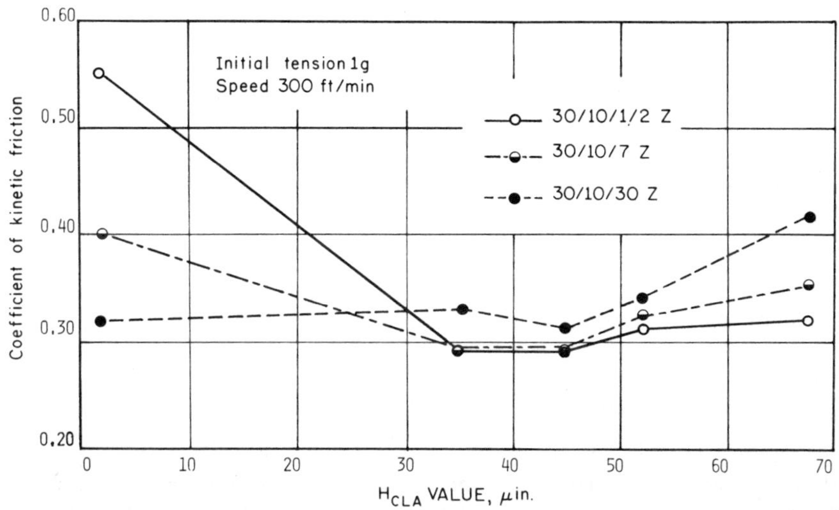

FIG. 23. Friction of nylon vs. guide surface roughness of alumina pins. Reprinted from ref. 41, p. 105, courtesy of Journal of the Textile Institute.

in lower friction on smooth pins. This effect is minimized when using rough pins.

c. Guide Diameter. Another variable affecting yarn friction that influences the area of contact is the guide diameter. Olsen [19] has demonstrated that on polished chrome surfaces (RMS-5) friction remains constant for guide diameters in the range from 1.0 to 0.2 in.

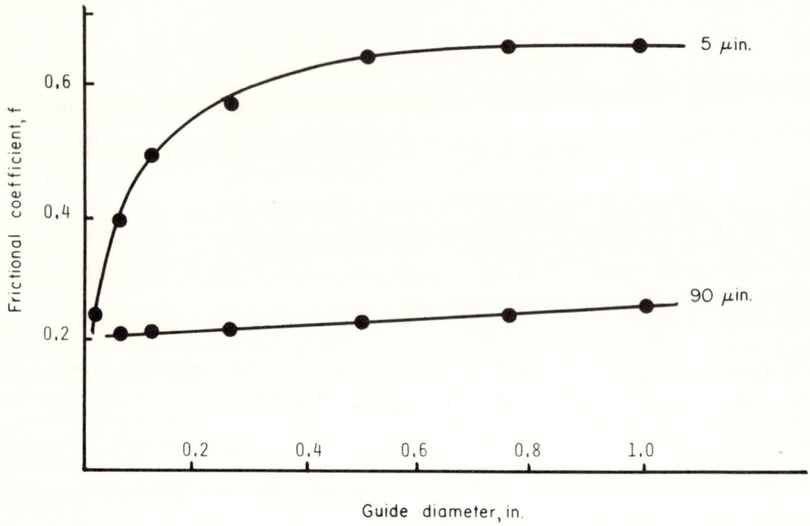

FIG. 24. Friction vs. guide diameter. Test conditions: yarn is semidull nylon 70/34 (duPont), finish is 0.8% wof, chrome pin 5AA and 90AA μin., speed 250 yd/min, T = 10 g, temperature 22°C, RH 65%. Reprinted from ref. 19, p. 36, courtesy of Textile Research Journal.

and then falls off rapidly as shown in Fig. 24. Similar results have been reported by Baird and Mieszkis [41] and Lyne [26]. This decrease in friction with decreasing diameter is attributed again to a decrease in area of contact. In contrast friction decreases only insignificantly on rough chrome surfaces (RMS-90) with decreasing guide diameter, because changes in contact area are minimized.

d. Pre-tension. Another factor that affects the area of contact and consequently greatly influences friction is the pre-tension of the yarn. The effect of pre-tension on friction at the F/F interface has already been demonstrated in Fig. 7 and additional data using a smooth chrome pin (RMS-4) at the F/M interface are illustrated in Fig. 25 for the low-speed region and in Fig. 26 for the high-speed region [40]. In the low-speed region stick-slip has been eliminated by following the procedure described in Section a on loop size. It is evident from Fig. 25 that friction increases with increasing pre-tension. In analogy with the explanation given for the F/F interface, the increase in friction with increasing pre-tension in the boundary region is attributed to increased pressure at the F/M interface with consequent increase in area of contact; an increase of the latter leading to increased fiber/guide wear. Similar results were observed in the high-speed region as shown in Fig. 26. Since one of the requirements for the use

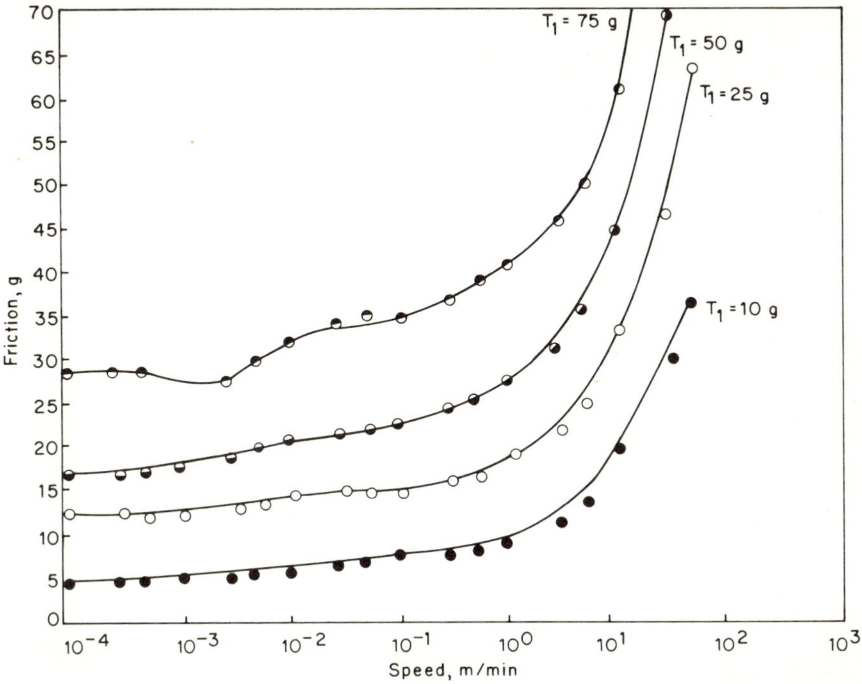

FIG. 25. Effect of pre-tension on friction at F/M interface. Test conditions: Chemstrand F meter, yarn is bright nylon 66 200/34 (duPont), finish is 1% wof Nopcostat 1296, chrome pin RMS-4, 0.5-in. diameter, contact angle 90°, temperature 70°F, RH 50%. Reprinted from ref. 40, p. 256, courtesy of Textile Research Journal.

of the capstan equation, i.e., the constancy in area of contact at the F/M interface, has not been met, the results in the high-speed region are also expressed in terms of the frictional force F, rather than in terms of the frictional coefficient f. In analogy with the explanation given for the boundary region, the increase in friction with increasing pre-tension in the hydrodynamic region is attributed to increased pressure at the F/M interface with consequent increase in area of contact. An increase of the latter leads to an increase in shear of the continuous fluid film at the F/M interface.

It must be emphasized that the pattern demonstrated in Fig. 26 holds true when using a rough pin (RMS-60) as shown in Fig. 27 and for tire yarns with denier 840 to 1,000 [40]. Friction vs. pre-tension plots are linear for each speed level both in the low- and high-speed region [40].

e. Denier. The effect of denier on friction of nylon and polyester, which varied widely according to denier of the yarn, is illustrated in

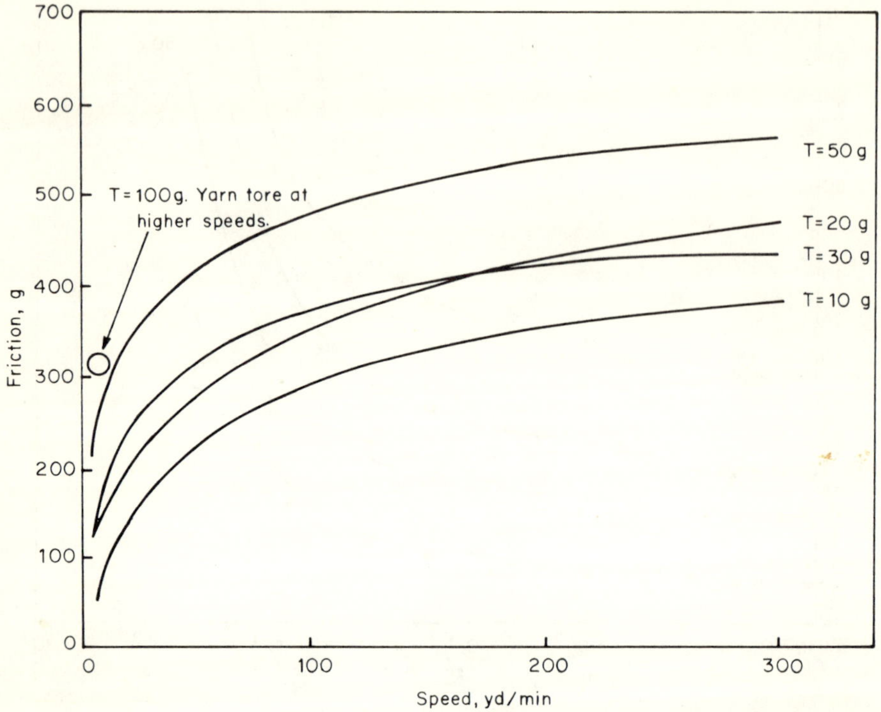

FIG. 26. Effect of pre-tension at the F/M interface. Test conditions: Rothschild F meter, yarn is bright nylon 66 200/34 (duPont), finish is 1% wof Nopcostat 2152-P, chrome pin RMS-4 (smooth), 0.5-in. diameter, contact angle 360°, temperature 70°F, RH 50%. Reprinted from ref. 40, p. 257, courtesy of Textile Research Journal.

Fig. 28 [40]. Since the capstan equation does not seem applicable for this comparison, the data are expressed in terms of frictional force, F. Friction increased significantly with increasing denier in the hydrodynamic region, which in line with the explanation given above for the effect of loop size, twist, guide diameter, and pre-tension, is attributed to an increase in area of contact at the F/M interface.

f. Moisture Regain. It is well known that the friction of cotton increases with increasing regain [7]. For comparison, frictional data for a man-made hydrophilic fiber such as rayon for a 24-hr regain are presented. The effect of relative humidity in the range from 12 to 69% on the friction of rayon yarn is illustrated in Fig. 29 [40]. In analogy with the findings reported for cotton [7], a significant increase in friction with increasing relative humidity has been observed in the hydrodynamic region. This increase in friction is attributed again to an

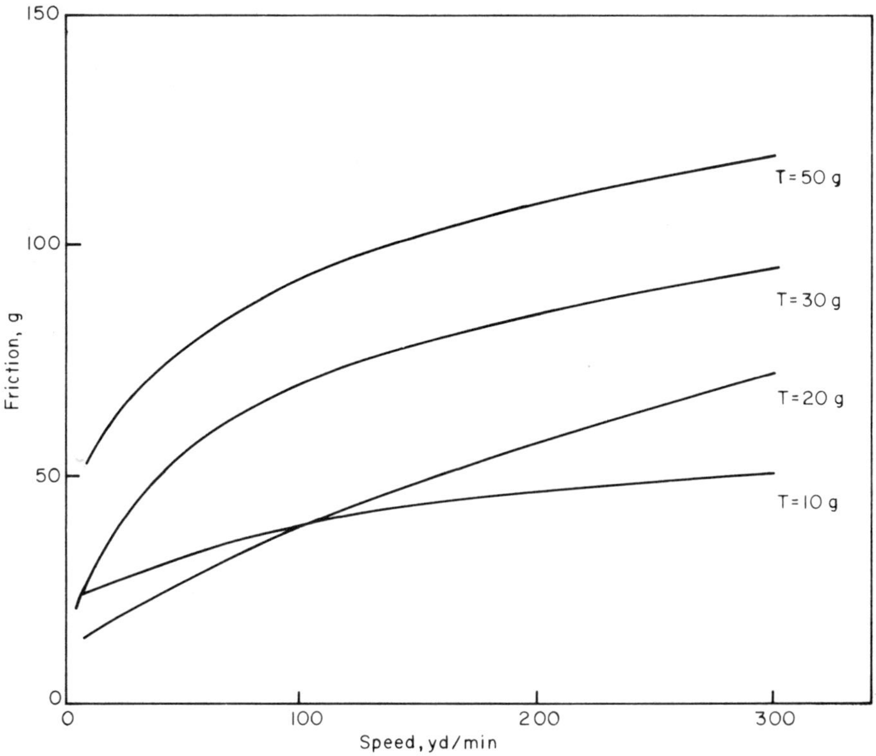

FIG. 27. Effect of pre-tension at the F/M interface. Test conditions: Rothschild F meter, yarn is bright nylon 66 200/34 (duPont), finish is 1% wof Nopcostat 2152-P, chrome pin RMS-60 (rough), 0.5-in. diameter, contact angle 360°, temperature 70°F, RH 50%. Reprinted from ref. 40, p. 256, courtesy of Textile Research Journal.

increase in area of contact at the F/M interface by the swollen rayon yarn.

6. Guide Material

Bowden and Tabor [1] postulated that the boundary friction of clean solid surfaces should be independent of the kind of material, provided that its shear strength is equal to or greater than that of the fiber and shearing does not take place at the junction interfaces. Frictional coefficients observed by Fort and Olsen [20] for clean fibers sliding over other solid surfaces support this concept and the pertinent data are listed in Table 2. F/F friction was compared with that of the yarn against $\frac{1}{4}$-in. diameter pins of chrome, "AlSiMag" 586, and

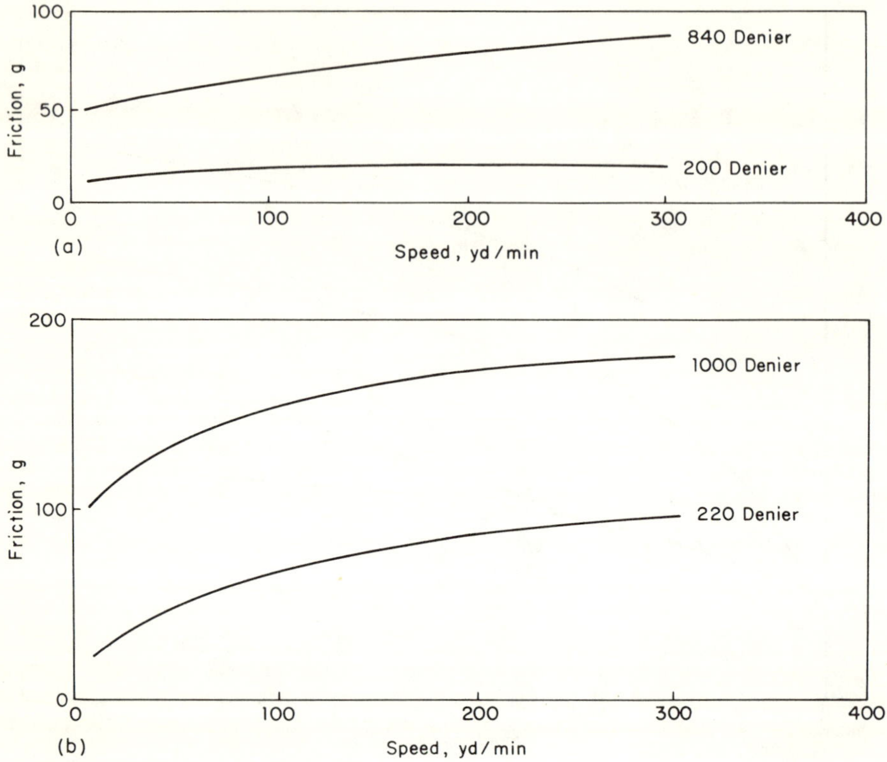

FIG. 28. Effect of denier on friction at the F/M interface. Test conditions: (a) Rothschild F meter, yarn is bright nylon 66, finish is 1% wof Nopcostat 1296, chrome pin RMS-60, 0.5-in. diameter, contact angle 360°, T = 0.05 g/denier, temperature 70° F, RH 50%. (b) Yarn is bright polyester, finish is 1% wof Nopco tire yarn finish. Reprinted from ref. 40, p. 258, courtesy of Textile Research Journal.

synthetic sapphire, which are conventional textile guide materials. Although the frictional coefficients for the different substrates are not identical for any given fiber, they agree well enough to indicate that the boundary friction is controlled by the shear strength of the fiber, which is the weaker material. Teflon exhibits a low friction irrespective of the type of solid surface; the shearing of junctions is undoubtedly taking place at the guide-to-polymer interface [13]. Thus, with the exception of Teflon, all yarns appear to follow the adhesion mechanism in friction.

The data of Table 2 have been reinforced by a recent study [39] of boundary friction with equal-denier yarns using equal diameter (0.5-

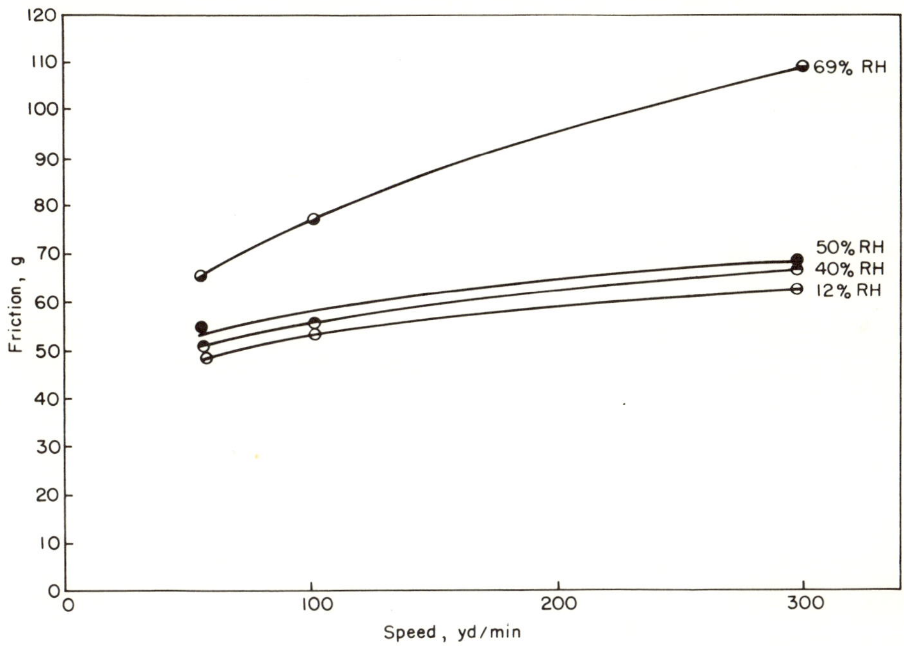

FIG. 29. Effect of relative humidity on friction of rayon. Test conditions: Rothschild F meter, yarn is rayon 300/44 (AVC), finish is Nopco 715, chrome pin RMS-60, 0.5-in. diameter, contact angle 360°, T = 15 g, temperature 70°F, 24-hr moisture regain. Reprinted from ref. 40, p. 258, courtesy of Textile Research Journal.

in.) and equal surface roughness (RMS-20) chrome and "AlSiMag" pins. It has been demonstrated again that boundary friction is independent of the guide material used, but is controlled by the shear strength of the fiber, which is the weaker material.

In the presence of lubricants the effect of the guide material on friction is overshadowed by the wetting properties of the lubricants. Since the guides are usually made from materials having high-energy surfaces, wetting by polar lubricants usually is satisfactory. In contrast satisfactory wetting of hydrophobic fibers requires judicious choice of lubricants as will be discussed in Section 9.

7. Charge

So far, our knowledge of the interrelation between friction and charges transferred to guide surfaces by moving yarns is limited [37]. In order to establish the existence of such an interrelation, both these parameters have been examined as a function of pre-tension over a

TABLE 2

Boundary Friction of Clean Fibers Against Different Substrates Expressed in Terms of the Frictional Coefficient (f)[a]

Fibers	Yarn-to-yarn	Chrome	"AlSiMag"[b] 586	Sapphire
70-34-0.5Z-200 nylon	0.63	0.56	0.60	0.55
150-40-2.5S bright rayon	0.72	0.80	0.81	0.70
300-80 bright acetate	0.68	0.70	0.75	0.71
75-30-0.3Z polyacrylonitrile	0.48	0.47	0.50	0.42
400-60 Teflon	0.10	0.10	0.10	0.10
250-50-0-55 Dacron[c]	0.70	0.73	0.71	0.73

[a]Reprinted from ref. 20, p. 1010, courtesy of Textile Research Journal.
[b]Registered trademark of American Lava Corporation.
[c]Registered trademark for DuPont's polyester fiber.

wide range and the pertinent results are given in Figs. 30 and 31 [39]. The data of Fig. 30 indicate that friction increases with increasing pre-tension in the high-speed region. This increase in friction with increasing pre-tension in the hydrodynamic region is attributed to increased pressure at the F/M interface with consequent increase in area of contact. An increase of the latter leads to an increase in shear of the continuous fluid film at the F/M interface. Similar results were observed with the corresponding charge generation as shown in Fig. 31. The charge generated and transferred to a chromium surface for two specific yarn lengths increases with increasing pre-tension. It is concluded from the approximately parallel behavior between friction and charge generation that the increase in friction with increasing pre-tension causes increased charge generation at the F/M interface.

Changes in charge generation may be brought about by other means than changes in pre-tension. In our program of screening the

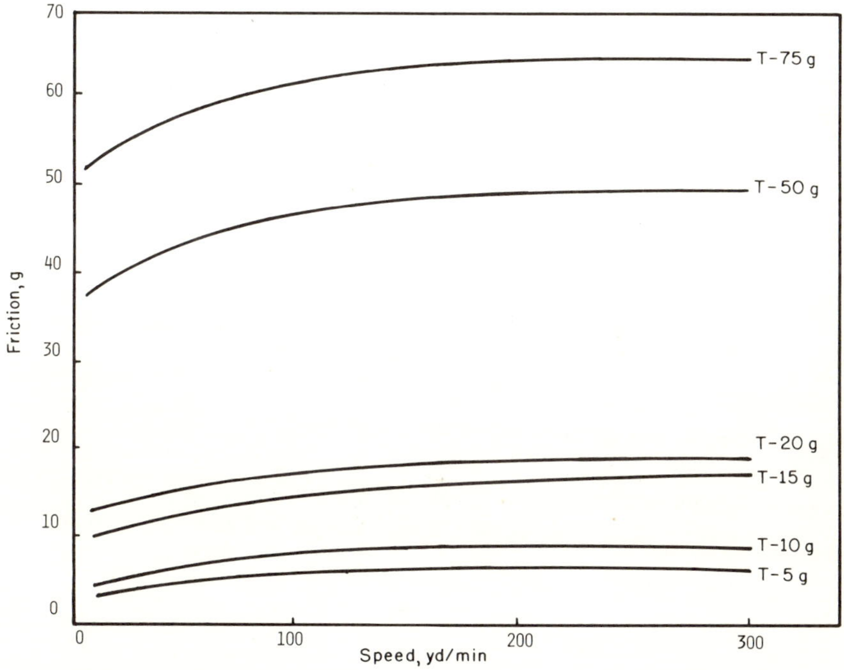

FIG. 30. Effect of pre-tension on friction. Test conditions: Rothschild F meter, yarn is bright nylon 66 200/34 (duPont), finish is 1% wof n-butylstearate, chrome pin RMS-60, 0.5-in. diameter, contact angle 360°, temperature 70°F, RH 50%. Reprinted from ref. 39, p. 761, courtesy of Textile Research Journal.

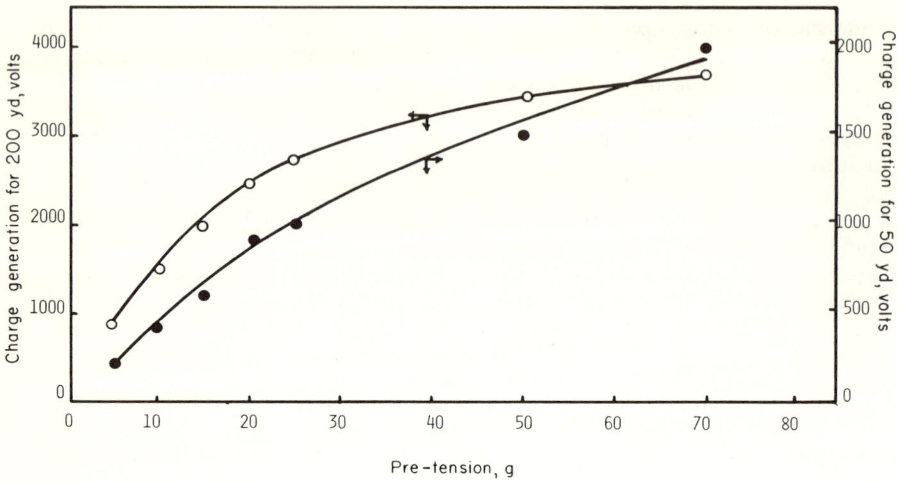

FIG. 31. Effect of pre-tension on charge generation. Test conditions: Rothschild F meter, yarn is bright nylon 66 200/34 (duPont), finish is 1% wof n-butylstearate, chrome surface, speed 300 yd/min, temperature 70°F, RH 50%. Reprinted from ref. 39, p. 762, courtesy of Textile Research Journal.

effectiveness of antistats in spin finishes, the frictional and antistatic properties of the pure lubricant are compared with mixtures of 10% antistat and 90% lubricant. Table 3 [39] illustrates the results of such a screening test in which the same parallel behavior between friction and antistatic properties was observed. Antistat I is ineffective and raises both the antistatic properties and friction of the lubricant, whereas antistat II is highly effective in improving the antistatic properties with a concomitant reduction in friction. Thus, increases in antistatic properties correspond to increases in friction, or vice versa.

It is concluded that the parallel behavior of friction and charge generated is the result of a close interrelation between these two parameters at the F/M interface without any clear indication of which one is the dependent variable.

8. Guide Wear

High friction does not necessarily lead to a high wear rate on guides, but depends on the type of friction mechanism [19]. The data of Fig. 32 demonstrate that the wear rate between dull Dacron 40/27 (4.5 tex) and a 0.375-in.-diameter chrome pin at point B with f = 0.25 and a guide surface roughness of 150 μin. is nine times that at point

TABLE 3

Frictional and Antistatic Properties of 10%/90% Mixtures of Antistats with Lubricants[a]

Finish	Fiber	Chrome pin RMS-60		Charge generation		Antistatic half-life (sec)
		S	f	Volts	Yards	
100% lubricant	Nylon 66 200/34	5.5	0.101	+30	50	230
		55	0.198	+30	50	
		300	0.290	+30	50	
				+125	200	
10% antistat I/90% lubricant	Nylon 66 200/34	5.5	0.160	+250	50	990
		55	0.255	+250	50	
		300	0.339	+250	50	
				+1,000	200	
10% antistat II/90% lubricant	Nylon 66 200/34	5.5	0.065	0	50	3.5
		55	0.144	0	50	
		300	0.246	0	50	
				0	200	

(continued)

TABLE 3 (continued)

Finish	Fiber	Chrome pin RMS-60		Charge generation		Antistatic half-life (sec)
		S	f	Volts	Yards	
100% lubricant	Polyester 220/50	5.5	0.147	+20	50	80
		55	0.283	+15	50	
		300	0.370	+10	50	
				+50	200	
10% antistat I/90% lubricant	Polyester 220/50	5.5	0.173	+20	50	13
		55	0.316	+15	50	
		300	0.387	+10	50	
				+50	200	
10% antistat II/90% lubricant	Polyester 220/50	5.5	0.121	0	50	1.7
		55	0.217	0	50	
		300	0.316	0	50	
				0	200	

[a]Reprinted from ref. 39, p. 762, courtesy of Textile Research Journal. Test conditions: Rothschild F meter, finish 1% wof, contact angle 360°, T = 0.05 g/denier, temperature 70°F, RH 50%.

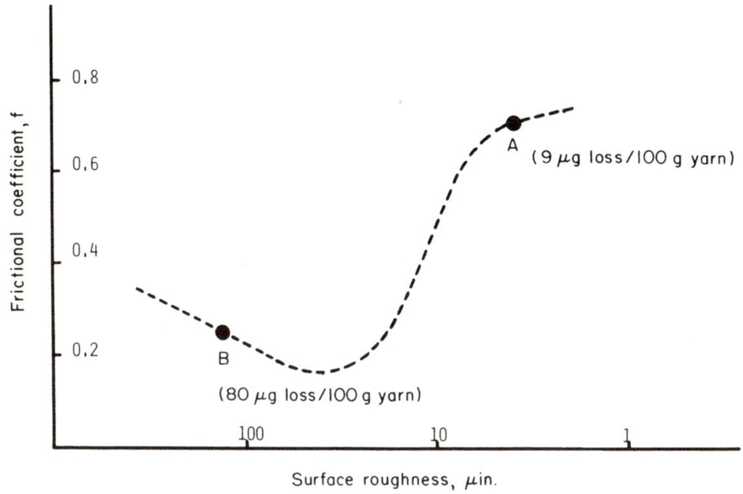

FIG. 32. Guide wear rate vs. type of friction. Test conditions: yarn is dull Dacron 40/27 (duPont), finish is 0.8% wof, chrome pin 0.375 in. diameter, speed 250 yd/min, T = 10 g, temperature 22°C, RH 65%. Reprinted from ref. 19, p. 36, courtesy of Textile Research Journal.

A with f = 0.65 and a guide surface roughness of 4 μin. In this comparison point B represents semiboundary conditions, whereas point A represents hydrodynamic conditions. These data support the concept that intermittent contacts between fiber and substrate exist in the boundary and semiboundary regions, whereas a continuous fluid film separates fiber from substrate in the hydrodynamic region.

9. Fiber Material and Wetting of Fibers

a. Introduction. So far this discussion of experimental results has dealt with factors related to instrumentation. The present and subsequent sections deal with factors related to the fiber material or to the physical properties of the lubricant.

Zisman and co-workers [42] have originated the concept of critical surface tension. According to this concept, a critical surface tension can be ascribed to any solid surface, and perfect equilibrium wetting requires a liquid surface tension below that of the critical surface tension of the substrate. An attempt was made to relate the frictional and equilibrium wetting properties of several fiber materials and the results are presented below. The results conform with the concept of critical surface tension.

b. Fiber Material and Equilibrium Wetting. It is generally accepted that different fiber materials require specific lubricants. Therefore, the results of a comparative study of the friction of three hydrophobic fiber materials of comparable 200-220 denier and pretension of T = 0.05 g/denier over a broad speed range are given in Figs. 33 to 35 [43]. The data of Figs. 33 and 34, which represent the use of two low-viscosity lubricants, such as n-butylstearate with η = 9.05 cP and Nopcostat* 1296 with η = 22.4 cP, demonstrate a wide range in friction, with friction increasing in the following order of fiber materials: nylon 66, polyester, and polypropylene. In contrast, the data of Fig. 35, which represent the use of a high-viscosity lubricant, such as Nopcostat PE-131 with η = 5,830 cP, demonstrate a high friction with a narrow spread in friction for all three fiber materials.

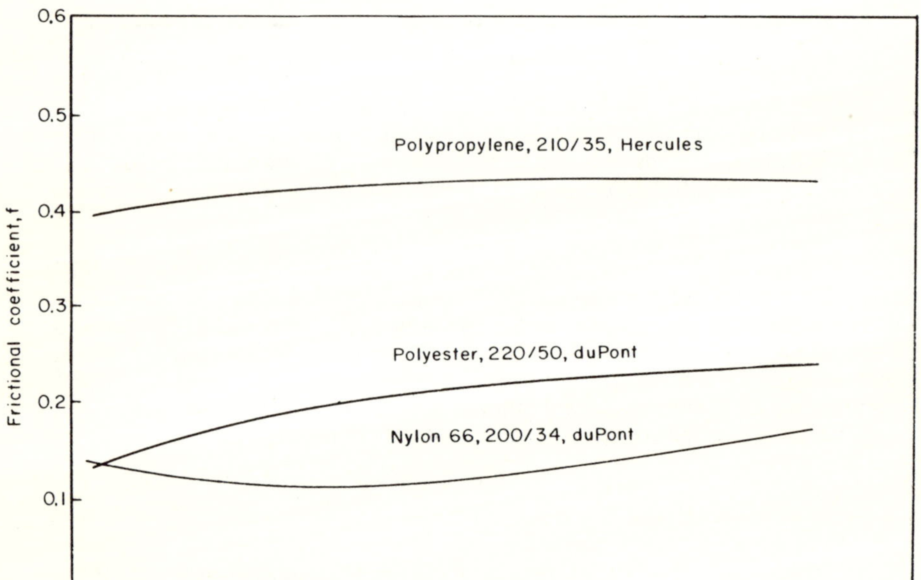

FIG. 33. Effect of fiber material at constant denier 200-220. Test conditions: Rothschild F meter, finish is 1% wof n-butylstearate (η = 9.05 cP), chrome pin RMS-60, 0.5-in. diameter, contact angle 360°, T = 0.05 g/denier, temperature 70° F, RH 50%. Reprinted from ref. 43, p. 343, courtesy of Textile Research Journal.

*Registered trademark for antistatic agent of Diamond Shamrock Corporation.

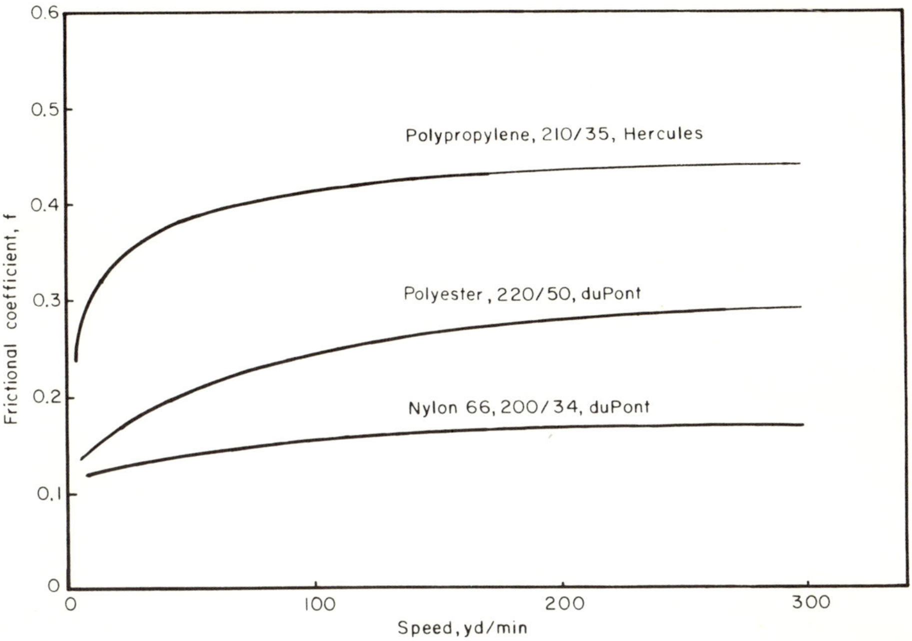

FIG. 34. Effect of fiber material at constant denier 200-220. Test conditions: Rothschild F meter, finish is 1% wof Nopcostat 1296 (η = 22.4 cP), chrome pin RMS-60, 0.5-in. diameter, contact angle 360°, T = 0.05 g/denier, temperature 70° F, RH 50%. Reprinted from ref. 43, p. 343, courtesy of Textile Research Journal.

In order to establish a link between frictional and equilibrium wetting properties, contact angles of the lubricants were measured on thin plastic films made from the same material as the fibers and these angles are listed in Table 4 [43]. Advancing contact angles in the range from 0-90° signify perfect wetting of the film at $\theta = 0°$ and no wetting at all at $\theta = 90°$. It follows from combining the data of Figs. 33, 34, and Table 4 that the almost perfect wetting of nylon 66 by n-butylstearate and Nopcostat 1296 is reflected in very low friction. Moreover, the progressive decrease in wetting by these two lubricants, in passing from nylon 66 to polyester and polypropylene, is reflected in consecutive increases in friction. In contrast to these good wetting results of n-butylstearate and Nopcostat 1296, the wetting of Nopcostat PE-131 was poor on all three polymer films. Consequently, the poor wetting results of Nopcostat PE-131 are reflected in high friction as shown in Fig. 35. Other factors also play a role, as shown in Section 10, in explaining the high friction of Nopcostat PE-131.

Furthermore, it may be deduced from the data of Table 4 that the equilibrium wetting properties conform with the concept of the critical

TABLE 4

Contact Angles of Fiber Lubricants or Antistats on Thin Plastic Sheets (Temperature 70° F)[a]

Polymer	Critical surface tension substrate γ_c (dyn/cm)	Contact angle, θ		
		n-Butylstearate (γ = 31.9 dyn/cm)	Nopcostat[b] 1296 (γ = 32.2 dyn/cm)	Nopcostat[b] PE-131 (γ = 51.5 dyn/cm)
Nylon 66	46 [42]	2	4	48
Polyester (Mylar[c])	43 [42]	6	6	44
Polypropylene	24 [44]	8	8	38
Teflon[d]	18.5 [42]	47	36	67

[a]Reprinted from ref. 43, p. 343, courtesy of Textile Research Journal.
[b]Registered trademark for antistatic agent of Diamond Shamrock Corporation.
[c]Registered trademark for DuPont's Polyester film.
[d]Registered trademark for DuPont's TFE-fluorocarbon film.

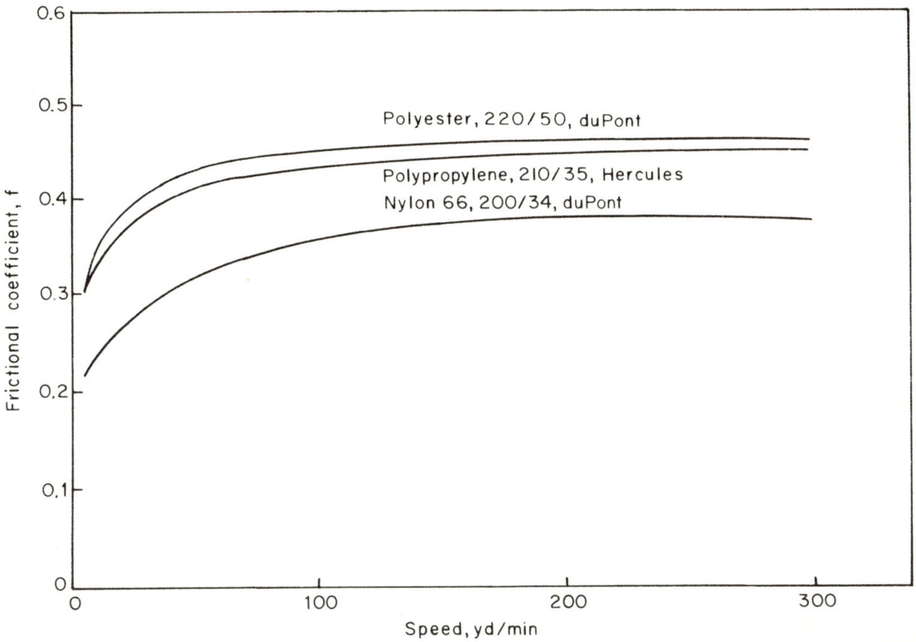

FIG. 35. Effect of fiber material at constant denier 200-220. Test conditions: Rothschild F meter, finish is 1% wof Nopcostat PE-131 (η = 5,830 cP), chrome pin RMS-60, 0.5-in. diameter, contact angle 360°, T = 0.05 g/denier, temperature 70°F, RH 50%. Reprinted from ref. 43, p. 343, courtesy of Textile Research Journal.

surface tension of the substrate. Or in other words, perfect wetting requires a liquid surface tension below that of the critical surface tension of the substrate, which is borne out by the authors' results. It is concluded from this discussion that good wetting of the fiber material is an important requirement for low friction of lubricated yarns over the entire speed range. The role of the other important parameter in fiber lubrication, i.e., lubricant viscosity, will be discussed in Section 10.

c. Fiber Material and Dynamic Wetting. The above correlation was based on equilibrium wetting data of thin plastic films made from the same material as the fibers. However, fiber lubrication involves the exposure of a moving surface to wetting by a liquid, a process in which no true equilibrium is reached. Therefore, a correlation between wetting and friction of lubricated yarns, which is based on dynamic wetting data, seems to be a superior method and is presented below. Moreover, in this correlation fiber surfaces were used for wetting and friction measurements.

TABLE 5

Dynamic Wetting of Multifilament Yarns of Comparable Denier (200-220)[a]

Liquid	Rise canceling velocity, V_c (cm/sec)			
	Nylon 66 200/34 (46)[b]	Polyester 220/50 (43)[b]	Polypropylene 210/35 (24)[b]	Teflon 200/30 (18.5)[b]
H_2O	0.45	0.01	0.00	0.00
Nopcolube HT	0.80	0.65	0.49	0.18
Isocetylstearate	2.65	2.30	2.25	1.60
Nopcolube HT-50	3.25	2.50	2.50	1.65
Mineral oil	3.20	3.05	2.90	2.20
Dimethyl polysiloxane (L-45 10 centistokes)	6.90	6.00	5.75	6.60
n-Butylstearate	9.20	7.30	-	7.10

[a]Reprinted from ref. 39, p. 763, courtesy of Textile Research Journal.
[b]Numbers in parentheses indicate critical surface tension, γ_c [42,44].

The dynamic wetting of nylon 66, polyester, polypropylene, and Teflon yarns of comparable denier (200-220) was examined with seven liquids and the results are expressed in terms of the rise canceling velocity V_c, as shown in Table 5 [39]. A high value of V_c signifies good wetting, and vice versa [45]. Since the object of this correlation was to examine variations in V_c of four yarns for each specific liquid, only the surface interactions were considered and the hydrodynamic interactions were eliminated. Likewise, variations in V_c as a function of yarn diameter were minimized by the use of yarns with comparable diameter. As anticipated, wetting of all four hydrophobic fibers by water was extremely poor. The data of Table 5 indicate that with a single exception the dynamic wetting expressed in terms of V_c decreases for each liquid with decreasing critical surface tension of the fiber. Thus, it has been demonstrated that Zisman's [42] concept on the dependence of equilibrium wetting of solid surfaces on the critical surface tension of the substrate is also applicable to dynamic wetting of yarns.

These dynamic wetting data of four different yarns, which are listed in Table 5, have also been correlated with the corresponding frictional data [39]. As an example, the frictional data using dimethyl polysiloxane as lubricant are presented in Fig. 36. A good correlation between friction and dynamic wetting is apparent for nylon, polyester, and polypropylene, which encompass the critical surface tension region from 46 to 24 dyn/cm. Thus, an increase in friction correlates with a decrease in rise canceling velocity V_c, or in other words low friction requires good wetting and vice versa. In contrast the Teflon data do not conform with this correlation. Teflon exhibits the lowest values of

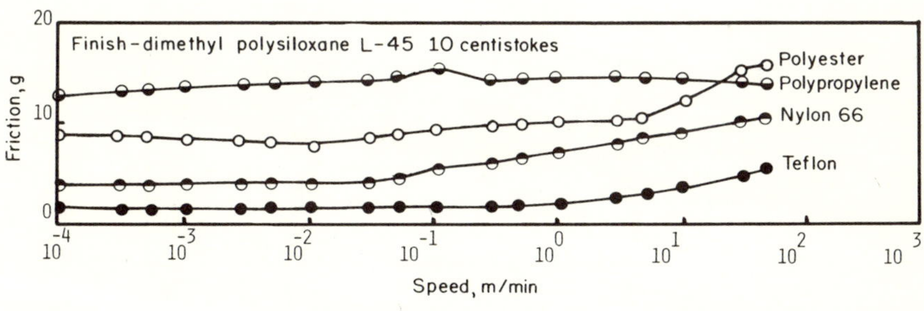

FIG. 36. Effect of critical surface tension of fiber materials on friction. Test conditions: Chemstrand F meter; yarn is Teflon 200/30 (duPont), nylon 66 (duPont), polyester 220/50 (duPont), polypropylene 210/35 (Hercules); chrome pin RMS-4, 0.5-in. diameter; contact angle 90°; T = 0.05 g/denier; temperature 70° F; RH 50%. Reprinted from ref. 39, p. 764, courtesy of Textile Research Journal.

the rise canceling velocity V_c, i.e., poorest wetting, but also contrary to expectations the lowest friction. This atypical frictional behavior is attributed to its very low critical surface tension of 18.5 dyn/cm, which results in a different shearing mechanism [13,20,43].

Figure 37 presents a comparison between the friction of nylon and Teflon yarn at the F/F interface as a function of pre-tension, i.e., the pressure existing at the F/F interface [39]. The object of this comparison is to relate the atypical frictional behavior of Teflon as shown in Fig. 36 to its low critical surface tension of 18.5 dyn/cm and to contrast it with that of nylon having a higher critical surface tension of 46 dyn/cm [42]. As anticipated friction increases with increasing

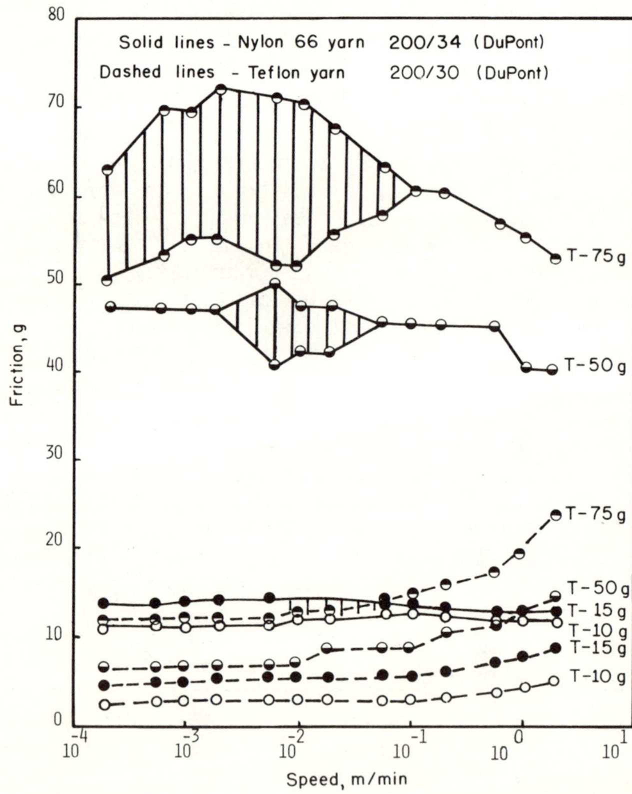

FIG. 37. Effect of critical surface tension of fiber materials on stick-slip. Test conditions: Chemstrand F meter, F/F - 2½ turns, 2.5 in., 5 pulleys, finish is 1% wof n-butylstearate, temperature 70°F, RH 50%. Reprinted from ref. 39, p. 764, courtesy of Textile Research Journal.

pre-tension for both types of yarns as shown in Fig. 37. However, a distinct difference in the magnitude of friction at corresponding pre-tension levels is noted with nylon yarn exceeding by far the frictional values of Teflon yarn. Similarly, a difference in the formation of stick-slip is evident. Nylon yarn exhibits stick-slip at pre-tension levels above 10 g, whereas Teflon yarn exhibits no stick-slip at any pre-tension level in the range from 10 to 75 g. It is concluded from the low frictional values and the complete absence of stick-slip formation with Teflon that no intrinsic adhesion occurs in friction of this low-energy fiber and as a result, shearing takes place primarily at the junction interfaces [13,20,43]. In contrast nylon with a higher surface energy exhibits intrinsic adhesion and, consequently, shearing takes place in the bulk phase with consequent significant stick-slip formation. It is worth noting that Tabor and co-workers differentiate between clean and slightly contaminated surfaces in their interpretation of the frictional behavior of Teflon [46,47].

10. Lubricant Viscosity

Another important parameter in fiber lubrication, the lubricant viscosity and its effect on friction at the F/M and F/F interfaces, is illustrated in Figs. 38 to 44 [43] for three fiber materials on a series of lubricants covering a wide viscosity range. The friction of each fiber material was examined both in the low- and high-speed region, i.e., under boundary, semiboundary, and hydrodynamic conditions. Figures 38 and 39 indicate that friction of nylon 66 yarn is independent of lubricant viscosity in the boundary region, whereas friction increases with increasing lubricant viscosity in the hydrodynamic region. The frictional results on polyester yarn follow the same pattern as shown in Figs. 40 and 41. With the exception of Nopcostat PE-131 the lubricants used in these frictional measurements exhibited very good wetting properties on nylon 66 and polyester (see Table 4). Thus, if we combine the findings on the effect of wetting and viscosity on friction, it is concluded that under conditions of good wetting, friction is independent of lubricant viscosity in the boundary region, but highly dependent on lubricant viscosity in the hydrodynamic region. The independence of friction from the effect of lubricant viscosity in the boundary region is attributed to the existence of numerous penetrations in the interfacial film as well as to direct fiber-to-metal contacts. The dependence of friction on lubricant viscosity in the hydrodynamic region is attributed to an increase in shear stress in the continuous liquid interfacial film with increasing viscosity.

In contrast the effect of lubricant viscosity on the friction of polypropylene yarn follows a different pattern, as shown in Figs. 42 and 43. Likewise, no dependence of friction on lubricant viscosity is apparent in the boundary region. However, no dependence of friction on lubricant viscosity was also observed in the hydrodynamic region.

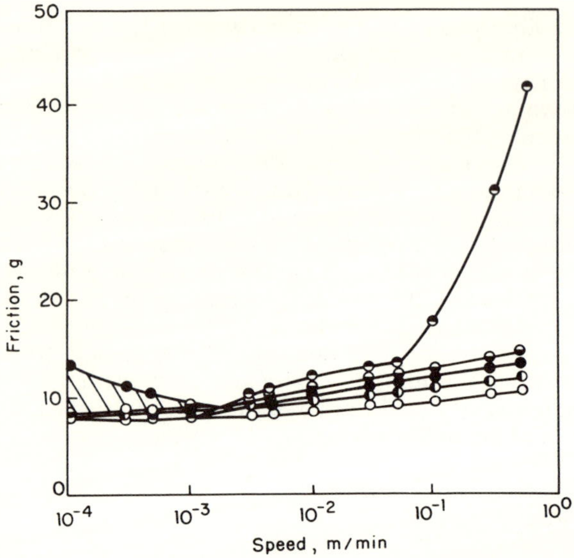

FIG. 38. Effect of lubricant viscosity on friction of bright nylon 66 200/34 (duPont). Test conditions: Chemstrand F meter, finish is 1% wof, chrome pin RMS-4, 0.5-in. diameter, contact angle 90°, T = 10 g, temperature 70°F, RH 50%. ◓ Nopcostat PE-131 (η = 5,830 cP), ◒ Nopcostat 1296 (η = 22.4 cP), ● Nopcostat 049-A (η = 37.1 cP), ◐ Nopcostat 2152-P (η = 96.7 cP), ○ n-butylstearate (η = 9.05 cP). Reprinted from ref. 43, p. 344, courtesy of Textile Research Journal.

These lubricants exhibited poor wetting on polypropylene, see Table 4. Thus, if we combine again the findings on the effect of wetting and viscosity on friction, it is concluded that under conditions of poor wetting, the effect of lubricant viscosity is nullified in the boundary as well as in the hydrodynamic region. Poor wetting of the sliding surfaces appears to prevent formation of a continuous fluid film even at the highest speed level.

Olsen [19] has demonstrated in a study of five Newtonian oils that a plot of friction vs. logarithm of viscosity of 2% lubricant on nylon follows a linear relation when passing either over a smooth or matte chrome pin. With the latter pin the effect of viscosity is still evident, but its influence is greatly reduced because the rougher guide surface increases the contribution of boundary conditions at the expense of hydrodynamic conditions.

Finally, results on the effect of lubricant viscosity on friction of polyester yarn at the F/F interface are presented in Fig. 44. In analogy with the results at the F/M interface, friction increases with

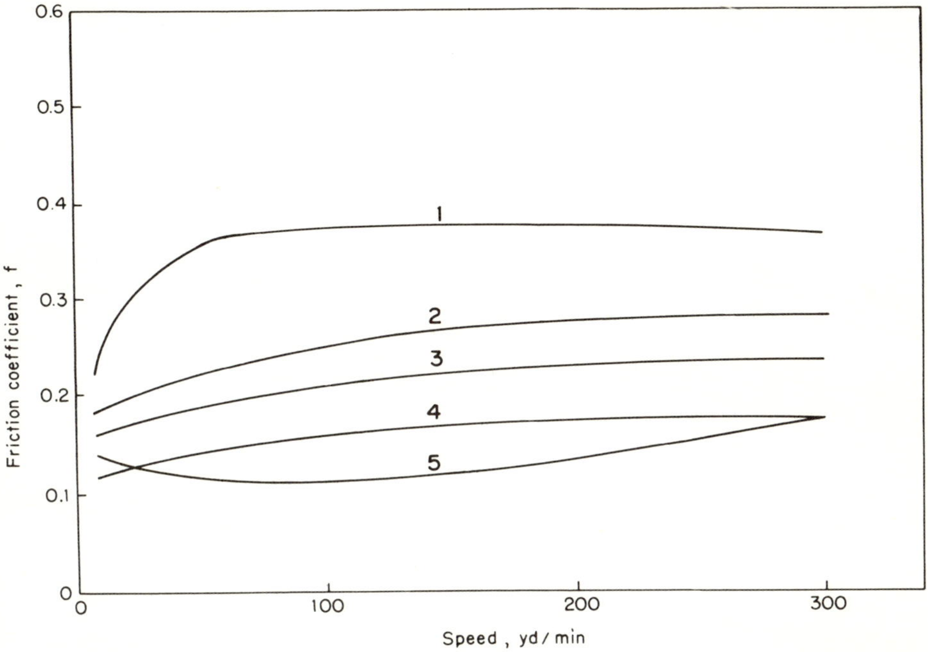

FIG. 39. Effect of lubricant viscosity on friction of bright nylon 66 200/34 (duPont). Test conditions: Rothschild F meter, finish is 1% wof, chrome pin RMS-60, 0.5-in. diameter, contact angle 360°, T = 10 g, temperature 70°F, RH 50%. (1) Nopcostat PE-131 (η = 5,830 cP), (2) Nopcostat 2152-P (η = 96.7 cP), (3) Nopcostat 049-A (η = 37.1 cP), (4) Nopcostat 1296 (η = 22.4 cP), (5) n-butylstearate (η = 9.05 cP). Reprinted from ref. 43, p. 344, courtesy of Textile Research Journal.

increasing lubricant viscosity in the hydrodynamic region, but is independent of lubricant viscosity in the boundary region. In summary it is concluded that the effect of lubricant viscosity on friction follows the same pattern at the F/M and F/F interfaces.

11. Fiber Luster

The intensity of fiber luster is dependent on the amount of TiO_2 added to the fiber melt. Three types of fibers are distinguished, namely, bright fibers with about 0.1% TiO_2, semidull fibers with about 0.5% TiO_2, and dull fibers with about 2.0% or more TiO_2. According to Scheier and Lyons [48] variations in TiO_2 content also affect the geometry of the fiber surface, namely, bright polyester fibers have a

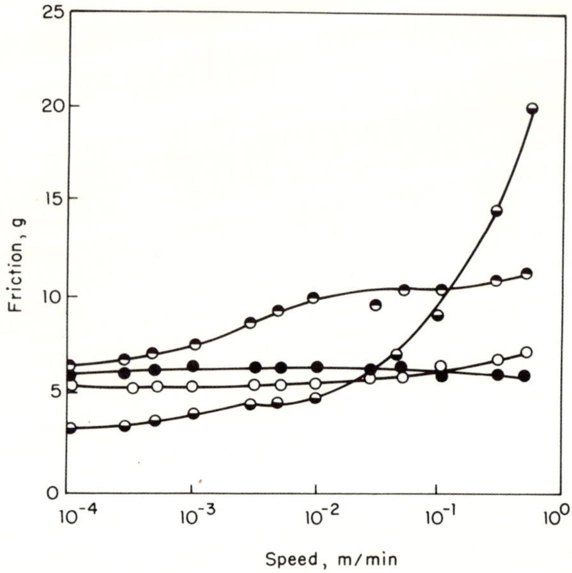

FIG. 40. Effect of lubricant viscosity on friction of bright polyester 220/50 (duPont). Test conditions: Chemstrand F meter, finish is 1% wof, chrome pin RMS-4, 0.5-in. diameter, contact angle 90°, T = 11 g, temperature 70°F, RH 50%. ◒ Nopcostat PE-131 (η = 5,830 cP), ◓ Nopco texturizing finish (η = 34.9 cP), ● Nopcostat 1296 (η = 22.4 cP), ○ n-butylstearate (η = 9.05 cP). Reprinted from ref. 43, p. 345, courtesy of Textile Research Journal.

smooth surface, whereas dull fibers have a rough surface. Furthermore, Grindstaff [49] has shown that rough polyester fibers wet better than smooth ones [50].

Differences in fiber luster have been related to yarn friction and the pertinent results are given in Figs. 45 to 48 [39]. Friction on a smooth pin (RMS-4) increased in the order: bright, semidull, and dull as shown in Fig. 45. Thus, friction increases with increasing fiber surface roughness in the boundary and semiboundary regions, which conforms with the findings of Fig. 13. This increase in friction in the boundary and semiboundary regions is attributed to increased fiber wear with increasing fiber surface roughness. In the hydrodynamic region the reverse relationship was observed for frictional results over a smooth pin (RMS-4) as shown in Fig. 46, which is in line with the results given in Fig. 12. The low friction of the dull yarn is attributed to a disruption of the continuous fluid film between yarn and guide by the rough surface resulting in a shift toward conditions simulating a semiboundary region with a consequent lowering of friction.

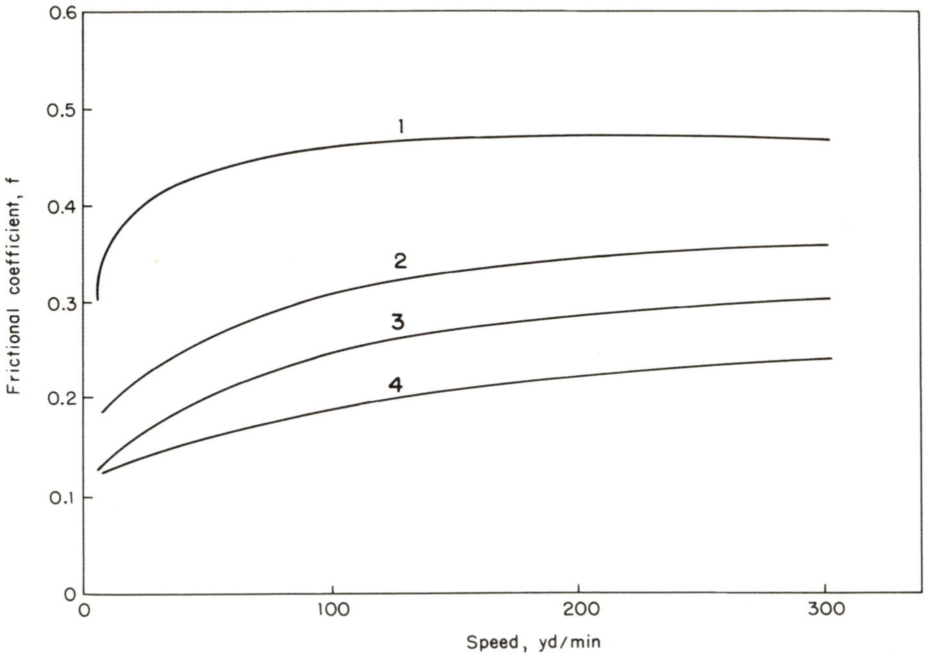

FIG. 41. Effect of lubricant viscosity on friction of bright polyester 220/50 (duPont). Test conditions: Rothschild F meter, finish is 1% wof, chrome pin RMS-60, 0.5-in. diameter, contact angle 360°, T = 11 g, temperature 70° F, RH 50%. (1) Nopcostat PE-131 (η = 5,830 cP), (2) Nopco texturizing finish (η = 34.9 cP), (3) Nopcostat 1296 (η = 22.4 cP), (4) n-butylstearate (η = 9.05 cP). Reprinted from ref. 43, p. 345, courtesy of Textile Research Journal.

In addition the data of Fig. 46 indicate that the discrepancy between the friction of bright and dull yarns in the hydrodynamic region is minimized when using a rough pin (RMS-60). It is concluded that the contribution of the guide surface roughness overshadows that of the fiber surface roughness. Baird and Mieszkis [41] have arrived at the same conclusion in their study of the friction of yarns as a function of TiO_2 content on alumina pins as illustrated in Fig. 47. The degree of smoothness of the alumina surfaces was measured with a Polysurf instrument [41a] and was expressed as H_{CLA} values in microinches.

Furthermore, the superior wetting of the rough surface of the dull yarn in contrast to the smooth surface of the bright yarn may also be considered a contributing factor to the lower friction of the dull yarn in the hydrodynamic region [49,50].

The discrepancy in fiber surface roughness is also reflected in the charge generated by passing a specific yarn length over a chromium

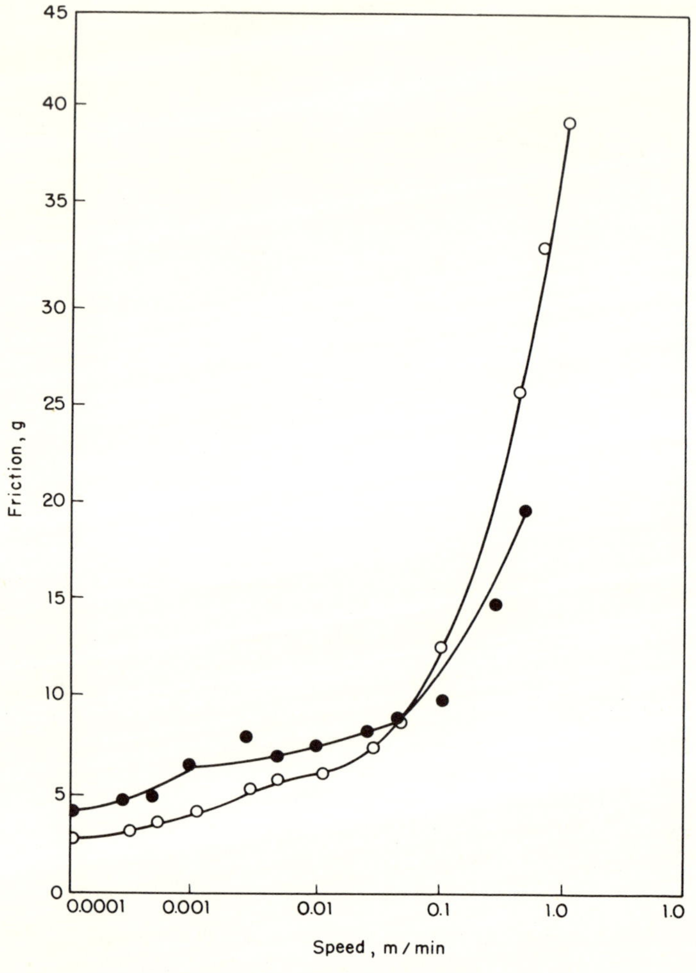

FIG. 42. Effect of lubricant viscosity on friction of bright polypropylene 210/35 (Hercules). Test conditions: Chemstrand F meter, finish is 1% wof, chrome pin RMS-4, 0.5-in. diameter, contact angle 90°, T = 10.5 g, temperature 70°F, RH 50%. ○ Nopcostat PE-131 (η = 5,830 cP), ● Nopcostat 1296 (η = 22.4 cP). Reprinted from ref. 43, p. 345, courtesy of Textile Research Journal.

surface as illustrated in Table 6 [39]. The charge generated increased in the following order: bright, semidull, to dull nylon. Thus, an increase in surface roughness resulted in an increase in charge generation and is more pronounced when the yarn passes a rough guide surface than when passing a smooth one.

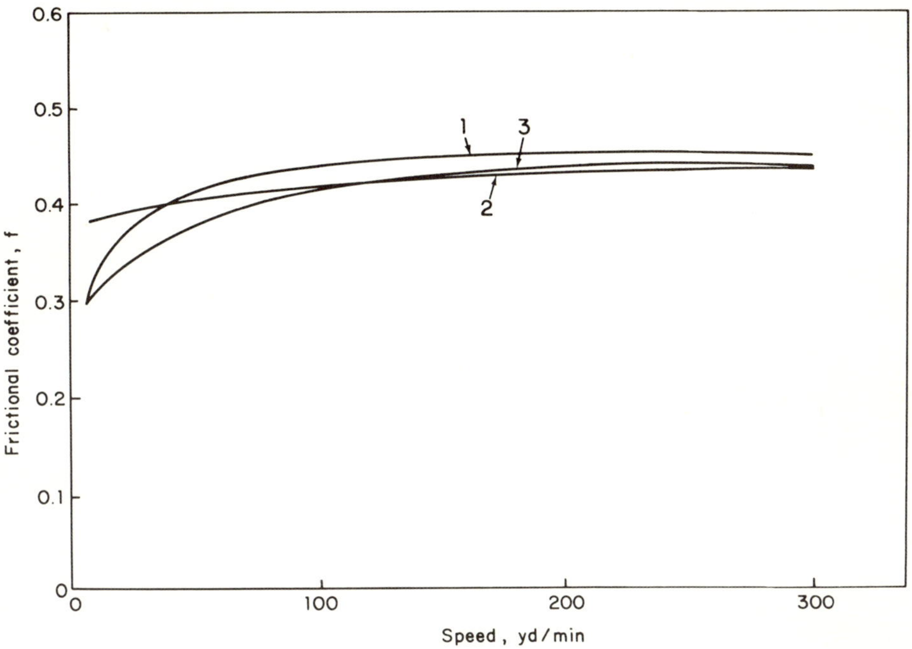

FIG. 43. Effect of lubricant viscosity on friction of bright polypropylene 210/35 (Hercules). Test conditions: Rothschild F meter, finish is 1% wof, chrome pin RMS-60, 0.5-in. diameter, contact angle 360°, T = 10.5 g, temperature 70° F, RH 50%. (1) Nopcostat PE-131 (η = 5,830 cP), (2) Nopcostat 1296 (η = 22.4 cP), (3) n-butylstearate (η = 9.05 cP). Reprinted from ref. 43, p. 346, courtesy of Textile Research Journal.

12. Glass Transitions

Park et al. [51] have demonstrated that plots of friction at the F/M interface vs. sliding speed pass through a maximum for multifilament yarns lubricated by copolyoxyalkylene monoethers. The close similarities between the frequency of maximum dynamic loss modulus and the speed of maximum friction in their molecular weight and temperature dependences support the postulate that the maximum friction is governed primarily by dynamic-loss characteristics of the polymeric lubricant; the relative position of the maximum in the friction vs. speed curves can be qualitatively predicted from the glass transition temperature of the polymeric lubricant.

An alternate explanation for the maxima observed in friction vs. sliding speed for multifilament yarns lubricated with a viscous phosphate ester has been given by Schick [52] as shown in Fig. 49. The

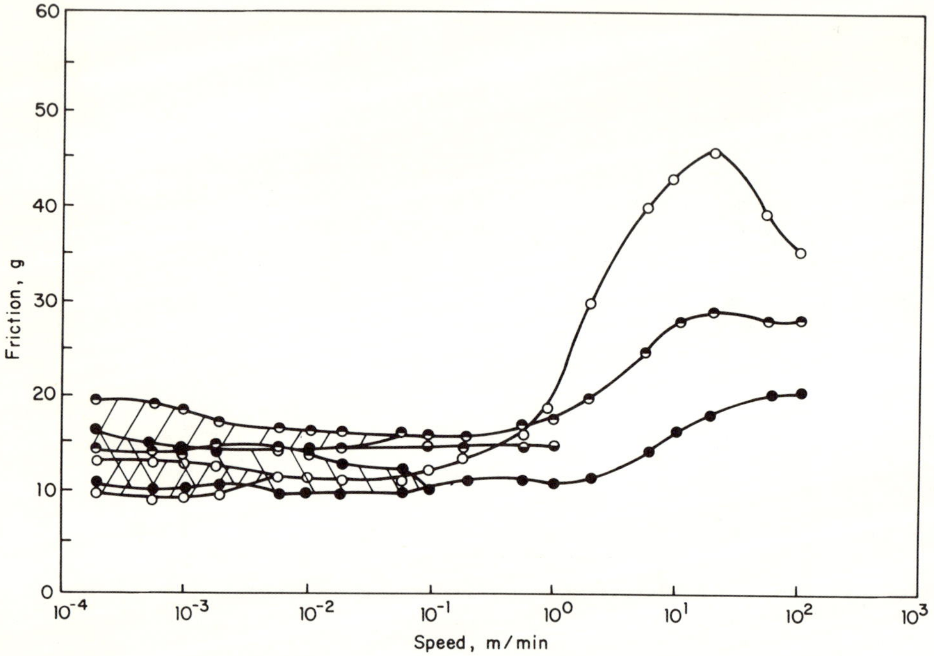

FIG. 44. Effect of lubricant viscosity on friction at the F/F interface. Test conditions: Chemstrand F meter, F/F $2\frac{1}{2}$ turns, 2.5 in., 5 pulleys, yarn is bright polyester 220/50 (duPont), finish is 1% wof, T = 11 g, temperature 70° F, RH 50%. ○ Nopcostat PE-131 (η = 5,830 cP), ◓ Nopco texturizing finish (η = 34.9 cP), ◒ n-butylstearate (η = 9.05 cP), ● Nopcostat 1296 (η = 22.4 cP). Reprinted from ref. 43, p. 346, courtesy of Textile Research Journal.

maximum is attributed to heat effects that result in a decrease in viscosity beyond the maximum. In addition morphological changes, i.e., shifts of the glass transition temperature, may be considered a contributing factor.

13. Lubricant Concentration

Olsen [19] has shown that a curve similar to the hypothetical curve of Fig. 1 can be developed by varying the quantity of lubricant applied from 0 to 2% as presented in Fig. 48. The minimum point in the curve coincides approximately with the concentration of lubricant theoretically required to form a continuous monomolecular film. The frictional coefficient f for clean, dry yarn is 0.6 and decreases to about 0.4 at 0.15% lubricant concentration. Friction increases again and

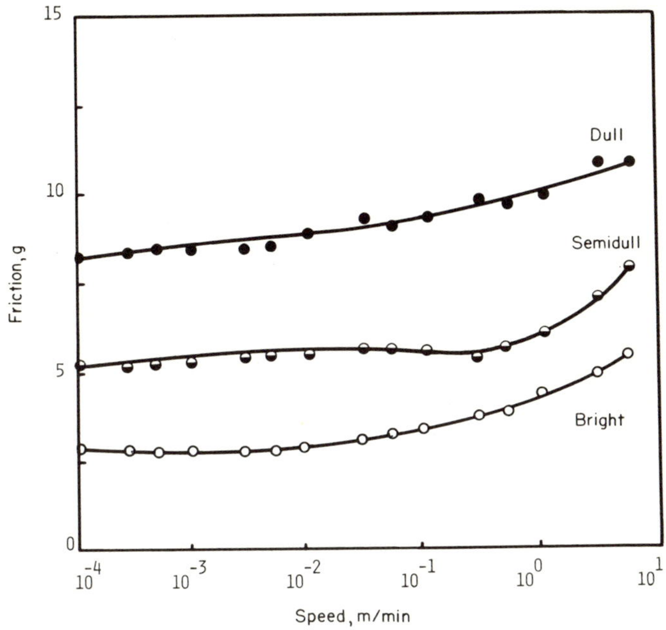

FIG. 45. Effect of fiber luster on friction. Test conditions: Chemstrand F meter, yarn is nylon 66 200/34 (duPont), finish is 1% wof n-butylstearate, chrome pin RMS-4, 0.5-in. diameter, contact angle 90°, T = 10 g, temperature 70° F, RH 50%. Reprinted from ref. 39, p. 759, courtesy of Textile Research Journal.

reaches a plateau region at 0.7% lubricant concentration. It is concluded that a lubricant concentration corresponding to the practical use level of 0.7% is required to obtain meaningful results in frictional tests [43].

IV. FRICTION AND LUBRICATION IN MULTICOMPONENT SYSTEMS

A. Model Systems

Formulations of fiber lubricants are in general multicomponent systems comprising at least one lubricant and one antistatic component. Results of a study of the frictional behavior of a two-component model system consisting of a volatile lubricant (A) and a nonvolatile antistat (B) are given below [52]. Component (A) is a polyoxyethylene fatty acid, whereas the antistatic component (B) is a phosphate ester of a

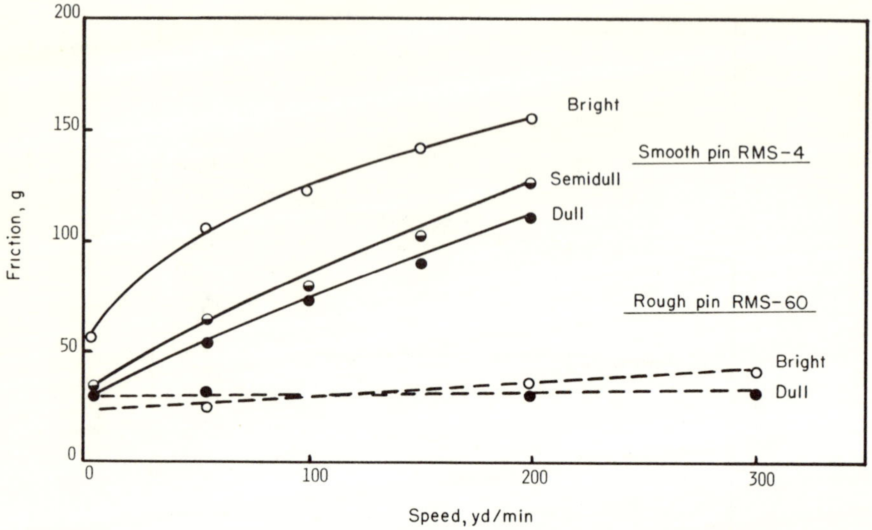

FIG. 46. Effect of fiber luster on friction. Test conditions: Rothschild F meter, yarn is nylon 66 200/34 (duPont), finish is 1% wof n-butylstearate, chrome pin 0.5-in. diameter, contact angle 360°, T = 10 g, temperature 70°F, RH 50%. Reprinted from ref. 39, p. 759, courtesy of Textile Research Journal.

polyoxyethylene aryl phenol. Thermal stabilities, i.e., volatilities, were determined on 2-g samples contained in an aluminum dish in a forced-air draft oven at 200°C for 1 hr and are expressed as percent weight loss. These amounted to 70.6% for component (A) and to 2.4% for component (B).

Frictional properties of this two-component system over the entire concentration range for nylon multifilament yarns at the F/M interface are illustrated in Figs. 49 and 50 [52]. The plots cover the entire speed range, which encompasses the boundary, semiboundary, and hydrodynamic regions. Figure 49 shows a family of curves for the speed range 10^{-4} to 50 m/min, which starts to diverge in the semiboundary region exhibiting the lowest friction for the lowest component (A), i.e., $\eta_{25.0°C}$ = 35 cP, and the highest friction for the high viscosity component (B), i.e., $\eta_{25.0°C}$ = 5,830 cP. The mixtures, as anticipated, yield intermediate values. These data support the findings of Section 10 that friction is independent of viscosity in the boundary region, but highly dependent on viscosity in the hydrodynamic region. A maximum was observed in this friction vs. speed plot for the high-viscosity component (B) at a speed of 10 m/min. The subsequent decrease in friction is attributed to a marked decrease in viscosity caused by the

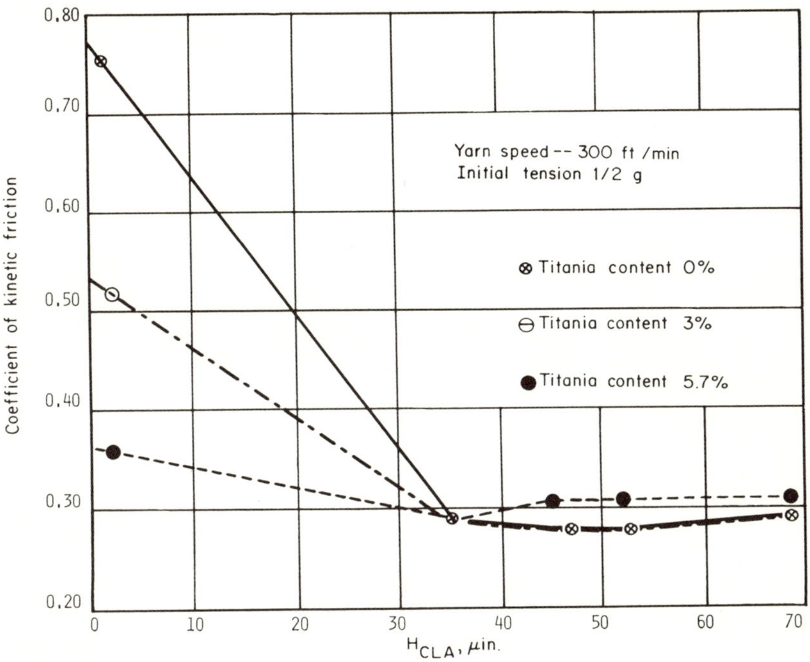

FIG. 47. Friction of nylon 30/10 on alumina pins as a function of titania. Reprinted from ref. 41, p. 107, courtesy of the Journal of the Textile Institute.

large heat developed when this highly viscous lubricant passes the smooth guide surface. In addition softening of the polymer with consequent morphological changes must be considered a contributing factor. The corresponding data for the higher speed region given in Fig. 50 indicate the same wide discrepancy in friction between the low viscosity component (A) and the high viscosity component (B), or in other words friction is highly dependent on viscosity in the hydrodynamic region.

A study of the temperature dependence of the frictional properties of this two-component system on nylon and polyester over the entire concentration range is illustrated in Fig. 51 [52]. All friction tests were carried out at a comparable speed of 55 yd/min using a heated chrome pin on the Rothschild F meter. Let us first examine the nylon plots of frictional coefficient vs. composition. With no exception, friction decreases when raising the temperature from 72 to 300°F. This phenomenon is attributed to a decrease in lubricant viscosity with increasing temperature. On further increase of the temperature, the friction of component (B) and 75%/25% mixture decreased, whereas that of the 50%/50% and 25%/75% mixtures and component (A) increased

TABLE 6

Effect of Fiber Surface Roughness on Charge Generation[a]

Luster	Charge generation			
	Smooth pin, RMS-4		Rough pin, RMS-60	
	Volts	Yards	Volts	Yards
Bright	-200	50	+250	50
	0	50	+240	50
	+500	50	+400	50
	+750	200	+1,000	200
Semidull	+1,500	50	+500	50
	+1,500	50	+700	50
	+1,250	50	+900	50
	+2,250	200	+3,500	200
Dull	+1,250	50	+4,000	50
	+1,250	50	+4,000	50
	+1,250	50	+4,000	50
	+2,250	200	Sparks	200

[a]Reprinted from ref. 39, p. 760, courtesy of Textile Research Journal. Test conditions: yarn was nylon 66 200/34 (duPont), finish 1% wof n-butylstearate, speed 300 yd/min, chrome pins 0.5-in. diameter, pre-tension 10 g, temperature 70°F, RH 50%.

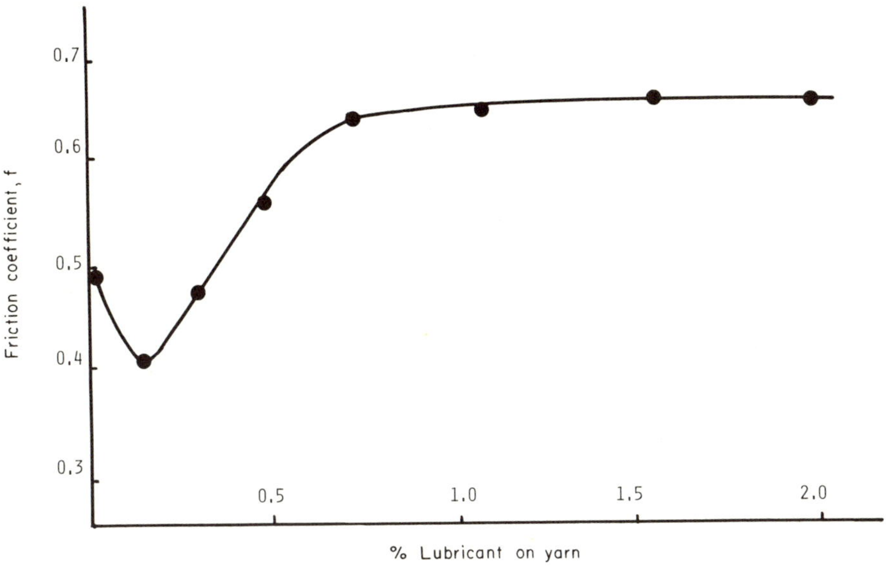

FIG. 48. Friction vs. percent lubricant on yarn. Test conditions: dull nylon 40/13 (duPont), chrome pin 5AA μin., speed 250 yd/min, T = 10 g, temperature 22°C, RH 65%. Reprinted from ref. 19, p. 36, courtesy of Textile Research Journal.

slightly. The behavior of component (B) and the 75%/25% mixture is attributed again to a decrease in lubricant viscosity with increasing temperature. In contrast, with the 50%/50%, 25%/75% mixtures and component (A) the decrease in viscosity is compensated by volatilization of the lubricant. It is concluded that the better heat stability of component (B) is reflected in these frictional measurements. In the corresponding polyester plots, friction decreases on raising the temperature from 72 to 300°F. Likewise, this phenomenon is attributed to a decrease in lubricant viscosity with increasing temperature. On further increase of temperature, friction of component (B) decreased, whereas that of mixtures and component (A) increased. In line with the explanation given for the nylon plots, the frictional behavior of component (B) is attributed to a decrease in lubricant viscosity, whereas with the mixtures and component (A), the decrease in viscosity has been overshadowed by the loss of the volatile lubricant. Thus, the better heat stability of component (B) is reflected again in these frictional measurements.

B. Formulations

The lubricant generally referred to as the spin finish is applied in the spinning or fiber-forming stage [53]. Its presence is required to

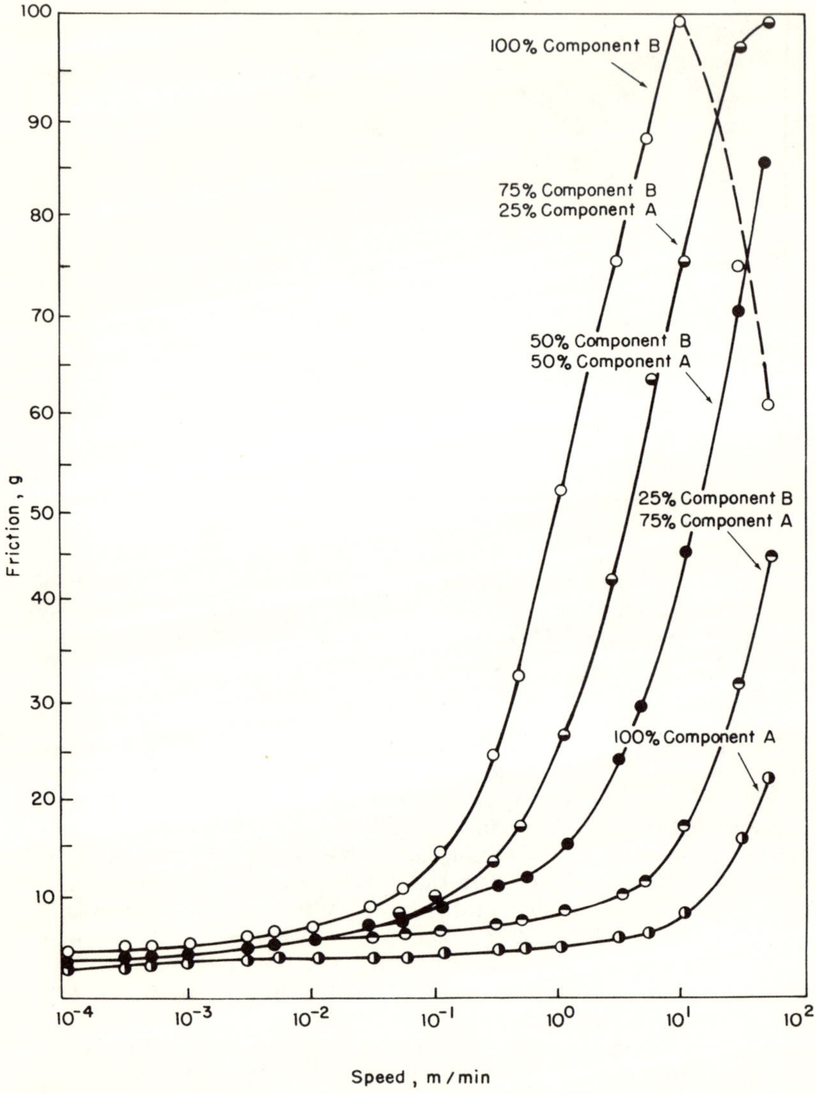

FIG. 49. Friction of two-component system. Test conditions: Chemstrand F meter, yarn is bright nylon 200/34 (duPont), finish is 1% wof, chrome pin RMS-4, 0.5-in. diameter, contact angle 90°, T = 10 g, temperature 70°F, RH 50%. Reprinted from ref. 52, p. 200, courtesy of <u>Textile Research Journal.</u>

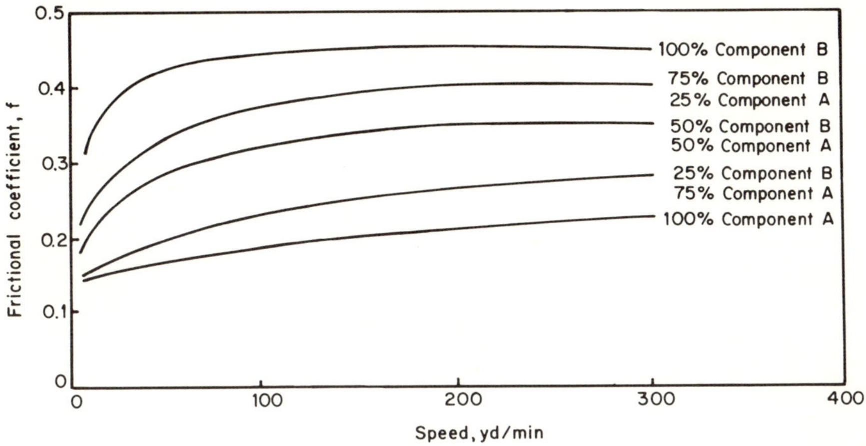

FIG. 50. Friction of two-component system. Test conditions: Rothschild F meter, yarn is bright nylon 200/34 (duPont), finish is 1% wof, chrome pin RMS-60, 0.5-in. diameter, contact angle 360°, T = 10 g, temperature 70°F, RH 50%. Reprinted from ref. 52, p. 200, courtesy of Textile Research Journal.

prevent damage to the filaments from their initial formation throughout the various steps of processing until the final product is made. Spin finishes are applied either neat or predominantly from a 10-25% aqueous emulsion by a lube roll or kiss roll application at a pick-up of 5-10% [54]. The latter method assures a more uniform application. Most spin finishes can simply be described as consisting of three major components, such as 40-60 parts lubricant, 5-20 parts antistat, and 30-50 parts emulsifier [55]. It is beyond the scope of this chapter, dealing with the fundamental aspects of friction and lubrication of synthetic fibers, to discuss the technology of spin-finish formulations and, therefore, the reader is referred to the copious patent literature on this subject.

V. CONCLUSIONS

It is concluded on the basis of all the experimental variables cited above that distinct differences in the frictional behavior of synthetic fibers exist under boundary, semiboundary, and hydrodynamic conditions. This division is based on the sharp transitions observed in the friction vs. yarn-speed plots. In these plots boundary friction is ascribed to the speed range from 10^{-4} to 10^{-1} m/min, semiboundary

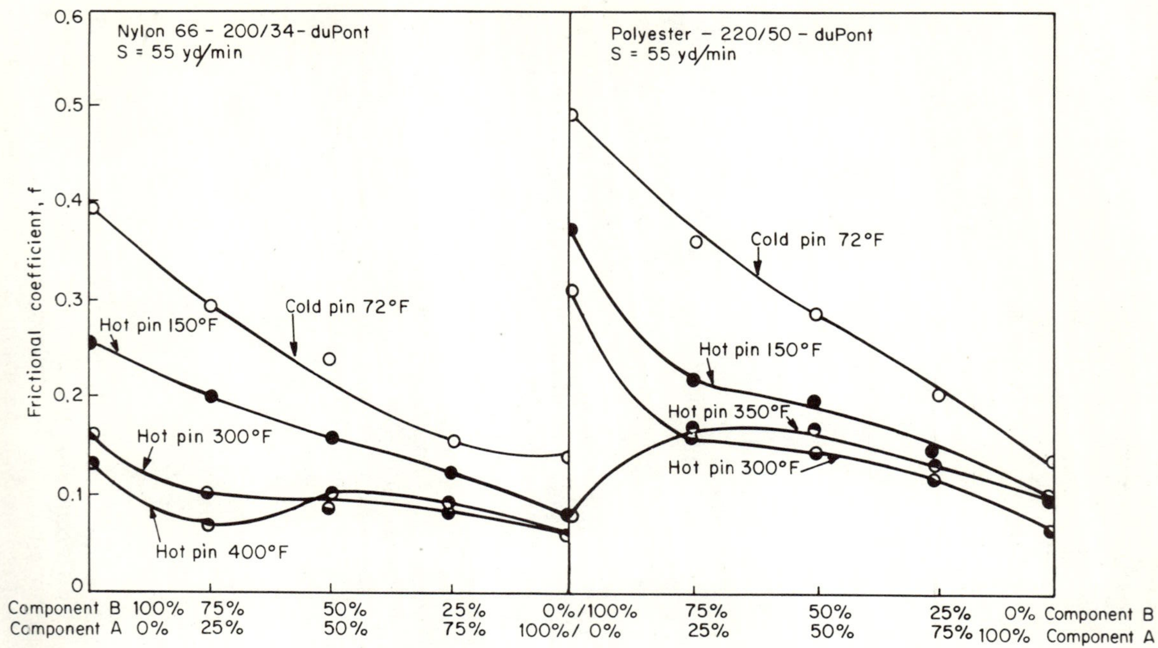

FIG. 51. Effect of temperature on friction of two-component system. Test conditions: Rothschild F meter, finish is 1% wof, chrome pin RMS-55 (heated), 0.5-in. diameter, contact angle 360°, T = 0.05 g/denier, temperature 70° F, RH 50%. Reprinted from ref. 52, p. 202, courtesy of Textile Research Journal.

friction to the speed range from 10^{-1} to 5.5 m/min, and hydrodynamic friction to speeds above 5.5 m/min. The data support the contention that intermittent contacts between fiber and substrate exist in the boundary region, whereas a continuous fluid film separates fiber from substrate in the hydrodynamic region. The dominant factors in boundary lubrication are the wetting and geometry of the sliding surfaces, whereas in hydrodynamic lubrication the viscosity of the lubricant and the wetting of the sliding surfaces are of equal importance. Therefore, it is recommended that yarn friction of single components should be tested in the low-speed region, whereas yarn friction of formulations, i.e., multicomponent systems, be tested in the high-speed region [25,43]. No significant difference is apparent between the frictional behavior at the F/M and F/F interfaces.

REFERENCES

1. F. P. Bowden and D. Tabor, The Friction and Lubrication of Solids, Oxford University Press, London, 1954.
2. F. P. Bowden and D. Tabor, The Friction and Lubrication of Solids, Part II, Oxford University Press, London, 1964.
3. E. Rabinowicz, Friction and Wear of Materials, Wiley, New York, 1965.
4. R. C. Bowers and C. M. Murphy, NRL Report 6466, Naval Research Laboratory, Washington, D.C., 1967.
5. D. Tabor, in Surface and Colloid Science, Vol. 5, E. Matijevic, ed., Wiley-Interscience, New York, 1972, pp. 245-312.
6. D. Tabor, Wear, 1, 5 (1957).
7. H. G. Howell, K. W. Mieszkis, and D. Tabor, Friction in Textiles, Butterworths, London, 1959.
8. G. Amontons, Memoires de L'Academie Royale, 1699, p. 206.
9. J. A. Greenwood and D. Tabor, Proc. Phys. Soc. (London), B68, 609 (1955).
10. M. W. Pascoe and D. Tabor, Proc. Roy. Soc. (London), A235, 210 (1956).
11. T. Fort, Jr., J. Phys. Chem., 66, 1136 (1962).
12. J. A. Chapman, M. W. Pascoe, and D. Tabor, J. Textile Inst., 46, P-3 (1955).
13. A. J. G. Allan, J. Polymer Sci., 24, 461 (1957).
14. R. C. Bowers, W. C. Clinton, and W. A. Zisman, Lubrication Eng., 9, 204 (1953).
15. R. C. Bowers, W. C. Clinton, and W. A. Zisman, Ind. Eng. Chem., 46, 2416 (1954).
16. W. W. Hansen and D. Tabor, Textile Res. J., 27, 300 (1957).
17. C. Schlatter, R. A. Olney, and B. N. Baer, Textile Res. J., 29, 200 (1959).
18. C. Schlatter and H. J. Demas, Textile Res. J., 32, 87 (1962).

19. J. S. Olsen, Textile Res. J., 39, 31 (1969).
20. T. Fort Jr. and J. S. Olsen, Textile Res. J., 31, 1007 (1961).
21. W. Wegener and B. Schuler, Ztschr. f. d. Ges. Textilind., 4, 250 (1964).
22. W. Wegener and B. Schuler, Ztschr. f. d. Ges. Textilind., 4, 362 (1964).
23. W. Wegener and B. Schuler, Ztschr. f. d. Ges. Textilind., 4, 458 (1964).
24. H. L. Roeder, J. Textile Inst., 44, T247 (1953).
25. M. J. Schick, Textile Res. J., 43, 103 (1973).
26. D. G. Lyne, J. Textile Inst., 46, 112 (1955).
27. L. W. Davis, U. S. Pat. 2,285,255 (1942).
28. C. Schlatter, U. S. Pat. 3,209,589 (1965).
29. H. G. Howell, K. W. Mieszkis, and D. Tabor, Friction in Textiles, Butterworths, London, 1959, p. 186.
30. M. J. Hammersley, J. Textile Inst., 64, 108 (1973).
31. B. G. Schuler, Melliand Textilberichte, 53, 1357 (1972).
32. Atlab Finish Applicator. Atlas Chemical Industries, Inc. Manufactured by Precision Machine & Development Corp., Newcastle, Del.
33. Rothschild Company, Zürich, Switzerland.
34. H. G. Howell, K. W. Mieszkis, and D. Tabor, Friction in Textiles, Butterworths, London, 1959, p. 47.
35. K. R. Makinson, J. Textile Inst., 61, 465 (1970).
36. F. H. Steiger, Textile Res. J., 28, 721 (1958).
37. D. A. Zaukelies, Textile Res. J., 29, 794 (1959).
38. W. Loebel, Faserforschung und Textiltechnik, 19, 110 (1968).
39. M. J. Schick, Textile Res. J., 44, 758 (1974).
40. M. J. Schick, Textile Res. J., 43, 254 (1973).
41. M. E. Baird and K. W. Mieszkis, J. Textile Inst., 46, P101 (1955).
41a. Reason, The Measurement of Surface Texture, Cleaver-Hume Press, 1950.
42. W. A. Zisman, Advances in Chemistry Series, No. 43, Chap. 1, "Contact Angle, Wettability, and Adhesion," American Chemical Society, Washington, D.C., 1964.
43. M. J. Schick, Textile Res. J., 43, 342 (1973).
44. A. Zettlemoyer, J. Colloid and Interface Sci., 28, 343 (1969).
45. R. V. Dyba and B. Miller, Textile Res. J., 40, 884 (1970).
46. K. R. Makinson and D. Tabor, Proc. Royal Soc. (London), A281, 49 (1964).
47. C. M. Pooley and D. Tabor, Proc. Royal Soc. (London), A329, 251 (1972).
48. S. C. Scheier and W. J. Lyons, Textile Res. J., 34, 410 (1964).
49. T. H. Grindstaff, Textile Res. J., 39, 958 (1969).
50. R. N. Wenzel, J. Phys. Chem., 53, 1466 (1949).
51. K. Park, C. G. Seefried, Jr., and G. M. Bryant, Textile Res. J., 44, 692 (1974).

52. M. J. Schick, Textile Res. J., 43, 198 (1973).
53. H. P. Stout, Wear, 15, 149 (1970).
54. J. E. Obetz, Fiber Producers Buyers Guide (April 1974), 123-127. Textile Asia, (Nov. 1973), 20-24.
55. J. P. Redston, W. F. Bernholz, and C. Schlatter, Textile Res. J., 43, 325 (1973).

Chapter 2

SURFACE PROPERTIES OF COTTON FIBERS

Kermit E. Duckett

Fiber Research Laboratory
University of Tennessee
Knoxville, Tennessee

I. INTRODUCTION

Surface properties of cotton play a very important role in textile technology. They, along with the easily measurable physical properties of length, strength, and fineness, are important to processing behavior. In addition to contributing to yarn uniformity and mill efficiency, surface characteristics influence dye accessibility and reactivity rate as well as fabric feel, or handle. Yet, because of the complexities involved, no single functional parameter is representative of all surface phenomena. The surface characterization that best describes spinning behavior may be completely unsuitable for describing fabric handle or dyeing behavior. Likewise, the primary determinant of luster may be completely unsuitable for describing electrostatic charge retention. However, all are legitimate representations of surface properties.

Many descriptions and methods of observation can be found and it is the purpose of this chapter to review and condense some of this information. In so far as is possible, a current broad account of the literature is presented. It is not intended to be complete. But it is intended to acquaint the reader with some of the characteristics of cotton fiber surfaces, ranging from the chemical composition of the cuticle and the fibrillar structure of the supporting walls to the more gross morphological features involving convolutions and frictional behavior. Wherever possible, processing and fabric behavior will be related to descriptions of specific surface phenomena.

II. FIBER STRUCTURE

The cross section of a cotton fiber consists of a number of concentric layers. They can be broadly classified as belonging to either the outermost region called the primary wall or the inner region designated as the secondary wall. However, this description is too general for surface-property discussion. The surface properties of cotton fibers are largely the result of a cellulosic structure plus a waxy material covering and impregnating this cellulosic structure. This surface layer constitutes approximately 5% of the total fiber weight. It is this outer region, in contact with the physical environment, that largely determines the processing behavior of the fiber. Nevertheless, the body of the fiber has a modifying effect on the periphery of the cotton hair and cannot be dismissed. Therefore, some description of fiber architecture is requisite for adequate interpretation of surface description and behavior.

A. General Morphology

A tubular growth, which encloses the living protoplast, is initially formed from the epidermal cells of the cotton seeds [1]. This initial layer, formed of both cellulosic and noncellulosic materials, is called the primary wall [2]. It determines, to a large degree, the length and circumference of the mature cotton fiber. Though the primary wall is generally considered to be the outgrowth of a single cell [3], there may be exceptions [4]. Secondary wall growth follows only after the completion of the primary wall extension and this secondary cellulose deposition determines the major wall thickness of the fiber.

The fiber morphology can be concisely summarized as follows [5]. Beginning with the outermost layer and continuing toward the center of the fiber, there are six distinct regions, which are illustrated in Fig. 1. The cuticle exists as the separable outer boundary and consists of a layer of wax and pectin materials that appears to be structureless. The surface shows deep rill-like folds. Below this layer is the primary wall, which consists primarily of a network of cellulose fibrils. These fibrils are long, thin crystalline structures of

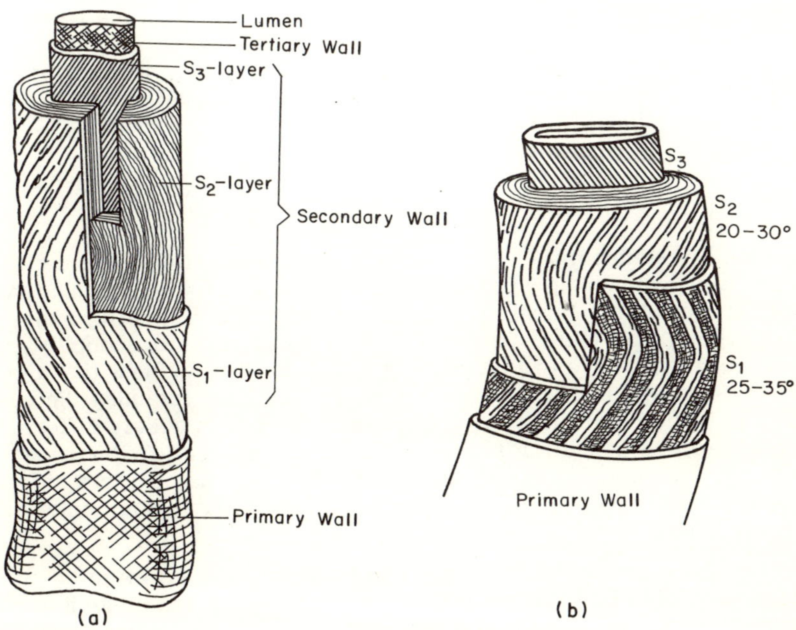

FIG. 1. Schematic drawings of the cotton fiber. (a) The layered structure, (b) the fibrillar orientation of the secondary wall.

cellulose molecules and the diameters of the crystallites are approximately 100 Å. The primary wall, together with the cuticle, constitutes the initial membrane of the single-celled cotton hair. Both of these are of vital concern to surface behavior and will be dealt with in greater detail throughout.

Below the primary wall is a thin (less than 0.1 μm thick) layer of the secondary wall (S1 layer). This layer is built up of closely packed parallel-lying fibrils having a spiral winding at 25-35° to the longitudinal axis of the fiber. There are no visibly porous zones in this region like those that may exist in other regions. Just beneath this thin layer is an additional secondary wall (S2 layer) that contains the majority of the cellulose in the fiber. This densely packed region of fibrils is several micrometers thick. The fibrils spiral about the axis of the fiber at angles up to about 25°. The spiral direction periodically reverses itself in this layer. Normal growth conditions generally produce variations in thickness of the concentric, diurnal layers. Continuing, a third layer (S3) of the secondary wall is distinguished. The thickness is, like that of the S1 layer, less than 0.1 μm. The S3 layer is similarly fibrous in character and presumably exists as a spiraled texture impregnated with noncellulosic substances. Small-angle X-ray scattering shows the primary and secondary walls to be threaded by a system of submicroscopic pores or capillaries [6]. In the innermost region, called the lumen, are deposited the remains of the cell contents. In general, this description of the breakdown of the fiber into layers [5] conforms to the fiber morphology visualized by many other investigators [7-13].

B. Cuticle and Chemical Composition

From the standpoint of chemical composition, cellulose is the predominant component. There are, in addition to the waxy substances of the cuticle, other noncellulosics in the form of ash, pectic, and proteinaceous materials. These are distributed inhomogeneously throughout the fiber, as illustrated in Table 1. It is evident that the chemicals comprising the primary wall (which here includes the cuticle) are found throughout the complete fiber, but in widely different percentages; the noncellulosic materials are confined for the most part to the lumen and primary wall [14]. The cuticle consists of wax and pectic substances, in about equal amounts. Nitrogenous material is also present.

The wax content appears to vary in quantity, as well as in chemical structure. There is apparently no relation between wax content and geographical origin. Nor is there a significant change in wax content with processing. Early work [15,16] on wax content shows the percentage, in terms of total fiber content, to vary approximately between 0.3 and 0.8%. The waxes are generally classed as alcohol or organic solvent extracted materials. Chemically, the composition of cotton

TABLE 1

Chemical Composition of Typical Cotton Fiber and Primary Wall [14]

Constituent	Percent of dry weight	
	Fiber	Primary wall
Cellulose	94.0	54
Protein	1.3	14
Pectic substance	1.2	9
Wax (alcohol solubles)	0.6	8
Ash	1.2	3
Other	-	4

waxes probably contains even-numbered-carbon primary alcohols ($C_nH_{2n+1}OH$) and normal fatty acids ($C_nH_{2n}O_2$) from C_{24} to C_{34}. Other substances such as glycerol, resinous materials, amyrin ($C_{30}H_{50}O$), and hydrocarbons, including probably heptacosane, have been reported in the wax content [17]. Amin and Truter [18] distinguished at least 16 components in lipid extractions: a complex mixture of waxes, fats, and resins. Included were saturated and unsaturated hydrocarbons, waxy esters, glycerides, long-chain alcohols, sterols, glycols, and acids. The most abundant group was that containing the aliphatic alcohols. To make matters additionally complex, the surface lipids undergo oxidation, generating many compounds containing hydroxyl and carbonyl groups. Polymerization may result from extensive auto-oxidation to form the resin fraction. There appears to be no significant interspecies differences, though the lipid content from the species G. barbadense tends to contain the lowest proportions of hydrocarbon.

A typical, mature cotton fiber contains between 0.6 and 1.2% pectic substances and most of these are localized in the cuticle and primary wall. They can more specifically be subdivided and classed as protopectins, pectins, and pectin acids. Though there are differences in the degree of extraction of these substances in neutral water and differences in the degree to which the acid groups are esterified with methanol, the name pectin is often used for the whole complex of pectin substances. All contain, as the primary constituent, chain molecules composed of d-galacturonic acid. They often occur as insoluble calcium, magnesium, or iron salts as well as soluble sodium, potassium, and ammonium salts [1]. Pectins are demonstrated by staining with ruthenium red and can be almost completely removed by kier boiling. On the basis of a limited number of samples there are indications that wax and pectate contents are related to relative surface areas of cottons [19].

There are additional constituents in the cotton fiber such as phosphorous compounds, organic salts other than pectic acids, vitamins, and pigments. As an example, the ash constituents of one variety, with their percentages of total ash content, are given in Table 2 [20]. These values are not constant for all varieties, however. Egyptian cottons consistently contain about twice as much phosphorus, as P_2O_5, as is found in American cottons [21]. The environment of the plant has a definite effect on the phosphorous content, though there are greater differences between Egyptian and American types than among similar varieties grown in different environments. Also, a high phosphorous content is apparently a property of the shorter cotton hairs. But on the average, the ash of cotton accounts for approximately 1.2% of the total fiber weight and is usually negligible after scouring and bleaching. Removal is an important step in processing because it is recognized that the calcium and magnesium mineral content of cotton fabric may affect dyeing [22]. The removal of these substances is easily accomplished by treating with dilute acetic acid, i.e., converting

TABLE 2

Percentage of Various Substances in the Total Ash Content of an Unknown Cotton Variety [20]

Ash constituent	%
K_2O	48.16
CaO	7.19
MgO	9.04
Na_2O	0.25
Fe_2O_3	0.20
Al_2O_3	0.31
SiO_2	0.76
SO_3	8.25
P_2O_5	4.22
Cl	1.22
CO_2	20.25
Total	99.85

calcium and magnesium to soluble acetates, and washing. Finally, there is a possibility that alkalinity of ash is related to fiber fineness [19].

The cotton fiber also contains approximately 1.3% protein and part of this is found in the primary wall in substantial amounts [14]. The rest is deposited almost totally in the lumen as desiccated protoplasm. Chemical identification is usually made through nitrogen-content determinations. From a physical standpoint, protein appears to have only a minor effect on dyeing properties but the high nitrogen content of finished fabrics due to poor scouring is said to favor damage by bacteria and molds [23]. Yellowing has been ascribed to proteins, but is probably due to the presence of oxycellulose arising from overbleaching [24]. As a side note, raw cottons do not differ as widely in nitrogen as in phosphorous content [25]. Egyptian cottons, however, contain appreciably more nitrogen than American varieties.

In connection with processing, significant difference is revealed between combed waste and combed sliver. This is presumably due to differences in the nitrogen content of immature and mature cotton. Nitrogen content tends to increase in amount with increasing fineness of the fiber. Since fineness and wall thickness are generally inversely related [26], there is further corroboration in that Guthrie et al. [20] find that the pectin, nitrogen, and sugar contents fall abruptly during a period of 10 to 15 days into the secondary-wall deposition period.

The primary wall, even though it consists primarily of cellulose, is impregnated with waxes, fatty materials, and pectinaceous materials found in the cuticle. As a result of this encrustation some investigators consider this layer and the cuticle a single layer. However, by the use of cellulose solvents, a distinct uniform waxy layer can be obtained [27], a finding that confirms the existence of a distinct separation between the two regions. A complete separation of the noncellulosic cuticle can be accomplished by a treatment with 30% H_2O_2 at moderate temperature or by an oxalate treatment or careful pressureless cooking in dilute soda lye [5].

III. MICROSCOPY

Though electron microscopy has contributed significantly to the structural and morphological understanding of the fiber, the light microscope was its predecessor by many years. However, the light microscope lacks the resolution and depth of field at high magnification. The resolution, largely a result of diffraction effects [28,29] is only approximately 1/100 of that of a transmission electron microscope, but the wide availability, relatively low cost, and ease of operation make the light microscope a most useful tool. Polarization, phase contrast, and interference accessories extend its capabilities multifold [30-35]. For example, all fibers are more or less doubly refractive, i.e., each fiber has more than one index of refraction. The fact that a fiber is a tubular outgrowth implies anisotropy in density and molecular orientation in at least two directions. Polarization and phase-contrast effects result from this anisotropy. Further extension of the light microscopic is obtained through the combined effects of fluorescence dichromism and polarization [34].

A. Optical Microscopy

The systematic approach to the study of the cotton fiber by application of the light microscope began with Balls [35,9,10] in the 1910's and continued strongly until the electron microscope became the primary source of fiber-surface information. Many early results remained questionable until substantiated by more modern techniques. From studies of the growing cotton plant in Egypt, Balls concluded that the single fiber was probably composed of concentric, diurnal growth layers, but the lack of sufficient resolution of the light microscope made observation impossible without some previous treatment. Axial concentric layering became visible in the swollen walls of the cotton fibers treated with CS_2 and NaOH. Transverse sectioning, better swelling techniques, and use of naphthamine as a stain supported the earlier findings on growth rings [11]. Balls concluded that there is only one, thin primary layer while the hair is growing in length [9].

Staining has contributed markedly to the contrast necessary for obtaining useful fiber information using the light microscope. Pectic content in the isolated primary wall is contrasted or rejected by staining with aqueous solutions of Nile blue sulfate, methylene blue, or ruthenium red. The reaction involved is between the staining compound and the carboxyl groups present in the pectin [36]. The proteinaceous constituents within the lumen stain for the same reason. But an iodine sulfuric acid technique, as well as the use of Simon's stain, gives color differences between the primary wall and lumen contents [37], which is useful on whole fibers. Wax complexes of the cuticle can be colored with such fat stains as Sudan black or oil red, but satisfactory intensity is often not achieved until the fibers have been treated in boiling ethyl alcohol.

Anderson and Kerr [3] have reviewed and extended the microscopic and staining work on the fiber surface. They conclude that fiber growth begins near the day of flowering and the primary wall of the young fiber possesses a coherent skeleton of cellulose from the first day of its appearance. This early-stage cellulose, however, does not react to the same microchemical tests that reveal mature cellulose. The absence of a blue color with chlorozinc iodide or with potassium triiodide and 70% sulfuric acid, the insolubility of the wall in fresh cuprammonia, and the apparent isotropy of the wall when viewed between crossed Nicols (polarizing microscopes) apparently led several investigators to conclude that the primary wall of the young fiber is free of cellulose. The shielding of the cellulose by larger amounts of wax and pectic substances was the probable cause of much confusion.

When young fibers are freed from pectic material, the coherent cellulose skeleton remains. This thin wall is doubly refractive in Canada balsam as well as in water and alcohol. In addition, it is soluble in fresh cuprammonia and gives characteristic reactions with chlorozinc iodide and with potassium triiodide and 70% sulfuric acid.

By staining young fibers with Congo red or chlorozinc iodide, the cellulose threads in the primary wall are made visible for examination under crossed Nicols [8]. Balls [10] first applied this method and observed two spiralling thread groups in one wall approximately 70° to the long axis of the fiber. A third layer of transverse cellulose strands is indicated as the microscope stage is rotated. Supporting visual evidence is made available by the position of the addition and subtraction colors produced when a gypsum plate Red I is inserted between crossed Nicols. Presence of the third layer is further suggested by a parallel extinction and by marked dichroism obtained when the long axis of the fiber is parallel to the polarizing Nicol [3].

Light microscopy has also been useful in studies on the secondary wall of the cotton fiber. At an early date, it was known that seconday-wall cellulose deposition gave a spiral structure. Balls, using polarized light, was able to observe regions in which the spiral reversed itself, i.e., points of reversal. He and Hancock [10,11] further noted a correlation between convolutions produced by the fiber upon drying out and the reversals of the spiralling cellulose fibrils in the secondary wall. This correlation implied a direct connection between internal structure and surface properties as represented by convolutions. Genetic and ordinary environmental influences do not affect the statistical peculiarities of the reversals. Nevertheless, the final mature length of the hair and the time taken in reaching that length, are related to the reversal distribution [12].

B. Electron Microscopy

With refined electron-microscopic techniques Roelofsen [38] has demonstrated that the outer region of the cellulosic primary wall contains layered fibrils roughly oriented in the axial direction. The appearance of the fibrils is illustrated in Fig. 2a. Often several fibrils are combined into bundles. Immediately beneath this, is another layer of fibrils that runs in an oblique direction, and still deeper are more fibrils having a predominantly transverse direction, i.e., perpendicular to the fiber axis. The innermost stratum is of considerable thickness and the fibrillar structure is more compact than on the

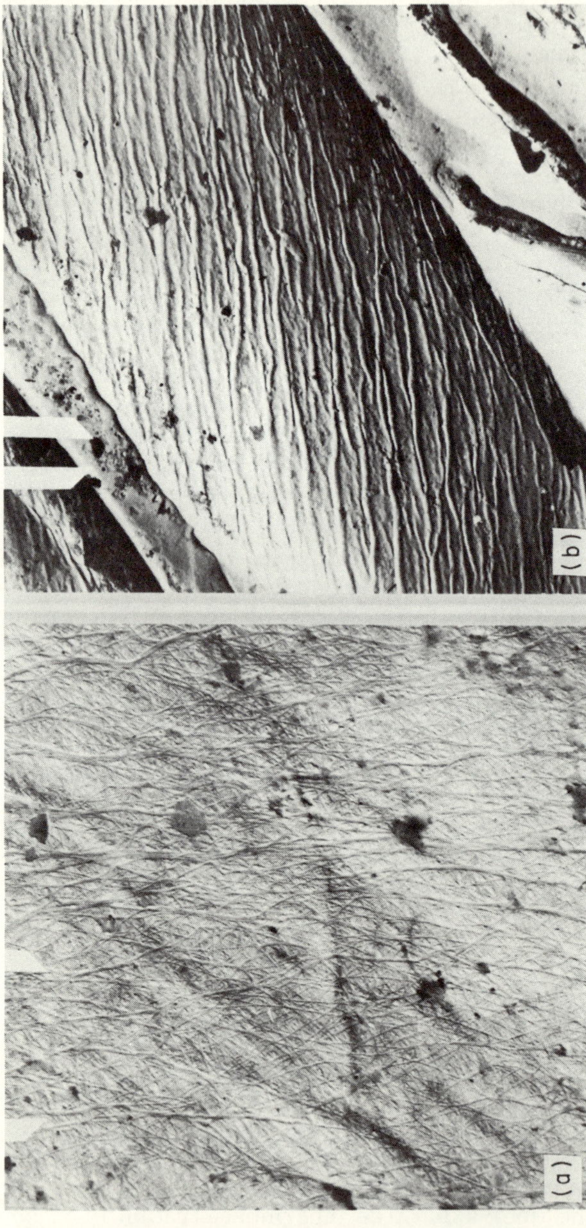

FIG. 2. Electron photomicrographs. (a) The primary wall, (b) the surface of a raw cotton fiber. (Courtesy of the U.S. Department of Agriculture.)

outside. The total thickness of the primary wall is near 0.1 μm. When viewed as a continuous cellulosic material throughout the primary wall thickness, there is a gradual change from a loose, more or less axial structure in the outer regions of the growing fiber to a compact transverse structure on the inside. This arrangement is supported by the positive birefringence observed at very early growth and the negative birefringence observed at a later date. In addition, interfibrillar spaces appear and are thought to result from separation by drawing or pulling. There is no indication of fibril fracture as a result of the drawing, which has led to the idea that the peripheral fibrils are reoriented from an initial transverse deposit as the single cell extends. The result would be the observed gradual change in fibril orientation. Rearrangement would account for the open spaces giving rise to a matted appearance of the noninnerwoven structure. Noncellulosics fill these spaces in the mature cotton fiber.

Overlaying this matted structure are the cuticle waxes that act as interfiber lubricants and aid in processing behavior. The surface of this waxy layer similarly shows irregularities, the presence of which affects certain fiber properties, though processing behavior may not be one. Light- and electron-microscopic techniques are both widely used in describing these geometric factors [39-42]. Using surface replica techniques, electron-microscope examination reveals a system of roughly parallel ridges and grooves spiralling about the fiber axis at angles ranging generally between 20 and 30°. The average height, as well as distance between the ridges, is approximately 0.5 μm and many ridge lengths are 10 or more μm. Figure 2b illustrates this surface phenomena.

All native cotton fibers show similar but not matching ridge-and-groove systems. There are statistically significant differences in the average ridge separation between certain samples. The patterns appear to predominate in thin-walled fibers. Thick-walled fibers show a smoother surface. In all cases, however, the surface of fibers from unopened bolls is relatively smooth, but assumes the roughness of a typical fiber on first drying. In fact, primary walls appear larger when they are mechanically separated than when they are an integral part of the fiber [36]. This larger size may represent the initial size of the fiber before secondary-wall deposition or fiber desiccation upon boll opening.

The spiral pattern is determined by the spiral arrangement of the cellulosic fibrillar groups lying immediately beneath the primary wall. Observations using the polarizing light microscope indicate similar axial positions in secondary-wall fibrillar reversal locations and reversals in the spiral direction of the ridge-and-groove systems. Initial drying of the cotton fiber results in a marked decrease in the cross-sectional area and such shrinking brings about folding and wrinkling of the cell wall. Increasing the roughness of the surface boundary follows naturally and wetting of the dried fiber does not restore the original smoothness. The increase in cross-sectional area when once dried cotton is wet amounts to about 30%. This is considerably less than the decrease accompanying the initial drying of the fiber [43,44].

IV. CONVOLUTIONS

When a fiber is viewed through a microscope, the most obvious characteristic is its generally flat band structure with corkscrew-like twisting. These twists are designated as convolutions [3,45]. They frequently alternate in direction of rotation and vary in pitch. They are not present during initial growth, but are formed as a result of fiber desiccation after boll opening. The extent is limited in very coarse cottons but increases significantly in average commercial varieties. Very thin-walled cottons generally are lacking in convolutions. Normally, the convolution frequency varies between 3.9 and 6.5 per mm and the number of reversals per mm ranges between 1.0 and 1.7. These dimensions cause no problems with microscopic resolution though detection is difficult with the unaided eye.

Convolution generation is intimately related to internal structure and secondary-wall thickness. The specific molecular mechanism that induces fiber twisting is not completely known. Denham [46] attributes normal convolutions to spiral structure of the wall layers and finds a relation between convolutions and the ratio of the ribbon width to wall thickness. Other investigators explain convolutions as a result of the secondary-wall layers having opposed spirals [1]. As has already been stated, convolution frequency has been correlated with fibrillar reversal regions within the secondary walls. Consequently, they are predetermined gross morphological features of the fiber and their twist is imparted only during desiccation [13].

Convolutions can be measured by optical microscopic methods. When defined as a twist of the fiber through 180°, Meredith's formula

$$\theta = \arctan(\pi/2)\ (\overline{D/c})$$

can be used to quantitatively describe the geometrical behavior [47]. In the formula, θ is the convolution angle, $\overline{D/c}$ is the mean value of D/c, D is the maximum cotton fiber width, and c is the pitch of the convolution. Figure 3 schematically illustrates the corkscrew-like twisting of the convolution structure as well as the geometrical symbolism used in Meredith's formula. Figure 4 shows an actual cotton fiber with the desiccated induced twist. From this microscopic photograph, it is quite obvious that the convolution angle varies widely and a large number of determinations are necessary for good statistical

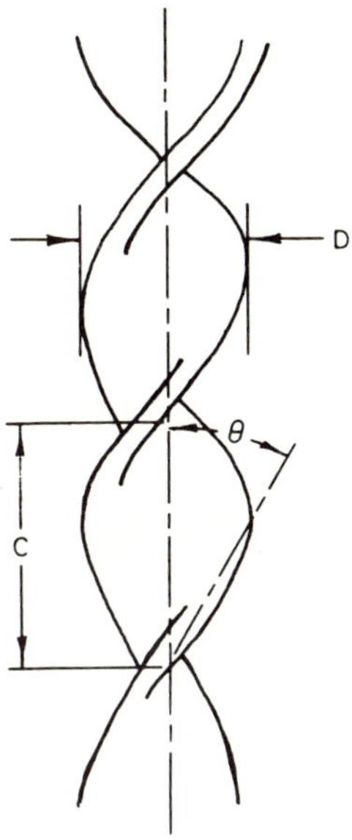

FIG. 3. Schematic diagram showing convolutions.

inference. There are more rapid techniques for obtaining a measure of convolution geometry and the results correlate well with the results obtained from Meredith's formula. One method simply requires the azimuthal monitoring of reflected light from the surface of a parallel array of tensioned fibers. An azimuthal angular determination is obtained from the anisotropic distribution of reflected light and is designated the 50% optical angle. This is closely analogous to the 40% X-ray angle determination method that measures the average angle of orientation of crystalline microfibrils away from the fiber axis direction [48,48a].

Table 3 lists the microscopically measured convolution angles (using Meredith's formula), the 40% X-ray angles, and the 50% optical angles obtained in 19 varieties of G. barbadense and G. hirsutum species. A linear regression analysis between the microscopically measured convolution angles and the 50% optical angles, given in Table 3, results in a high correlation coefficient ($r = 0.89$), which is not unexpected since both measurements describe surface geometry. Figure 5, on the other hand, illustrates a relation between the internal structure of the secondary wall and the gross surface properties characterized by the convolutions. Here the optical angle, and hence the convolution angle, is positively correlated with the 40% X-ray angle. However, there are differences between cotton species as implied by separate regression lines. The correlation coefficients

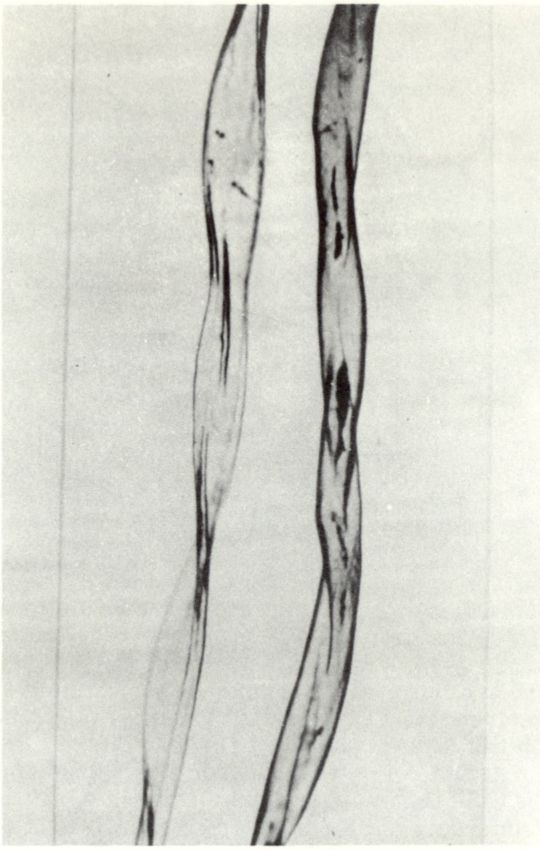

FIG. 4. Microscopic view of raw cotton fibers.

r are given for each species separately and both species combined. On the average, G. barbadense varieties contain fewer convolutions than the G. hirsutum varieties. Similar differences exist among other species [47].

Physical explanations for the differences in the regression lines of Fig. 5 are still subject to speculation. There may be actual secondary-wall differences in fibrillar orientation between species in the undried cotton hair and the degree of convolution generation in one species may require a different degree of fibrillar reorientation than that in another species. On the other hand, some investigators indicate that the fibrillar spiral angle within the secondary wall of an unconvoluted cotton fiber is constant at 21.25° and independent of genetic variety [49-51]. This theory of constancy in angle is still

TABLE 3

Cotton Fiber Properties [48]

Variety	Species	40% X-ray angle (degrees)	Microscopically measured convolution angle (degrees)	50% optical angle (degrees)
De Ridder Red	G. hirsutum	34.0	17.0 ± 4.1	24.41 ± 1.82
Cr-4-C	"	27.7	15.0 ± 2.4	21.23 ± 1.59
Lengupa	G. barbadense	39.0	18.0 ± 4.3	23.86 ± 0.94
NR-AHA-C	G. hirsutum	29.8	15.5 ± 3.4	20.67 ± 1.43
Paymaster	"	39.6	17.8 ± 4.1	23.58 ± 1.89
SL(SLS) 1000	"	41.8	18.3 ± 4.4	24.78 ± 1.39
Lankart	"	38.3	19.5 ± 5.1	24.23 ± 1.18
A4-42-176	"	29.6	16.3 ± 3.3	21.23 ± 1.00
Bobshaw 1-A	"	32.5	17.5 ± 4.5	24.55 ± 1.75
Thef 96	"	26.2	14.7 ± 3.4	20.40 ± 1.56
A4-42-176	"	33.7	16.95± 3.9	22.20 ± 1.37
HA-46-124	"	26.3	14.9 ± 3.4	20.81 ± 1.29
DPL 15	"	37.3	17.2 ± 3.9	23.54 ± 1.05
Peruvian Tanguis	G. barbadense	36.0	17.3 ± 4.0	22.29 ± 1.57
Acala 1517C	G. hirsutum	33.7	16.6 ± 2.9	21.97 ± 1.48
Coastland RNXET	G. barbadense	35.0	15.8 ± 3.8	20.82 ± 1.19
Pima S-1	"	31.8	14.9 ± 3.7	18.37 ± 0.87
Amsak	"	30.8	14.3 ± 2.9	18.73 ± 1.44
Old Pima	"	35.2	14.5 ± 2.7	20.03 ± 0.96

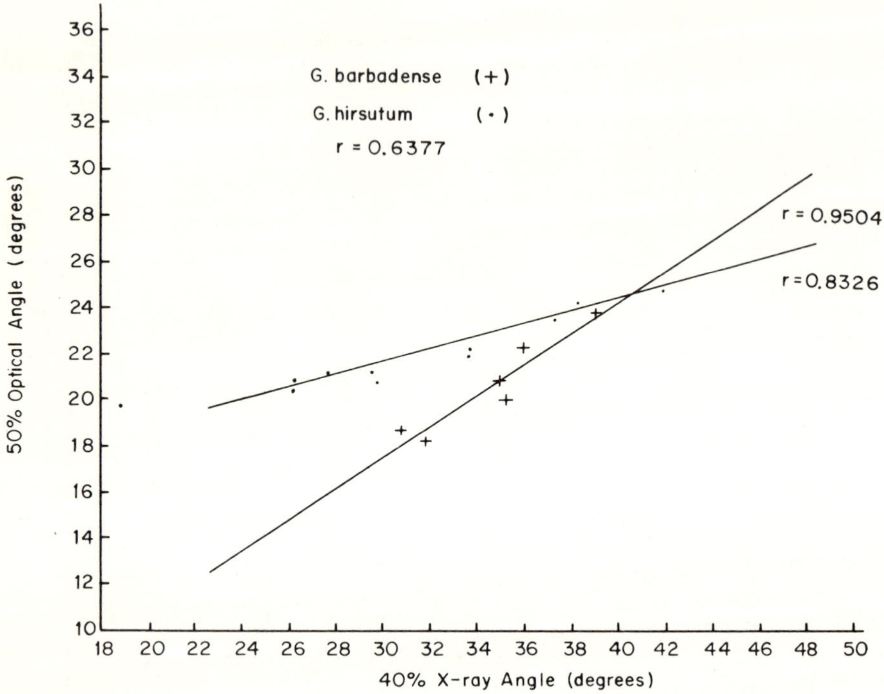

FIG. 5. 40% X-ray angle vs. 50% optical angle. (Reprinted from ref. 48, p. 265, courtesy of Textile Research Journal.)

somewhat open to debate. But if correct, it would imply that changes in X-ray angle are largely the result of degree of induced convolutions [52]. Readjustment of the fibrils within the cotton hair as a result of drying would contribute similarly. Differences in slopes might imply undried fiber density differences between species. Fiber fineness and convolution frequency add additional complexities to the interpretation.

Fiber convolutions also correlate negatively with such physical measurements as bundle tenacity, which indirectly implies a connection between convolutions and internal structure [53]. Radial stress distributions are modified during actual production of convolutions, even though the points of reversal within the secondary wall are of importance [54]. Convolutions may play a more direct role in yarn strength than they do in bundle tenacity, however. Yarns involve considerable interfiber interaction and this necessarily dictates surface interaction. Convolutions, being a surface characteristic, may be extremely important.

The effect of convolutions on the luster of cotton fibers has been recognized [55,56] though it is probably not as significant an effect as fineness. As a result of mercerization, the effect of decrystallization tends to overshadow the accompanying deconvolution process resulting from fiber swelling. In effect, the contribution of circularity of fiber to luster may be rather limited.

V. EFFECTS OF CHEMICAL TREATMENTS

Cotton fibers, yarn, and fabrics are often subjected to chemical treatments for various purposes. Such treatments may involve different phases of finishing, and may affect either the internal structure of the fiber, the surface of the fiber, or both [57]. Purification is one process that substantially involves the surface and deserves some attention.

In the preparation of cotton for textile finishing, the purification process generally includes scouring, i.e., boiling in an alkali, followed by an oxidation bleaching [58]. The main purposes of purification is the removal of the noncellulosics of the cuticle as well as those embedded in the primary wall and, thereby, the achievement of a more rapid wetting of the fabric with subsequent treating liquids. The overall result is a more uniform treatment and, hence, homogeneous product. The degree of success is easily measured microscopically by observing the failure of fibers to react to stains, an indication of wax and pectic content.

The amount of wax and pectic content removed depends not only on the chemical concentrations, but also on treatment times and temperatures. The commercial scouring methods apparently do not remove enough of the noncellulosic materials to reveal clearly the underlying cellulose structure. Fibrillar patterns of the primary wall can be observed, however, if all the noncellulosics are first removed by suitable solvents such as 95% ethanol followed by scouring with hot dilute alkali [39]. Further treatment with hot monoethanolamine fully exposes the interlaced arrangement of fibrils [7]. Other chemical treatments will similarly contribute to fiber purification and the primary wall then becomes the fiber surface.

Another commercial modification is the mercerization process. Mercerization enhances dyeability and luster of cotton materials. It consists of immersion of the fabric in 18 to 25% sodium hydroxide at ambient conditions. The cellulose of the fiber is almost instantaneously swollen and, after the sodium hydroxide is washed out, its effect on the fiber remains. The cellulose is so modified that the absorption and adsorption properties of the fiber are greatly improved. The specific surface areas available for chemical reactivity are increased and the heats of wetting [59] probably undergo change. When dyed in electrochemical solutions [60] the resulting colors are then greatly improved. The luster of yarns and fabrics is also improved by mercerization, but this change is accompanied by alterations in the crystalline and supermolecular structure of the cellulose [61]. The actual surface of mercerized fibers is not a great deal smoother than that of native fibers. Mercerization removes many of the gross concavities of the fiber surface, but the submicroscopic roughness remains. This result must be attributed to the relatively large difference between the perimeters of the primary wall and the secondary thickening of the fiber, differences that are not overcome by the swelling and shrinkage of these fiber components during mercerization.

Permanent surface coatings or additive finishes can be applied to fibers and fabrics to produce such desirable characteristics as soil and abrasion resistance, frictional behavior modification for increased processing efficiency, etc. The effect of such coatings on the fiber is, for the most part, limited to the surface, though penetration does exist and does affect nonsurface properties such as fabric tearing strength [62,63]. Nevertheless, additive finishes have wide commercial application. For example, soil resistance is enhanced by subjecting the fabric to a water suspension of carboxymethyl cellulose; frictional behavior is modified by colloidal silica or starch deposition; and water repellency and luster are enhanced by a coating of synthetic polymeric substances [44,58].

The effect of the specific additives on cotton surfaces may be determined by optical and electron microscopy [64]. The surfaces show varying changes in appearance. Low-level deposits of carboxymethyl cellulose additive on unscoured fabrics fill in the small crevices, but leave the overall topography of the surface unchanged. Increased deposition imparts a smooth finish to a fabric and retards soiling. Starch simply forms irregular flat deposits of various thickness. Deposited colloidal silica, which affects the soiling tendency as well as modifies the frictional characteristics [65], shows an abrasive surface of particulates, and it is this hard granular surface that apparently resists the attachment of soil particles [66,67]. For most synthetic polymeric substances, the hard spherical particulates collect in the grooves of the unscoured surface; the soft particulates coalesce on the surface, obliterating the details of the uncoated surface. Other finishes may form a smooth layer over the original topography.

Chemical modification of the internal structure (i.e., crease recovery through chemical linking of cellulose chains) would appear to be of minor importance to surface modification, unless the addition of new groups in the inner regions of the fiber cause physical enlargement of the cross section of the fiber. Such additions of groups would cause changes in the folds (ridge-and-groove system) on the fiber surface. Damage to the primary wall may even result, as is evident with certain formaldehyde treatments [68]. On the other hand, acetylation, partial cyanoethylation, and partial carboxymethylation result in a smoother surface than does either scouring or mercerization of fibers. In the first three processes the frequency and height of the ridge system is reduced. Wrinkle-resistance treatments with polymerization chemicals such as urea-formaldehyde and methylol melamine cause surface modifications also.

VI. FRICTION

The surface properties significantly influence the technological processing of fibers. In particular, fibers in contact are often made to move relative to one another, creating frictional forces that are of great importance [69-71]. This early recognition of frictional forces has led to the accumulation of much empirical data that characterize the interaction of fibers under numerous situations. The theory is somewhat lacking, which is not surprising in light of the complexities and especially the variability of cotton fiber properties. Nevertheless, theories are essential for adequate explanations of processing mechanisms.

As has already been noted, the surface of the raw cotton fiber consists of a waxy substance having a rill-like appearance. The amount of waxy substance varies somewhat with variety and the rill-like appearance is a result of secondary wall shrinkage after boll opening. Superimposed on this topography are the acknowledged crimp and convolution formations, as well as the cross-sectional shape and overall fiber size. These similarly depend on both growing and drying conditions. All of these intrinsic surface factors, as well as those associated with the internal structure, determine the frictional behavior of fibers. This combination of characteristics is often determined by some type of frictional measurement. In most situations, the "force of friction" or simply "friction" is measured and is defined as the resistance to the motion that is developed when one body slides over the surface of another. These measurements are dependent on experimental variables such as velocity, fiber tension, normal forces between contacting fibers, chemical treatments, and mechanical modifications. All of these tend to make interpretation of results difficult.

Nevertheless, some simplifying definitions and statements can be made and the reader is referred to Chapter 1 for additional discussions on friction. Static friction opposes the displacement of a body at rest against a second surface. This is contrasted with kinetic friction, which represents the opposition forces to the forces producing uniform motion. Quantitative description is often given by the coefficient of friction

$$\mu = \frac{F}{R}$$

where R is the normal load between the object and the restraining surface and F is the frictional force acting parallel to the contacting surfaces. If F is the force just necessary to initiate motion of the object, then μ is defined as the coefficient of static motion. This particular relationship is frequently known as Amontons' law [72]. With a moving object, the relation defines the coefficient of kinetic friction. Essentially, the relationship states that the frictional force is linearly proportional to the normal load so that the coefficient of friction remains constant for two macroscopically contacting bodies. It is also generally assumed that the frictional force is independent of the contact area. This is reasonably valid on the macroscopic scale but the independence of frictional force from contact area can deviate strongly at the microscopic level, i.e., level of fiber-to-fiber interaction.

Apart from the implied complexities of the cotton-fiber surface geometry, there appear to be at least three mechanisms responsible for friction; surface roughness, adhesion, and plowing [73]. The surface roughness resulting primarily from asperities, surface cracking, convolutions, and crimp produces frictional resistances by mechanical locking. This mechanical locking brings into play molecular forces between the two contacting surfaces and forms minute cohesive and adhesive bonds at regions randomly spaced over the surface. Continued relative motion requires additional forces to rupture or shear these contact junctions.

The process of sliding by shear is generally interpreted as a time-dependent phenomenon, where the time of contact at the interface is an important factor in determining the degree of cohesion or adhesion that occurs [74]. This time dependency largely accounts for the differences between static and kinetic coefficient of friction. In the case of kinetic behavior, each contact region receives energy that depends both on the yield strength of the material and the velocity of sliding.

The frictional behavior is dominated by the properties of the small regions of contact. For raw cotton, this would imply small regions of waxy materials in contact, though there may be an interface layer of contamination. By analogy, measurements have been made of the force of friction F between a $\frac{1}{2}$-in. diameter stainless steel ball sliding at constant speed while pressing against wax-coated glass plates [75]. For a series of monolayers of fatty acids and fatty alcohols, the coefficient of kinetic friction was less than 0.25. There was a strong negative influence on the coefficient of friction due to the number of carbon atoms in the straight molecular chain. The level-out value was near 0.05 for both fatty acids and fatty alcohols. This level-out value should correspond to the fatty acids and alcohols consituting the wax of the cotton fiber. Whether these results can be specifically carried over to the frictional behavior of cotton fibers remains to be seen. Determinations of the coefficient of friction on fibers are generally substantially larger. For example, early work by Morrow [76] lists coefficients of friction for raw dry cotton around 0.25. Soda-boiled cottons are somewhat higher and wet cottons even higher in friction. These values are comparable with those of many other investigators.

A. Fiber Pairs and Tufts

The strength of twist yarns varies between 15 and 75% of the tensile strength of the fibers in the cross section. This large variation in transfer of fiber strength to yarn strength illustrates the importance in the clinging power of single fibers and the cohesive forces between fibers. Clinging power is recognized to be a function of convolutions and crimp as well as the coefficient of friction. In fact, fiber configurations may be the major effect, with interfiber friction playing a lesser role in sliver strength [77]. One method of measuring clinging power is to monitor the force necessary to withdraw a single fiber from between two layers of combed cotton lint pressed together. Adderley [78] has shown by this technique that convolutions are significantly related to clinging power. Maximum clinging power is obtained when the convolutions of contacting fibers produce a "best fit," i.e., contribute to maximum area of contact. Uniformity of convolution distribution is also a contributing factor.

Closely related to the work of Adderley is the fiber twist method of Hood [79]. Surface interaction between pairs of fibers was observed by twisting two fibers together for a known number of turns and moving one fiber relative to the other. The resisting frictional forces are again related to the clinging power via convolutions. A comparison of several varieties show that the clinging power is dependent on (in addition to convolution frequency and regularity) the ratio of convolutions between fibers and the reversals of the convolutions. These are additional parameters necessary for a "best fit" situation.

An additional relationship observed using the fiber twist method was a decreasing clinging power with increasing humidity, which might imply increased lubrication with increased moisture content; but such a conclusion contradicts some of the frictional measurements to be discussed later in the chapter. Though all varieties do not respond in the same manner, the clinging power probably decreases as a result of a change in the convolution frequency. It has been observed that the wettability properties of waxes are apparently functions of the number of carbon atoms in the straight chains of the alcohols and fatty acids comprising cotton wax [75]. Though the cotton wax chemistry may vary somewhat between varieties, the difference probably shows little influence on clinging-power variations.

In an attempt to obtain a more representative determination of the coefficient of friction, Mercer and Makinson [80] used crossed-fiber pairs and the stick-slip principle of Bowden and Leben [69]. Knowing the average normal forces and the force for pulling one fiber across the other, these authors could obtain the coefficient of friction from Amonton's law. The results gave a value of 0.29 between cotton fibers taken from sewing thread and 0.57 between fibers taken from raw cotton. The former reflects the effects caused by processing, i.e., a reduction in crimp and surface abnormalities. The latter value more likely reflects clinging power. Even more significant, though, was the observation of a drop in the coefficient of friction with increasing normal load. This is also observed in the fiber-twist results of Hood [79] and raises questions of the accuracy of Amontons' law, when applied to fibers.

The relation $F = aR^n$ between the frictional force F and normal reaction R is found to hold for most fibers [81]. It has been proposed that it replace Amontons' law. The friction index n lies between about 0.80 and unity. Only for unity do the two relations agree. Since Amontons' law presupposes independence of the apparent area of contact, the replacement relation implies areal changes with changes in normal loading [82,83].

The choice of $F = aR^n$ and the failure of Amontons' law is explained on the basis of viscoelastic considerations [84,85]. When two surfaces are pressed together with a force R, contacts are formed only at the tips of a number of irregular asperities. As a consequence, the pressures acting on these small real areas of contact can be sufficiently high for the yield point of the material to be exceeded. Plastic flow occurs, thus extending the area until the pressure has dropped to the yield value P_y and the load R is supported elastically. The real area of contact, A, is related to the load by

$$A = \frac{R}{P_y}.$$

Because of the high pressures, the surfaces may be welded together at the points of contact and subsequent motion of one surface with respect to the other requires a tangential force F to shear the area A of contact of the material having shear strength S at the junction. The parameters are related by

$$F = SA.$$

If the two relations are combined, Amontons' law is given by

$$F = (S/P_y)R.$$

This is a special case of the more general form

$$F = aR^n,$$

holding when n = 1 and a = S/P_y = constant. The equivalency is linked with the notion of plastic deformation of the contact area. Experimentally, deviations from the special case are more common. The extent is largely determined by the degree of elasticity or plasticity and n usually varies between 2/3 and 1. The coefficient a is generally negatively correlated with the exponent n [83,86]. And both are dependent on the molecular cohesion of the fiber material, as well as possibly on the mechanical properties of the material and the nature of the deformation.

Several other relations between frictional force and loading have been proposed. For two fibers pressed together under a load w and having partly plastic and partly elastic behavior, Gralen et al. [87] have proposed the relation

$$F = \alpha w + \beta \ell r.$$

The length of the contact line ℓ and the fiber radius r shows areal implications. The factors α and β are constants related to the plastic and elastic properties of the fiber material. duBois [82], using fringe samples, has found a quadratic formula to be most representative.

A comprehensive extension of the use of the stick-slip processes from crossed-fiber pairs is given by Belser and Taylor [88]. With analog instrumentation, data plots of the frictional forces encountered in traversing one fiber across another at right angles are obtained [89]. An example of such a plot is shown in Fig. 6. Both kinetic and static coefficients of friction are determined from this plot. The coefficient of static friction is arbitrarily determined by using the ten highest peaks from an analog plot; the coefficient of kinetic friction is determined by averaging the data plot over a selected length.

Experimentally, successive runs on the same fiber pairs result in a decrease in both coefficients of friction, which indicates that surface polishing through processing would produce reduced frictional resistance. Heating the cotton results in only a slight increase in coefficient of friction. A marked increase in static friction is observed with increasing traversing velocity, though the kinetic friction coefficients remain relatively constant over the same range. With normal loading increases, the coefficients of friction decrease slightly after an initial peaking near 5g normal force [Fig. 7a]. Fig. 7b shows that the coefficients of friction are highly sensitive to relative humidity, especially above 60%. The ratio, μ_s/μ_k, however, remains

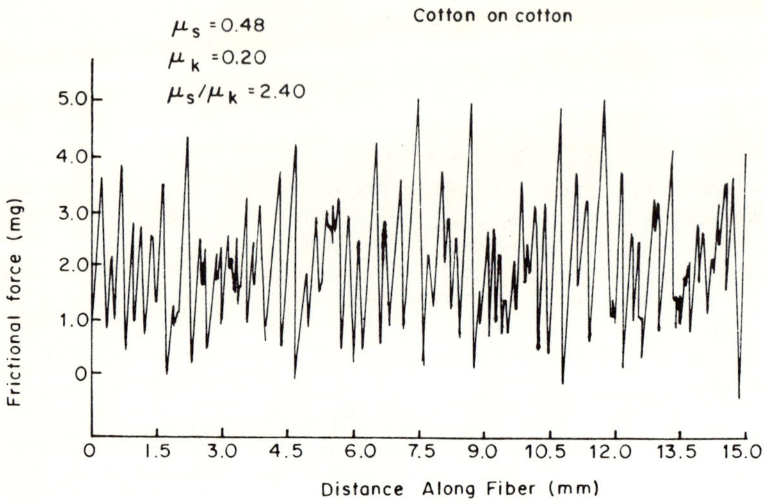

FIG. 6. Schematic showing stick-slip mechanism. (Reprinted from ref. 88, p. 116, courtesy of U.S. Department of Agriculture.)

reasonably constant and best represents the frictional measurement most independent of atmospheric moisture. The sensitivity of either the static or kinetic friction coefficient to experimental conditions suggests caution when comparing values between different investigators.

Intrinsic fiber characteristics have an additional influence on frictional measurements. These are most easily isolated and controlled using synthetic fibers because surface characteristics in cottons vary widely along even a single fiber. Nevertheless, certain generalizations can be made concerning frictional parameters of cotton. The linear density of the fiber affects the coefficient of friction; the correlation is positive. Values for μ_k of 0.289, 0.225, and 0.173 were recorded for Micronaire values of 4.37, 3.7, and 2.8 μg per inch, respectively [88,90,91]. This is an areal influence and is supported by the work of Pascoe and Tabor [92]. Using crossed nylon cylinders, they obtained the relation

$$\mu = c_1 s\, w^{-0.26} D^{0.52}$$

where w is the normal force, s is the adjusted shear value for nylon, c_1 is a constant, and D is the diameter of the fiber. The relation clearly shows that an increase in fiber diameter, and hence linear density, predicts an increase in coefficient of friction.

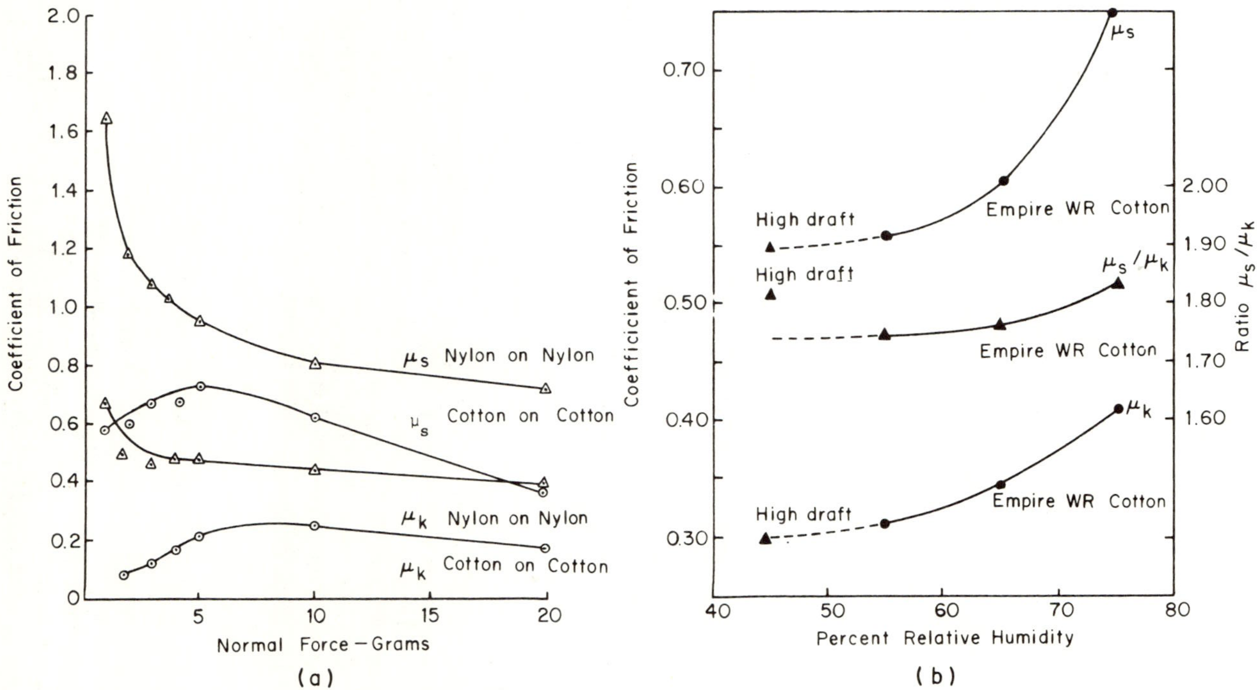

FIG. 7. Coefficients of friction. (a) Variation of coefficients of kinetic and static friction of cotton and nylon at low normal force; (b) coefficients of friction vs. relative humidity for Empire WR cotton (hand ginned). (Reprinted from ref. 88, pages 88 and 95, courtesy of U.S. Department of Agriculture.)

Differences between natural and dewaxed cotton are observed, though they are not greatly different. For an Empire WR cotton of 1-in. staple length, a coefficient of kinetic friction of 0.43 was obtained on natural cotton and a value of 0.41 was obtained on dewaxed cotton. Corresponding coefficients of static friction were 0.88 and 0.72 [88,93]. These values may be subject to excessive instrumental error.

Insofar as processing is concerned, any mechanism that reduces the variations in longitudinal shape or polishes the surface [94] probably produces lower values for coefficients of friction. Specific processing, which results in increased surface roughness, amplifies the stick-slip process, resulting in increased coefficients of static friction. This may include fractures of the fiber surface as well as surface additives.

In an attempt to bridge the gap between separate individual fiber interactions and cohesive behavior operative in raw-cotton processing, Viswanathan [86,95] has analyzed the frictional interaction between fiber fringes. The experimental results show that Amontons' law again does not hold. The expression $F = \alpha R^n$ explains the behavior of the fibers. In addition, α and n were again found to be highly negatively correlated, a result that could be expected from multiple-asperity-contact theories.

The fringe measurements show that fiber characteristics of length, fineness, and maturity have a decreasing influence on the frictional behavior as the applied normal load increases. Nevertheless, cottons with a low linear density (increasing fineness) and high immaturity generally show lower interfiber friction than mature, coarse cottons. Long and fine cottons give lower friction values than short, coarse ones [96].

Recognizing the increasing importance of blends in textile processing, Viswanathan has obtained results on friction between fringes of Karnak (or Kalyan) cotton and of polyester (terene), Viscose, and tufcel (polynosic) fibers [97]. These results are given in Table 4. The coefficients of friction for all fringe pairs show the expected decrease with increase in normal load. The exponent n varies between 0.59 and 0.76 and indicates the degree of failure of Amontons' law.

B. Yarn and Roving

The effect of fiber friction within rovings and yarns can be expressed reasonably well by means of "minimum twist of cohesion" [98,99]. In this context, the "minimum twist of cohesion" is defined as the number of turns per unit length existing in the yarn at the moment it breaks as a result of fiber slippage when it is subjected simultaneously to tension and untwisting.

TABLE 4

Coefficients of Friction μ between Fiber Fringes at Varying Applied Normal Load R and Values for a and n in the Expressions $\mu = aF^{n-1}$ or $F = aR^n$ [97]

Fiber	μ R = 30	R = 75	R = 150	R = 225	R = 300	a	n
Karnak	0.595	0.443	0.324	0.304	0.275	1.88	0.66
Kalyan	0.678	0.552	0.386	0.336	0.303	2.44	0.64
Terene	0.813	0.594	0.393	0.360	0.330	3.29	0.59
Viscose	0.546	0.389	0.296	0.253	0.242	1.87	0.64
Tufcel	0.463	0.384	0.285	0.260	0.241	1.30	0.70
Karnak-Terene	0.506	0.481	0.358	0.306	0.284	1.37	0.73
Karnak-Viscose	0.436	0.428	0.315	0.275	0.263	2.26	0.60
Karnak-Tufcel	0.459	0.456	0.337	0.295	0.278	1.14	0.76
Kalyan-Terene	0.535	0.499	0.357	0.298	0.277	1.66	0.69
Kalyan-Viscose	0.470	0.424	0.316	0.279	0.256	1.28	0.72
Kalyan-Tufcel	0.484	0.429	0.319	0.283	0.255	1.36	0.71

The measuring technique, using a torsiometer, consists of fastening a length of yarn between a variable tensioning clamp and a second clamp which can be turned to impart an untwisting in the yarn. At some degree of untwist, there is a sharp drop in yarn tension as a consequence of yarn breakage through fiber slippage. There is little, if any, fiber breakage. The "minimum twist of cohesion" is simply obtained by subtracting the decrease in twists from the original twist in the yarn. In addition to "minimum twist of cohesion" (turns per m), the results can be expressed as a twist coefficient. The twist coefficient is related to the inverse metric yarn count system and to turns per meter.

This technique has allowed a study of the principal properties of the minimum twist of cohesion (residual twist) in connection with fiber and yarn properties. The minimum twist of cohesion is related to both fiber parameters (length and fineness) as well as yarn parameters (count and initial twist) [100]. Modification of the fiber surface, either by chemical or mechanical means, influences the minimum twist of cohesion [101]. Consequently, this measurement is a surface characteristic of the fibers within the roving or yarn. Roving structure and individual fiber properties similarly effect the residual twist.

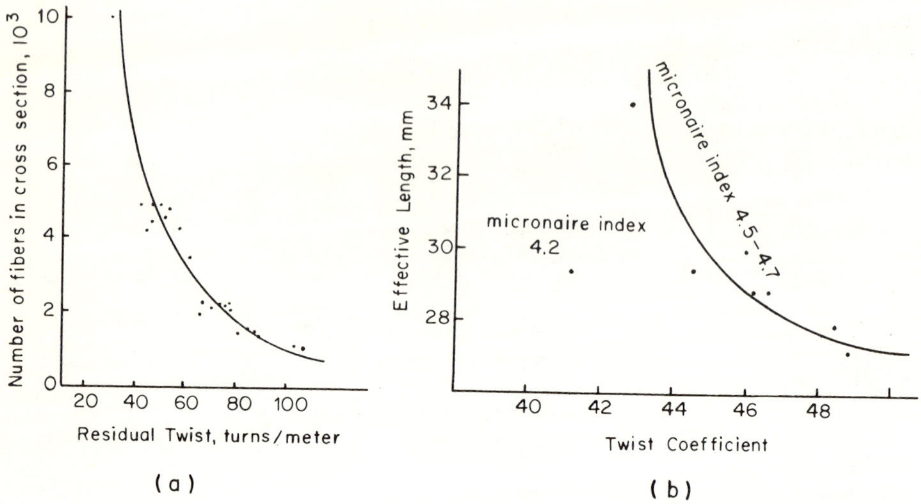

FIG. 8. Residual twist curves. (a) Effect of cross section on residual twist, (b) effect of fiber length on residual twist. (Reprinted from ref. 99, p. 223, courtesy of Textile Research Journal.)

As is shown in Fig. 8a, residual twist is highly responsive to yarn cross section. In fact, the minimum twist of cohesion increases proportionately to the square root of the number of fibers N_0, which is in qualitative agreement with the area dependence of frictional forces. For rovings, it is in qualitative agreement with the theoretical work of Hannah [102], provided that the fiber fineness is fixed. Also, the minimum twist of cohesion is roughly independent of the original yarn twist, when this is not too high.

Fiber properties strongly affect the cohesion measurements. The minimum twist of cohesion is dependent on the mean fiber length. This is to be expected since it is well recognized that longer fibers tend to produce stronger yarns. The relation between the coefficient of minimum twist of cohesion, τ_t, and the mean length ℓ is curvilinear. For cotton

$$\tau_t = 0.13\ell^2 - 8.42\ell + 175.79.$$

Fig. 8b shows a curvilinear plot of fiber length versus coefficient of minimum twist of cohesion. As is indicated, there should be a series of curves, each classed according to fineness or micronaire index. The effect of micronaire index on the coefficient of minimum twist of cohesion is somewhat greater than the effect of length. In general, for a given roving weight, the minimum twist of cohesion is positively correlated to micronaire index and negatively correlated to fiber length [99]. A quantitative extension of this behavior allows further conclusions to be drawn concerning roving and yarn strength. Roving strength F for given fiber characteristics within a limited range, increases as the fourth power of twist t and varies inversely to the 2.3 power of inverse metric count of the roving, N_m,

$$F = \frac{107 \times 10^{-6} t^4}{N_m^{2.3}}.$$

Twist-strength plots for rovings have been obtained and extended in range by other investigators using different techniques [103,104]. In general they agree and show that long, fine cottons require less twist to produce a given strength than do short, coarse cottons (Fig. 9a). Roving size, fiber length and fineness, as well as position on the roving bobbin and method of manufacture, are parameters in any strength relation.

Increasing the roving twist continues to increase the strength until a point is reached where the strand no longer breaks by fiber slippage. Both fiber strength and frictional characteristics come into play in a complex manner [105-107], which is characterized in Fig. 9b. Curves A and B represent two fibrous strands of the same cotton twisted by two separate methods. Note that Fig. 9a represents the portion below about 3 turns per in. (TPI) in Fig. 9b. It has been suggested that the point of inflection is where the fibers first begin to break [104].

C. Multiple Card Web Specimens

In an attempt to eliminate the large variability that arises from measurements on too small a specimen and to reduce the number of tests necessary for statistical significance, attention has been directed to 50-g multiple card web specimens [108]. The method used involves the measurement of the loss of mechanical energy as the fiber assembly, with its many points of fiber-to-fiber contacts, is subjected to an oscillatory shear action. The instrument is essentially an oscillating pendulum coupled to plates between which the specimen is placed. The energy losses are responsive primarily to surface interactions, but it is recognized also that some of the losses are the result of internal losses due to the deformations the fibers undergo during oscillatory motion. This latter effect is found to be very small. The method is capable of displaying small differences in cottons as well as distinguishing between processing procedures [108-114b].

The measurements are also sensitive to many factors such as static friction, fiber surface polishing, fiber orientation, humidity, relaxation, fiber maturity, fiber surface treatments, and growth conditions. The mechanism causing the energy loss has been treated

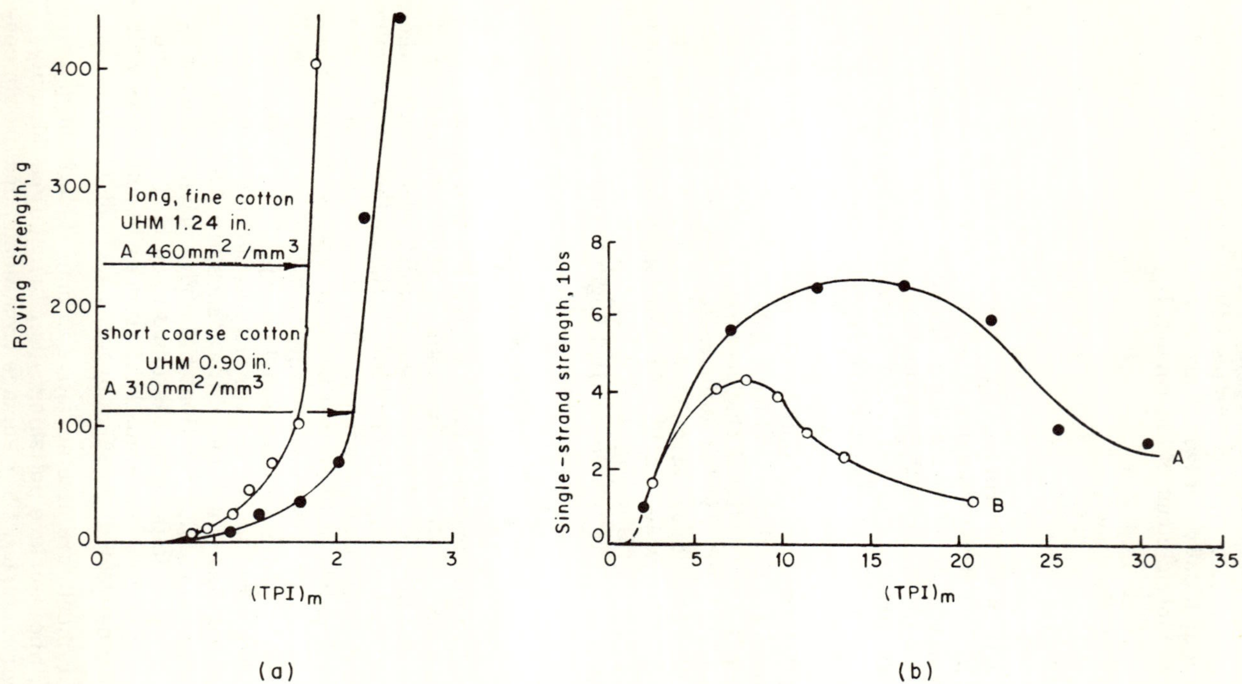

FIG. 9. Twist-strength curves. (a) Twist strength curves showing the effect of fiber properties on the curves for 2.0-hank roving; (b) twist-strength curves developed over a wide range of twist. (Reprinted from ref. 103, pp. 487-488, courtesy of Textile Research Journal.)

theoretically and analyzed experimentally. The relation between the amplitude and the number of swing is [113]:

$$\left(\sqrt{A_1} - \sqrt{A_{n+1}}\right)\sqrt{T} = \left[\frac{8}{15K}\left(\frac{r}{R}\right)^{3/2}\right]\left[L + \frac{L_0}{2}\right]\left[\nu\mu_k\sqrt{\frac{k_{\parallel}}{F_s}}\right]n$$

where A_1 is the beginning amplitude of the instrument's pendulum, A_{n+1} is the amplitude at the end of the nth swing, and $n = 1,2,3,\ldots$ The first factor on the right hand side of the equation is instrumental; the second factor depends on the specimen weight L_0 and the load L on its top; the third factor is a function of lap structure, the coefficient of kinetic friction μ_k, the static friction force F_s at a contact point, and the tensile modulus of a pair of contact points, $k_{\parallel}$. The frictional characterization of the fiber assembly is determined by the third factor above and this term has been designated the coefficient of energy dissipation,

$$h = \nu\mu_k\sqrt{\frac{k_{\parallel}}{F_s}}.$$

The fundamental relationship is reduced to

$$\left(\sqrt{A_1} - \sqrt{A_{n+1}}\right)\sqrt{T} = chn$$

where c combines instrumental and loading terms. Further simplification combines c and h into H or

$$\left(\sqrt{A_1} - \sqrt{A_{n+1}}\right)\sqrt{T} = Hn$$

and H then becomes the frictional parameter for a specific instrument. Experimentally, the value H is constant up to five swings; the variability is low. Table 5 lists the results obtained from four specimens of six cotton varieties each run twice. Figure 10 shows a plot of $\left(\sqrt{A_1} - \sqrt{A_{n+1}}\right)\sqrt{T}$ vs. n for the same six cotton varieties. The slopes of each represent H values listed in Table 5. These slopes are apparently quite well representative of scroop or handle.

A number of practical conclusions have been drawn from these studies. The energy losses or corresponding coefficient of energy dissipation do vary among bales, varieties, and regions. The values are reduced when the carded web is heated above the melting point of wax but increases again at higher temperatures. The measurements

TABLE 5

Energy Loss Values of Cotton Fibers [114]

Variety	Designation	Slope H (cm)	Coefficient of variation (%)
Lankart-57	SR-6	0.629	1.18
Deltapine smooth-leaf	SR-2	0.566	0.94
Stoneville 7A	SR-4	0.555	0.99
Acala 4-42	SR-8	0.552	1.21
Cal 7-8	SR-7	0.540	2.00
Pima S-2	SR-13	0.520	0.98

are responsive slightly to humidity between 30 and 65% RH; they increase rapidly with moisture above 65% RH. Differences have been observed between spinning wastes and roving being spun; the spinning wastes generally dissipate more energy than the roving.

VII. ABRASION

Abrasion, primarily the result of mechanical effects, manifests itself first in surface damage. While it is most often associated with surface-to-surface interaction, abrasion can also result from fiber flexure. Internal structure, as reflected by stiffness, affects abrasion as well as surface characteristics [115]. The combination of effects has been recognized as a major cause of fabric or yarn loss of strength or color retention (frosting). There are standardized methods for determining the resistance of yarns and fabrics to flexing and abrasion [116-118]. The reader is particularly referred to Chapter 5 for a discussion of abrasion of synthetic fibers.

There have been attempts to modify the fiber chemically to resist abrasion enhancement [119]. The situation is rather complex and the results depend to a large extent on the method of generating abrasion [120,121]. In an effort to relate untreated and chemically modified cotton fibers to consumer problems, for example, abrasion damage due to machine washing or drying was observed with both the scanning and transmission electron microscope. Two kinds of fiber damage became apparent: fiber breakage and fibrillation. Wet abrasion tends to result in splintering or fibrillation whereas dry fibers tend to break

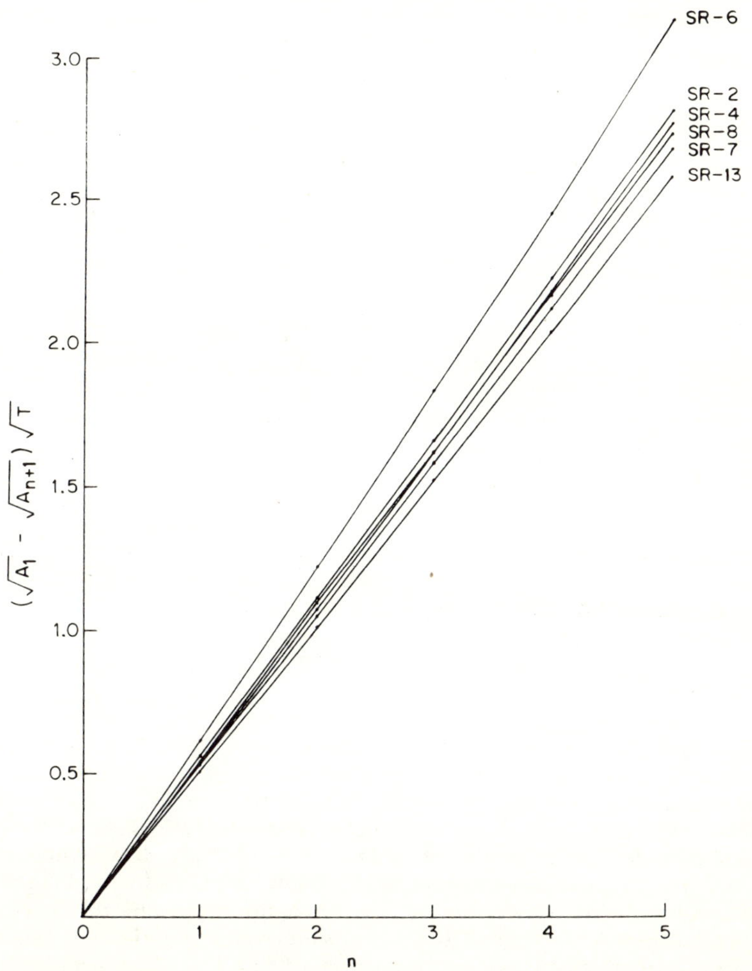

FIG. 10. $\left(\sqrt{A}_1 - \sqrt{A_{n+1}}\right)\sqrt{T}$ (cm) vs. n for cotton fibers. (Reprinted from ref. 114, p. 582, courtesy of Textile Research Journal.)

or crack on abrasion. Greater differences were observed between washer- and dryer-abraded samples than between treated and untreated samples abraded by the same method [122]. Abrasion resistance and frictional characteristics resulting from differing degrees of chemical modifications, i.e., acetylation, carboxymethylation, cyanoethylation, are not necessarily positively correlated [101].

VIII. EFFECTS OF GAS PLASMA RADIATION

Exposing cotton fiber to electrical glow discharge causes changes in the fiber characteristics [123-126]. Significant alterations are found in the cotton wax, wetting rate, and alkali absorption of irradiated cotton fiber. Gas plasma radiation causes the surface waxes to melt and redistribute. The corrugated surface features (Fig. 2b), apparent in electron micrographs of untreated cotton fiber, are removed and broken surface areas appear. Analyses show only a slight loss of wax content from the irradiated fiber. The melting range of the waxes is, however, lowered approximately 10%.

The rate of water absorbency is considerably increased. The alkali absorption, i.e., alkali-centrifuge value, is increased nearly 100%. This is indicative of induced fiber damage and increased internal accessibility. However, there is generally an increase in breaking strength of yarns, contradictory to the expected result of fiber damage. The interfiber surface friction is known to increase as a result of gas discharge treatment. This is the likely reason for the observed yarn strength increase.

IX. STATIC ELECTRIFICATION AND ELECTRICAL CONDUCTIVITY

Problems arising from the static electrification of textile materials are encountered in mill processing and fabric handling. The accumulation of static charges during cotton fiber processing poses less of a problem than it does with many synthetics. See Chapter 6. Blends show a behavior somewhere in between [127]. Nevertheless, there are occasions when consideration must be given to the effect charge accumulation may have on yarn irregularity, ends down, etc. There are, conversely, occasions when electrostatic behavior is an asset, i.e., electrostatic separation of short and long fibers [128].

The production of static charges on a fiber, yarn, or fabric results from mechanical contact of the surface of that textile substance with the surface of another substance. Under the right conditions, mere separation of the contacting surfaces produces charges of opposite sign on the two materials. A number of possible processes can be suggested that would result in charge transfer at mechanical contact [129,130]. At the same time, there remains considerable ignorance of the mechanisms of charge transport in polymeric materials.

The nature, sign, and magnitude of the transferred charge are subject to several influences. The physical and chemical nature of the contacting surfaces are determined by the nature of the bulk materials, the surface contamination, and the presence of local neutralizing ions. The contact pressures and kinetic friction behavior introduces physical and thermal effects. Ambient and time-dependent conditions must be taken into account.

The determination of charge carriers, their origin, and their mode of transfer are required for satisfactorily formulating a charging mechanism. Qualitative simplification has been suggested in the form of triboelectric series. In this series, materials are listed according to the sign of the charge developed on a given surface of one by contact with another. A portion of one such series [131] is shown in Table 6. Other series have been suggested [132] and the ranking of similar substances is not always identical. The order, as illustrated in Table 6, is such that the upper of a pair of substances in the list will charge positively and the lower, negatively. Note that cotton is just above midway in this grouping of textile materials.

Though such a series is helpful, there are several limitations. Some groups of materials cannot be fitted into a linear series at all [133,134]. It is likely that there is not an exchange of a single type of charge carrier common to all the pairs of materials. Metal-to-metal contacts are thought to involve only electrons as charge carriers; the direction of transfer depends on the relative positions of the electron conduction bands. Identical arguments do not necessarily extend to polymers and, hence, cotton. Ionic and charged material particles may also act as carriers, though their effect may be limited [135].

TABLE 6

Triboelectric Series [131]

Positive
Wool
Nylon
Viscose
Cotton
Silk
Cellulose acetate
Polyvinyl alcohol
Dacron
Orlon
Polyvinyl chloride
Dynel
Velon
Polyethylene
Teflon
Negative

The dielectric constant, moisture content, and even yield strengths partially determine the positions of a substance in a triboelectric series. Interpretation of cotton behavior is increasingly complicated by it variable waxy surface substances. A review of the status of electrification in polymers is given by Morris [130].

Charge accumulation can be largely controlled by adjustment of relative humidity or the use of suitable static eliminators. Static charge retention is largely the result of low conductivity [136]. Often it is simply necessary to increase the conductivity and drain away unacceptable accumulated charge. Many textile fibers are hygroscopic and their electrical conductivity varies according to the moisture content. In fact, the electrical conductivity doubles its magnitude for about every 4% rise in relative humidity [137]. One theory is that the conduction of textile fibers is due to the conductivity of water condensed in the intermicellar regions. This implies primarily electronic conduction behavior. Ionic conduction is more likely, in view of the large differences in activation energy between fiber-bound water and pure liquid water. The activation energy of cotton at a regain of 6.4% is 0.94 eV and the activation energy of water is 0.44 eV [138]. This indicates that the conductivity of a cotton-water system must be of a different nature from the conductivity in pure liquid water. The fact that the activation energy of ice is 1.25 eV would suggest that the absorbed water is in a more strongly oriented form than is liquid water. But the conductivity in a cotton-water system is by no means identical with the conductivity in ice, since ice shows plarization. Perhaps absorbed water molecules produce impurity centers giving rise to ionic conduction. A review of the literature pertaining to electrical resistance of textile materials has been presented by Hearle [139].

Static eliminators are of two types [140]. One employs radioactive materials, the other a high-frequency corona discharge. In the vicinity of accumulated charge, both of these static eliminators ionize the air and permit rapid dissipation of the static charges.

Application of a suitable finish can also be beneficial, though no one finish or antistatic ingredient is effective in preventing static under all conditions. One of the groups of compounds usually consists of cationic surface-active substances that form a conductive film on the surface to carry static charges away. Other effective compounds contain polyethylene glycol chains. The effectiveness of antistats frequently increases with relative humidity.

X. SPECTROSCOPY AND FLUORESCENCE

A. Infrared Spectra from Cotton Fiber Surfaces

FMIR (Frustrated Multiple Internal Reflectance) and ATR (Attenuated Total Reflectance) infrared spectroscopy permits the identification of surface coatings on fibers without significant interference of internal structure. It also permits spectroscopic analyses of samples too opaque for ordinary transmission measurements. And spectra obtained do not contain the background that would otherwise be introduced by solvent and embedding materials required for transmission-sample preparation. But the spectra obtained from transmission spectroscopy generally show much greater resolution and the use of FMIR or ATR has not gained wide acceptance. Experimentally, the technique is more applicable to rather soft materials. Cotton does not fit well in the category. Unusually satisfactory FMIR spectra are quite limited, but have been obtained [141]. Harrick [142] has authored a comprehensive book on the field of internal reflection spectroscopy and the reader is referred to chapter 8 for a more detailed discussion of ATR Infrared Spectroscopy.

B. X-Ray Fluorescence

X-ray fluorescence techniques have been applied to the qualitative and quantitative elemental analyses of chemically modified cotton textiles [143,144]. The determination of specific elements not found in cellulose provides an effective quality control of chemically modifying and add-on reactants. For example, these techniques are applicable to problems of subsequent loss of fungicidal agents containing such elements as Cd, Zn, Se, Cl, S, and P.

For purposes of classification, elements introduced into or onto cotton fall into two broad categories: those of atomic number less than 22 and those of atomic number greater than 22. The lighter elements, including Cl, S, P, and Si, are often found in finishing procedures that confer flame resistance and water repellency. The heavier elements are used to impart flame resistance (Sb, Br) and protection against degradation by outdoor exposure (Cu, Hg, Pb, Cd, Zn, Se). They are used for catalytic purposes; they show up in certain classes of dyes (Cr, Ti, Fe) [143].

There are definite advantages in applying X-ray fluorescence techniques to the identification of nonmetals. Spectrochemical determinations are often difficult or time consuming in comparison to identification by the X-ray method. X-ray fluorescence methods, nevertheless, have limitations. Among other things, there is limited sensitivity for elements lighter than silicon [145]. For the heavier elements, quantitative accuracy can be handled at levels near 0.01%. For the lighter elements, 0.1% levels can be obtained.

REFERENCES

1. P. A. Roelofsen, The Plant Cell Wall, Gebr. Borntraeger Verlag, Berlin, 1959.

2. I. W. Bailey and T. Kerr, J. Arnold Aboratum, 15, 327 (1934).
3. D. B. Anderson and T. Kerr, Ind. Eng. Chem., 30, 48 (1938).
4. P. Kassenbeck, Ciba Review No. 4, 14 (1962).
5. G. Jayme and K. Balser, Melliand Textilberichte, 51, 3 (1970).
6. N. E. Dweltz, G. K. Ambady, and T. Radhakrishnan, The Proceedings of the Indian Academy of Sciences, 68, 26 (1966).
7. M. L. Rollins and V. W. Tripp, Textile Res. J., 24, 345 (1954).
8. V. W. Tripp and M. L. Rollins, Anal. Chem., 24, 1721 (1952).
9. W. L. Balls, Proc. Roy. Soc. (London), B90, 542 (1919).
10. W. L. Balls, Proc. Roy. Soc. (London), B95, 72 (1923).
11. W. L. Balls and H. A. Hancock, Proc. Roy. Soc. (London), B93, 426 (1922).
12. W. L. Balls and H. A. Hancock, Proc. Roy. Soc. (London), B99, 130 (1926).
13. S. M. Betrabet, J. Sci. Ind. Res., 26, 201 (1967).
14. V. W. Tripp, A. T. Moore, and M. L. Rollins, Textile Res. J., 21, 886 (1951).
15. L. V. Lecomber and M. E. Probert, J. Textile Inst., 16, T338 (1925).
16. P. H. Clifford, L. Higginbotham, and R. G. Fargher, J. Textile Inst., 15, T120 (1924).
17. A. C. Chilnall, S. H. Piper, A. Pollard, E. P. Williams, and P. N. Sahai, Biochem. J., 28, 2189 (1934).
18. S. A. Amin and E. V. Truter, J. Sci. Fd. Agric., 23, 39 (1972).
19. R. F. Nickerson, T. D. Fontaine, and C. B. Leape, Textile Res. J., 11, 154 (1941).
20. J. D. Guthrie, C. L. Hoffpauir, M. F. Stansbury, and W. A. Reeves, USDA Bulletin AIC-61, Agricultural Research Service, 1949.
21. A. Geake, J. Textile Inst., 15, T81 (1924).
22. Anonymous, Am. Dyestuff Reporter, 36, 104 (1947).
23. S. R. Trotman, Soc. Chem. Ind. J., 28, 1237 (1909).
24. J. C. Hebden, Ind. Eng. Chem., 6, 714 (1914).
25. B. P. Ridge, J. Textile Inst., 15, T94 (1924).
26. K. L. Hertel and C. J. Craven, Textile Res. J., 21, 765 (1951).
27. M. L. Rollins, Forest Prod. J., 18, 91 (1968).
28. J. B. Airy, Cambridge Phil. Trans., 5, 283 (1834).
29. A. F. Hallimond, The Polarizing Microscope, Vickers, York, England, 1970.
30. E. R. Schwartz, Textiles and the Microscope, McGraw-Hill, New York and London, 1934.
31. A. N. J. Heyn, Fiber Microscopy, Wiley-Interscience, New York, 1954.
32. N. H. Hartshorne, Nature, 184, 179 (1959).
33. B. Friedman, American Laboratory, 3:4, 25 (1971).
34. J. Isings, TNO-Nieuws, 20, 940 (1965).
35. W. L. Balls, The Development and Properties of Raw Cotton, Black, London, 1915.
36. M. L. Rollins and V. W. Tripp, Forest Prod. J., 11, 493 (1961).
37. A. T. Moore, Stain Tech., 28, 149 (1953).

38. P. A. Roelofsen and A. L. Houwink, Acta Botani Neerl, 2, 218 (1953).
39. V. W. Tripp, A. T. Moore, and M. L. Rollins, Textile Res. J., 24, 956 (1954).
40. V. W. Tripp, R. L. Clayton, and B. R. Porter, Textile Res. J., 27, 340 (1957).
41. L. L. Muller and M. L. Rollins, The Microscope, 20, 143 (1972).
42. A. M. Cannizzaro, W. R. Goynes, M. L. Rollins, and R. J. Berni, Textile Res. J., 38, 842 (1968).
43. V. W. Tripp, A. T. Moore, and M. L. Rollins, Textile Res. J., 27, 419 (1957).
44. V. W. Tripp, A. T. Moore, and M. L. Rollins, Textile Res. J., 27, 427 (1957).
45. E. E. Berkley, Textile Res. J., 18, 71 (1948).
46. H. J. Denham, J. Textile Inst., 14, T86 (1923).
47. R. Meredith, J. Textile Inst., 42, T291 (1951).
48. K. E. Duckett and C. C. Cheng, Textile Res. J., 42, 263 (1972).
48a. C. C. Cheng and K. E. Duckett, Textile Res. J., 44, 650 (1974).
49. J. J. Hebert, Textile Res. J., 37, 57 (1967).
50. J. J. Hebert, R. Giardina, D. Mitcham, and M. L. Rollins, Textile Res. J., 40, 126 (1970).
51. N. Morosoff and P. Ingram, Textile Res. J., 40, 250 (1970).
52. A. Peterlin and P. Ingram, Textile Res. J., 40, 345 (1970).
53. K. E. Duckett and R. S. Krowicki, Trans Soc. Rheology, 15:2, 261 (1971).
54. J. C. Warwicker, S. C. Simmons, and P. Hallam, Textile Res. J., 40, 1051 (1970).
55. J. Prakash, P. G. Oka, and R. L. N. Iyengar, Textile Res. J., 38, 52 (1968).
56. A. Adderley, J. Textile Inst., 15, T195 (1924).
57. J. R. Modi, S. S. Trivedi, and P. C. Mehta, J. Applied Polymer Science, 7, 15 (1963).
58. M. L. Rollins, A. T. Moore, and I. V. deGruy, Proceedings of the American Association of Textile Chemists and Colorists, Atlanta, Ga., June, 1963, p. 19.
59. S. R. S. Iyer and N. T. Baddi, Cellulose Chem. and Tech., 3, 561 (1969).
60. S. R. S. Iyer, G. Srinivasan, and N. T. Baddi, Textile Res. J., 38, 693 (1968).
61. S. S. Trivedi and A. G. Chitale, J. Textile Inst., 50, T390 (1959).
62. M. L. Rollins, W. R. Goynes, and R. J. Brysson, Applied Polymer Symposium #16, 35 (1971).
63. N. J. Abbott, T. E. Lannefeld, L. Barrish, and R. J. Brysson, 1st Symposium International de la Recherche Textile Cotonniere, Paris, 1969, p. 419.
64. M. L. Rollins, I. V. deGruy, A. M. Cannizzaro, and J. H. Carra, Norelco Reporter, 13, 119 (1966).
65. M. L. Rollins, I. V. deGruy, and V. W. Tripp, Symposium on Microscopy, Special Technical Publication #257, Am. Soc. for Testing Materials, 1959.

66. B. R. Porter, C. L. Peacock, V. W. Tripp, and M. L. Rollins, Textile Res. J., 27, 833 (1957).
67. V. W. Tripp, A. T. Moore, B. R. Porter, and M. L. Rollins, Textile Res. J., 28, 447 (1958).
68. B. Singh, C. L. Jain, A. Pande, V. B. Chipalkatti, and R. S. Parikh, Textile Res. J., 40, 940 (1970).
69. F. P. Bowden and L. Leben, Proc. Roy. Soc., 169, 371 (1939).
70. E. R. Kaswell, Textile Fibers, Yarns and Fabrics, Reinhold, New York, 1953.
71. G. W. Bartlett, T. M. Smith, and H. A. Thompson, Textile Res. J., 23, 647 (1953).
72. G. Amontons, Mem. Acad. Roy., 206 (1699).
73. S. Baker, Textile Res. J., 21, 453 (1951).
74. E. Rabinowicz, in Friction and Wear (Robert Davies, ed.), Elsevier, Amsterdam, London, New York, Princeton, 1959, p. 149.
75. W. A. Zisman, in Friction and Wear (Robert Davies, ed.), Elsevier, Company, Amsterdam, London, New York, Princeton, 1959, p. 110.
76. J. A. Morrow, J. Textile Inst., 22, T425 (1931).
77. I. H. Mehta and S. Y. Nanal, Textile Res. J., 40, 1131 (1970).
78. A. Adderley, J. Textile Inst., 13, 249 (1922).
79. B. G. Hood, Textile Res. J., 23, 495 (1953).
80. E. H. Mercer and K. R. Makinson, J. Textile Inst., 38, T227 (1947).
81. H. G. Howell and J. Mazur, J. Textile Inst., 44, T59 (1953).
82. W. F. duBois, Textile Res. J., 29, 451 (1959).
83. W. F. duBois, J. Textile Inst., 57, T262 (1966).
84. J. F. Archard, Proc. Roy. Society (London), A243, 190 (1957).
85. H. G. Howell, Textile Res. J., 23, 589 (1953).
86. A. Viswanathan, J. Textile Inst., 57, T30 (1966); 57, T263 (1966).
87. N. Gralen, B. Olofsson, and J. Lindberg, Textile Res. J., 23, 623 (1953).
88. R. B. Belser and J. L. Taylor, Frictional Properties of Cotton Fibers, USDA, Grant No. 12-14-100-7661 (72), February 1968.
89. R. B. Belser and J. L. Taylor, 1st Symposium International de la Recherche Textile Cotonniere, Paris, 1969, p. 419.
90. H. R. Levy, M.S. Thesis, Ga. Tech., Atlanta, Ga., 1966.
91. L. B. Whitworth, M.S. Thesis, Ga. Tech., Atlanta, Ga., 1967.
92. M. W. Pascoe and D. Tabor, Research (London), 8, S15 (1955).
93. T. E. McBride, M.S. Thesis, Ga. Tech., Atlanta, Ga., 1965.
94. R. S. Krowicki, M.S. Thesis, Univ. of Tenn., Knoxville, Tenn., 1960.
95. A. Viswanathan, Proceedings of the Fifth Joint Technological Conference, SITRA, Coimbatore, India, Dec. 1963.
96. C. Nanjundayya, J. Sci. Ind. Res., 17A, 412 (1958).
97. A. Viswanathan, private communication, 1972; J. Textile Inst., 64, 553 (1973).
98. A. Barella, P. Miro, and R. Crespo, Textile Res. J., 30, 633 (1960).

99. A. Barella and A. Sust, Textile Res. J., 32, 217 (1962).
100. A. Barella, J. Textile Inst., 43, P734 (1952).
101. A. Barella, A. Sust, P. Miro, and S. Anguera, J. Textile Inst., 54, T138 (1963).
102. M. Hannah, J. Textile Inst., 41, T57 (1950).
103. C. B. Landstreet, P. R. Ewald, and J. Simpson, Textile Res. J., 27, 486 (1957).
104. J. Simpson, C. B. Landstreet, B. Schnabel, and H. Hutchens, Textile Quality Control Papers, 3, 70 (1956).
105. R. R. Sullivan, J. Appl. Phys., 13, 157 (1942).
106. C. B. Landstreet, P. R. Ewald, K. L. Hertel, and C. S. Craven, Textile World, 104, 106 (1954).
107. L. C. Weiss, R. H. Tsoi, J. E. Sands, and J. N. Grant, Textile Res. J., 34, 605 (1964).
108. K. L. Hertel, Proc. 1967 Cotton Research Clinic, Pinehurst, N.C., 1967, p. 24.
109. K. L. Hertel, Textile Res. J., 37, 433 (1967).
110. K. L. Hertel and R. Lawson, Proc. 1968 Cotton Research Clinic, Pinehurst, N.C., 1968, p.48.
111. K. L. Hertel, K. E. Duckett, and R. Lawson, 1st Symposium International de la Recherche Textile Cotonniere, Paris, 1969, p. 139.
112. K. L. Hertel and R. Lawson, Textile Res. J., 40, 472 (1970).
113. C. C. Cheng and K. E. Duckett, Textile Res. J., 42, 51 (1972).
114. C. C. Cheng and K. E. Duckett, Textile Res. J., 42, 580 (1972).
114a. C. C. Cheng and K. E. Duckett, Textile Res. J., 43, 527 (1973).
114b. K. E. Duckett and C. C. Cheng, Textile Res. J., 44, 365 (1974).
115. R. J. Harper, E. J. Blanchard, J. T. Lofton, and G. A. Gautreaux, Textile Res. J., 37, 233 (1967).
116. American Society for Testing and Materials, Part 24, D1175-71 (1971).
117. American Society for Testing and Materials, Part 24, D1379-64 (1971).
118. American Society for Testing and Materials, Part 24, D2814-69T (1971).
119. J. N. Grant, F. R. Andrews, L. C. Weiss, and C. B. Hassenboehler Jr., Textile Res. J., 38, 217 (1968).
120. I. V. deGruy, J. H. Carra, V. W. Tripp, and M. L. Rollins, Textile Res. J., 32, 873 (1962).
121. M. L. Rollins, I. V. deGruy, T. P. Hensarling, and J. H. Carra, Textile Res. J., 40, 903 (1970).
122. W. R. Goynes and M. L. Rollins, Textile Res. J., 41, 226 (1971).
123. R. B. Stone, USDA Publication ARS 42-37, 1959.
124. R. B. Stone Jr., and J. R. Barrett Jr., Transactions of the ASAE, 5, 46 (1962).
125. J. L. Kotter and A. Baril, Textile Res. J., 39, 385 (1969).
126. W. J. Thorsen, Textile Res. J., 41, 321 (1971).
127. M. J. Williams, D. R. Lawson, and M. L. Staples, Can. Textile J., 88, 63 (1971).

128. J. J. Lefranca Jr., M. Mayer Jr., and H. W. Weller Jr., Textile World, 116, 58 (1966).
129. M. Hayek and F. C. Chromey, American Dyestuff Reporter, 40, P164 (1951).
130. W. T. Morris, Plastics and Polymers, 38, 41 (1970).
131. S. P. Hersh and D. J. Montgomery, Textile Res. J., 25, 279 (1955).
132. D. J. Lehmicke, American Dyestuff Reporter, 38, P853 (1949).
133. P. E. Shaw and C. S. Jex, Proc. Roy. Soc. (London), A118, 97 (1928).
134. P. E. Shaw and C. S. Jex, Proc. Roy. Soc. (London), A118, 108 (1928).
135. S. P. Hersh and D. J. Montgomery, Textile Res. J., 26, 903 (1956).
136. H. W. Weller and A. Baril, Textile Res. J., 41, 945 (1971).
137. F. P. Slater, J. Textile Inst., 16, P53 (1925).
138. S. Baxter, Trans. Faraday Soc., 39, 207 (1943).
139. J. W. S. Hearle, J. Textile Inst., 43, P194 (1952).
140. P. S. H. Henry, British J. Appl. Phys., 4, S78 (1953).
141. T. S. Herman, J. Appl. Poly. Sci., 9, 3953 (1965).
142. N. J. Harrick, Internal Reflection Spectroscopy, Wiley-Interscience, New York, 1967.
143. V. W. Tripp, B. Piccolo, D. Mitcham, and R. T. O'Connor, Textile Res. J., 34, 773 (1964).
144. B. Piccolo, D. Mitcham, and R. T. O'Connor, Appl. Spectroscopy, 22, 502 (1968).
145. H. A. Liebhafsky, H. G. Pfeiffer, G. H. Winslow, and P. D. Zemany, X-Ray Absorption and Emission in Analytical Chemistry, Wiley, New York, 1960.

Chapter 3

SURFACE PROPERTIES OF WOOL FIBERS

K. Rachel Makinson

Division of Textile Physics
Commonwealth Scientific and Industrial Research Organization
Sydney, Australia

I. INTRODUCTION

The direction of research efforts into the various aspects of the surface properties of wool fibers has been markedly different from that in the study of most other materials. This is partly because the character of the fiber surface makes the customary measurements inappropriate or difficult, and partly because certain unusual properties of the surface and near-surface layers have considerable practical and commercial interest, and have therefore attracted much attention. To take some examples: it would be inappropriate at present to concentrate a great deal of effort on studies of the tertiary arrangement of molecules in the surface layers, since wool fibers have a histological substructure down to near-molecular dimensions which, while correlating with chemical differences, outweighs in importance any small differences in tertiary structure. Furthermore, it is difficult and of limited value to measure adsorption behavior because the bulk phase can absorb most of the substances that it can adsorb. As a third example: much effort has been concentrated on studies of frictional properties in the wet state because these have an importance which they do not have for any other material, being the basic cause of felting and felting shrinkage, and having to be modified if wool fabrics are to be shrinkproofed.

In this chapter, therefore, surface structure and the frictional properties of wet fibers will be described at some length, while some other surface properties will be mentioned only briefly.

II. STRUCTURE OF THE CUTICLE

A. General Structure of the Wool Fiber

Wool fibers have the form, approximately, of slightly elliptical cylinders (axial ratio usually between 1.1 and 1.3), with mean diameters for different types ranging from 15 μm to more than 50 μm; the range from 20 to 25 μm is typical for merino wool, which is most used for apparel. The fibers are composed principally of keratin, a protein or complex of proteins characterized by the presence of a considerable amount of the amino acid cystine, HOOC—CH(NH_2)—CH_2—S—S—CH_2—CH(NH_2)—COOH. The disulfide bond in this amino acid residue forms crosslinkages between different protein chains or different parts of any one chain, thereby making keratin chemically more stable and physically less soluble than most proteins. Keratin also contains many amino acid residues with polar (acidic or basic) side chains, and many hydrogen atoms lying in such positions that they can form hydrogen bonds within and between the chains. Salt links between the acidic (COO^-) and basic (NH_3^+) terminal groups on the side chains help the hydrogen bonds to stabilize the structure of dry keratin, but both these types of bond are broken to an increasing extent as the keratin absorbs water, which it can do up to a maximum of approximately 34% of its dry weight. The disulfide bonds are unaffected by water. Hydrophobic bonding between nonpolar chains also occurs, but only, of course, at high water contents.

Each wool fiber has complex histological, chemical, and crystallographic substructures. We are concerned here only with the structure of the near-surface layers, i.e., with the cuticle of the fiber. The typical external appearance is shown in Fig. 1, and the basic histological structure is shown diagrammatically in Fig. 2. The cuticle is made up of flattened overlapping cells that are cemented firmly to each other and to the underlying material. The cuticle encloses a cortex made up of spindle-shaped cells, also strongly cemented together. The thickness of the cuticle in merino wool varies from about 0.5 to 1.5 μm. The length of the exposed part of the cell is, on the average, about 12-15 μm. It has been generally accepted that the cuticle cells in merino wool overlap for only about one-sixth of their total length [4], but there is some recent evidence [5] that the degree of axial overlap may vary around the circumference and may be much greater on one side than on the other. In the circumferential direction, in merino wool, there are usually only one or two cells, with little overlap, in the outermost layer.

The finer kinds of wool fiber, such as merino, are waved or crimped, with up to 10 crimps/cm. In such wools the cortex is bilaterally differentiated, chemically and physically. The smaller moiety, which runs along the inside of the crimp and is called the paracortex, is chemically more resistant than the remainder, which runs along the outside of the crimp and is called the orthocortex. The paracortex is more highly crosslinked by disulfide bonds than the orthocortex, and the two regions differ in their histological and crystallographic substructure. There is some evidence, not yet generally accepted or explained, that there may also be chemical differences between the cuticular cells on the inside and outside of the crimp [6-10].

B. The Scales

It can be seen from Figs. 1 and 2 that the overlapping cuticle cells form a stepped structure, like a tiled or shingled roof. This is

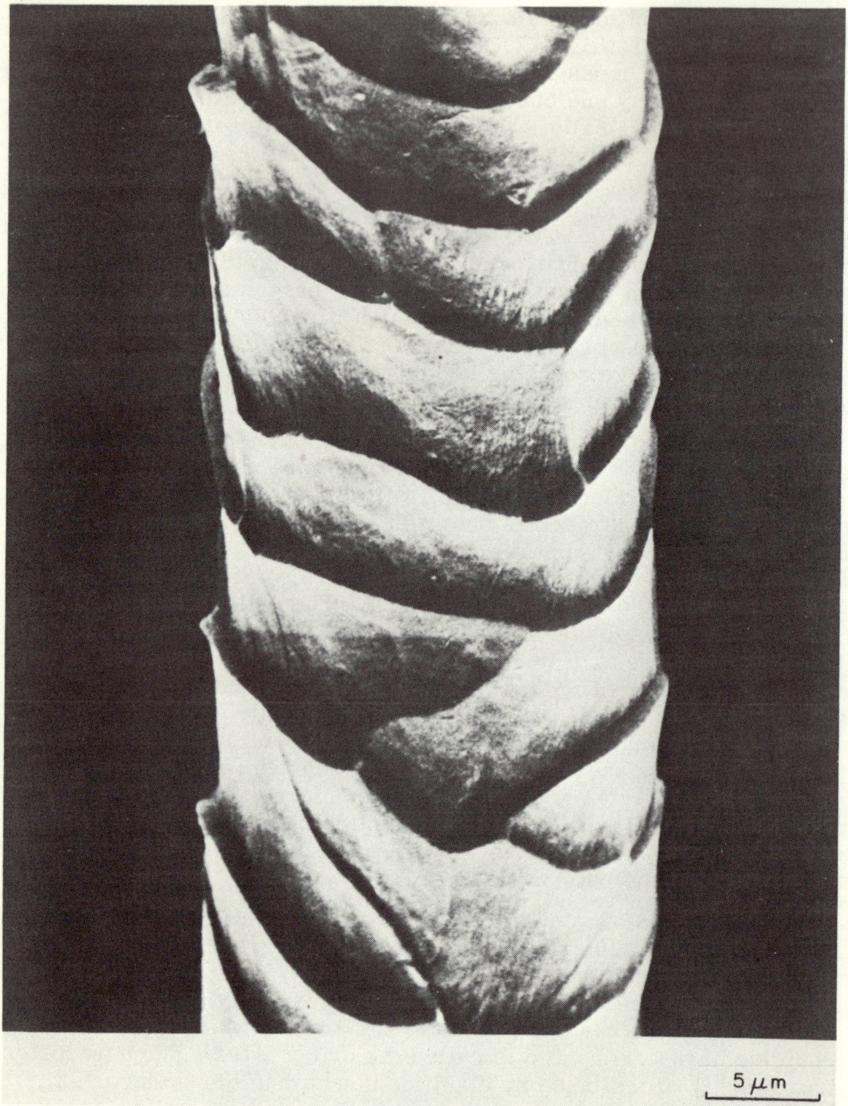

FIG. 1. Scanning electron micrograph of a merino wool fiber. Reproduced from ref. 1, Fig. 2, courtesy of Textile Research Institute.

usually called a scale structure, from its resemblance to the scaly skin of many reptiles. The steep face of each scale is toward the distal or tip end of the fiber. A number of workers have shown that there are also "false" scale edges (F in Figs. 2 and 4) between the "true"

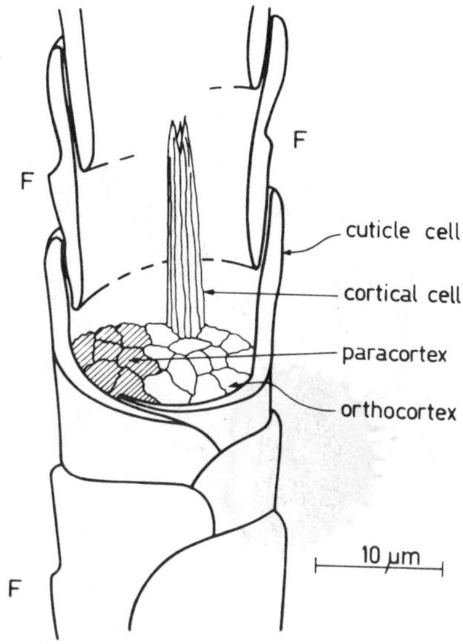

FIG. 2. Diagram of the histological structure of a wool fiber. F is a "false scale edge." Based on ref. 2 per ref. 3.

edges that correspond to the ends of the cells; in merino wool these false scale edges make up between one quarter and one half of the total number. Their presence has been established by electron microscopy of longitudinal sections of the fiber [5,11-13], by optical and electron microscopy of isolated cuticle cells [14], by optical microscopy of whole fibers [8], and by use of the Allwörden reaction, a technique of osmotically swelling the cuticular cells after degrading their protein content with chlorine [5,14,15]. It is now believed that all the scale edges, "true" and "false," are molded on the cuticle during growth of the fiber by the serrated inner root sheath of the follicle from which the fiber grows. In merino wool there is synchronization between the periodicity of the inner root sheath and double the periodicity of the cuticle cells, so that nearly all distal ends of cuticle cells correspond to scale edges, although the converse is not true. In some types of animal hairs there is no relation at all between the scale edges and the boundaries of the cuticle cells [14].

C. Fine Structure of the Cuticle Cells

Electron microscopy reveals that each cell in Fig. 2 has a substructure; that of the cuticle cells is shown in Figs. 3 and 4.

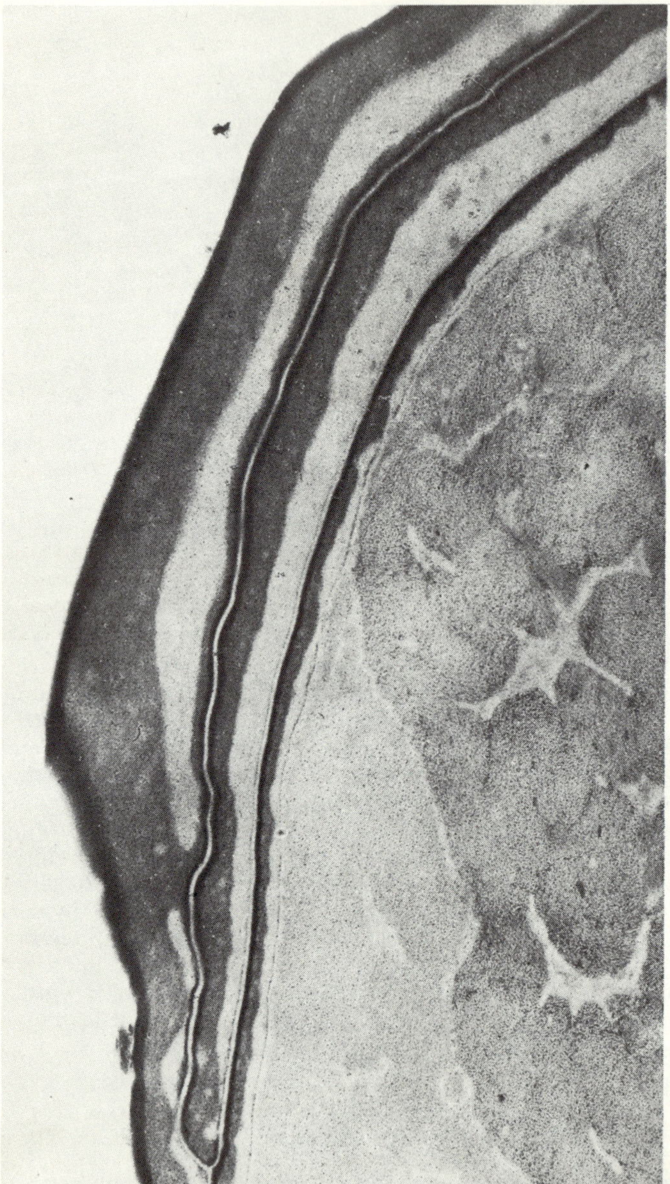

FIG. 3. Electron micrograph of a transverse section of a wool fiber, stained with mercury [16] to show structure of cuticle cells. Compare layers with those in Fig. 4. Micrograph courtesy of the Textile Physics Laboratory, University of Leeds, England.

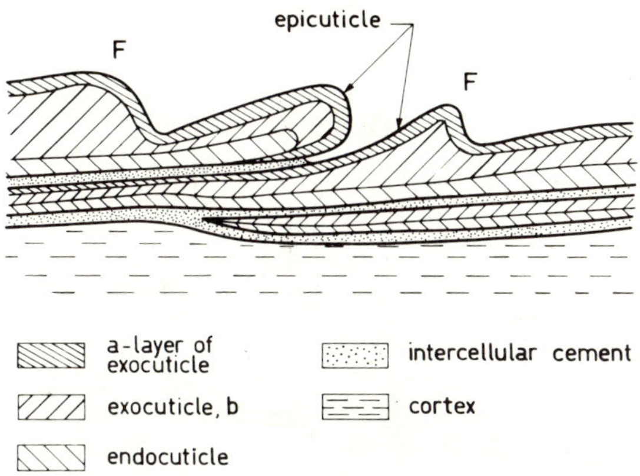

FIG. 4. Diagram of the structure of cuticle cells, in longitudinal section (not drawn to scale). F is a "false scale edge."

1. The Epicuticle

Every cell in the fiber is enclosed in a thin membrane, the thickness of which, for cuticle cells, has been variously estimated as 50-100 Å and 20-40 Å [17-19]. The former value seems more probable, as it corresponds to the usual thickness of plasma membranes (those membranes forming the outer envelopes of cells); the lower values may be due to splitting of the membranes, or loss of some component lipids, during isolation. Further work is needed to resolve this question. The parts of the plasma membranes of the cuticle cells that lie on the outside of the fiber are collectively called the epicuticle. This name is a relic of the history of the discovery of this membrane [17, 18,20-26]; many, though not all, workers believed that it formed a continuous film over the outer surface of the fiber. Recent work [14, 15,27] has shown conclusively that the epicuticle is discontinuous and is formed by the external part of the plasma membrane of each cell, these membranes being firmly cemented together, where they touch, by the intercellular cement (to be discussed subsequently in Section 2).

Cell membranes from cuticle and cortex together comprise 2-3% of the weight of the fiber [28]. This figure includes the lipid component but excludes the intercellular cement. The parts that are collectively called the epicuticle form only a small part of this. Isolation and analysis of the epicuticle is therefore difficult; it is likely either to be contaminated with material from other parts of the cuticle, or to have been damaged by the isolation procedures. It has at various times been considered to consist predominantly of lipids [29,30] or, more specifically, sterols [31], or to be proteinaceous with some carbohydrate content [32]. Most workers now agree that it is largely proteinaceous [33-38], as are biological membranes in general. The proportion of lipid content is uncertain, as it may be removed during the preparative procedures. The protein component is keratinous, containing about 12% cystine [28,37]; there is nothing in the composition of this component, taken in isolation from the lipid, to explain the extreme chemical inertness of the membrane, which can be damaged by extraction of the lipid component with strong organic solvents, but unless so damaged is resistant to alkalies, acids, oxidizing and reducing agents, and enzymes [5,27,39]. Such inertness is characteristic of biological membranes in general; the explanation is to be found in their molecular organization and the interaction between proteins and lipids. The information available in this field comes from the study of various other types of cell membrane, not that of wool cells in particular.

There is such activity at present in the field of membrane structure that it is impossible to present a comprehensive survey of current theories in the confines of this chapter. Figure 5a shows what may be considered the traditional picture [40,41], perhaps the simplest of the many structures that have been proposed. This structure is known as the unit membrane. It consists of a bimolecular lipid leaflet sandwiched between protein layers (which are keratinized in the case of plasma membranes). However, this structure is now considered to be thermodynamically unstable [42]. A currently favored structure is that shown in Figs. 5b and 5c [42-44]. In this model, globular protein molecules are embedded in a bimolecular lipid leaflet. All structures based on such a bimolecular leaflet are expected to be capable of splitting into two thinner sheets (along the direction of the arrow in Fig. 5b), and this in fact has been demonstrated recently [43-45] by the freeze-etching technique [46,47]. Another structure that has been proposed is one in which globular lipid micelles are embedded in protein [48,49]. Some proposed structures allow for extensive hydrophobic bonding between proteins and lipids, which would account for the high stability of membranes. The structure shown in Figs. 5b and 5c is one example; another is a two-layer structure of globular protein having the interstices filled with lipid molecules so oriented as to form small areas of bimolecular lipid leaflet [50]. There is some evidence that membranes may change their structure reversibly when exposed to changing conditions [51]. Many of the proposed structures are discussed in refs. 52-56.

2. The Intercellular Cement

When two cells in a tissue are juxtaposed, junctions of various degrees of closeness are formed over limited discrete areas of their membranes; at most places these membranes are not in direct contact but are held together by a layer of protein which is usually called

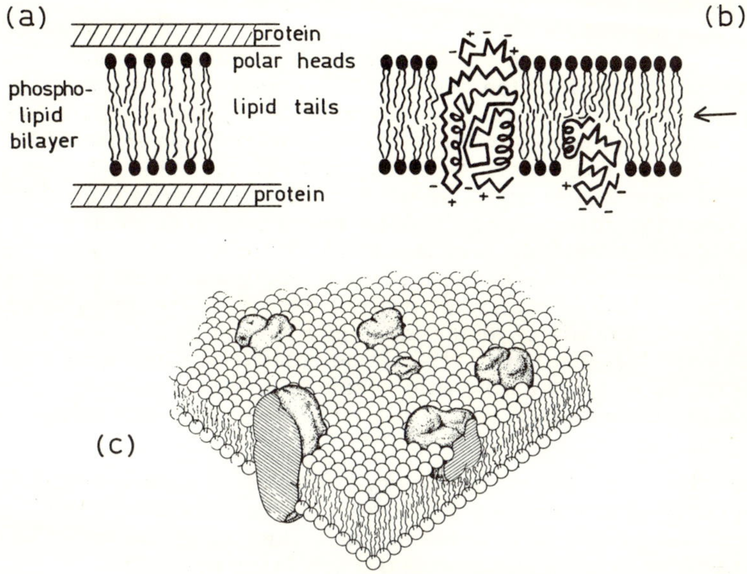

FIG. 5. Diagrams of two models proposed for plasma cell membranes. (a) The unit membrane [40,41]; (b) and (c) the fluid mosaic model [42]. In (b) and (c) the globular protein molecules, which have polar and nonpolar regions, are shown embedded in the phospholipid matrix, which has a bilayer structure. Figure 5a based on and Figs. 5b and 5c reproduced from ref. 42, courtesy of the authors and the publishers. Copyright 1972 by the American Association for the Advancement of Science.

intercellular cement. The term "cell-membrane complex" is applied to the complete structure consisting of two neighboring membranes and the cement between them. Such cement is estimated to make up about 1% of the weight of the wool fiber [28]. Its thickness is usually about 150 Å, varying from place to place, as it fills all gaps between cells. It is considered to be nonkeratinous, having an extremely low cystine content [57-60]; it is consequently labile, being dissolved by acids, alkalies, oxidizing and reducing agents, and enzymes [61,62].

3. The Exocuticle

Below the epicuticle, and adhering firmly to it, lies a thicker layer of protein that is also essentially keratinous in nature [Figs. 3 and 4]. Some of the limited number of electron micrographs of longitudinal sections that have been published appear to show this material extending around the "true" scale edge [12,63]; and the electron micrograph of a transverse section reproduced in Fig. 3 shows a thin layer of it in the under part of the outermost cuticle cell. In many cuticle cells, however, there appears to be little or no exocuticle in the under part. The exocuticle also fills the "false" scale edges [12]. It makes up approximately 5% of the weight of the wool fiber, and is approximately 0.3 μm thick, with considerable variation. It has been shown [64-69] to consist of two sublayers, the outer, or a layer, having a higher content of cystine than the inner, or b layer. The a layer is 0.05-0.1 μm thick. Since the cystine content of the whole cuticle is about 16% [70] and that of the endocuticle (see Section 4, following) is very low indeed, that of the exocuticle and of the a layer in particular must be exceptionally high. This is confirmed by recent work by Bradbury and Ley [71], who find that 20% of the amino acid residues in the whole exocuticle (a and b layers) are half-cystine residues. Correspondingly, the proportion of polar amino acid residues in the exocuticle is lower than that in the epicuticle, which in turn is lower than that in the whole fiber.

It is difficult to determine the crystallographic nature of these small quantities of material, but Rogers [65] has concluded that the protein of the exocuticle is predominantly amorphous, as would be expected because protein chains containing a large proportion of cystine residues cannot readily form an ordered array [3]. It is not known whether the protein consists of globular molecules or of chains folded at random.

4. The Endocuticle

Below, or inside, the exocuticle lies the endocuticle (Figs. 3 and 4), a layer of nonkeratinous protein material. Like the exocuticle, it constitutes about 5% of the weight of the whole fiber and is about 0.3 μm thick on the average, but the relative proportions of exo- and endocuticle vary widely between cells. Because of its extremely low cystine content of about 3% [71] it is relatively labile and can be extracted by digestion with proteolytic enzymes [62]. It is also dissolved by acids [72]. Analysis shows that its content of polar amino acid residues is high [71].

D. The Allwörden and Herbig Reactions and the Effect of Chlorine on the Cuticle Cells

The description given above of the fine structure of the cuticle of wool fibers is not based solely on electron microscopy and chemical analysis; such studies have proceeded in parallel with studies of the Allwörden reaction, a striking phenomenon that has proved to be of considerable value in structural investigations.

In 1916 von Allwörden [21] reported that when a wool fiber is immersed in chlorine water, characteristic protuberances (Auswüchsen), clearly visible under a microscope, are formed on its surface, each protuberance covering one or, less often, two scales. Later workers showed that the protuberances were really sacs [24] containing a liquid [22] enclosed by a thin membrane [23]. Subsequently this membrane was isolated and examined by electron microscopy by Lindberg et al. [17,18] and by Schuringa and Algera [25], and was named the epicuticular membrane or epicuticle [39]. Its more recent history has been discussed above.

The rate of development of Allwörden sacs on undamaged wool fibers is remarkably constant, provided the conditions are kept reasonably constant [10,73]. The sacs reach their maximum size about 20 sec after exposure of the fiber to saturated chlorine water at room temperature. Figure 6 shows a merino wool fiber with sacs well developed.

The Allwörden sacs collapse when punctured [22] and when exposed to concentrated salt solution [24,74]. These observations, along with all subsequent ones, confirm the hypothesis originally put forward by Stirm and Colle [75] and supported by later workers [18,23,37,76], that the sacs are formed by osmotic swelling, by water, of degraded material of high molecular weight beneath a semipermeable membrane (the epicuticle).

The Allwörden reaction can be produced on isolated cuticle cells [10,27,74]; this circumstance makes it appear certain that the epicuticle consists of the external parts of the cuticle cell membranes, and is not a continuous film.

It is generally though not universally accepted that the principal effect which chlorine has on keratin is to break the disulfide bonds, converting some of them to sulfonic acid groups, $-SO_3^-$ (see refs. 73 and 77). It is also accepted that there is some peptide-bond breakage at the same time; most workers consider this to be less important, but both types of attack may be necessary for the Allwörden reaction (and for shrinkproofing, see Section E below). The sulfonic acid groups solubilize the degraded protein material, promoting osmotic swelling in water. The part of the cuticle cell richest in disulfide bonds is the exocuticle, which is thickest at the scale edges and fills the outer half of the cuticle cells. This conclusion correlates well with the observation that Allwörden sacs begin to form at the scale edges and are usually formed only on the outer side of the cells, even when the cells are released from the fiber. However, there has been one report [74] of sac formation on both sides of isolated cells. If there is, as has been suggested [63] a very thin layer of exocuticle, probably an a layer, on the inner side of the cell, the distribution of swelling in the Allwörden reaction could well be determined by a rather critical balance between the levels of degradation of the exocuticle and endocuticle.

Absence of the Allwörden reaction in saturated chlorine water, or delay in development of the sacs, indicates that the cuticle has previously been modified in some way. Known or suggested possibilities

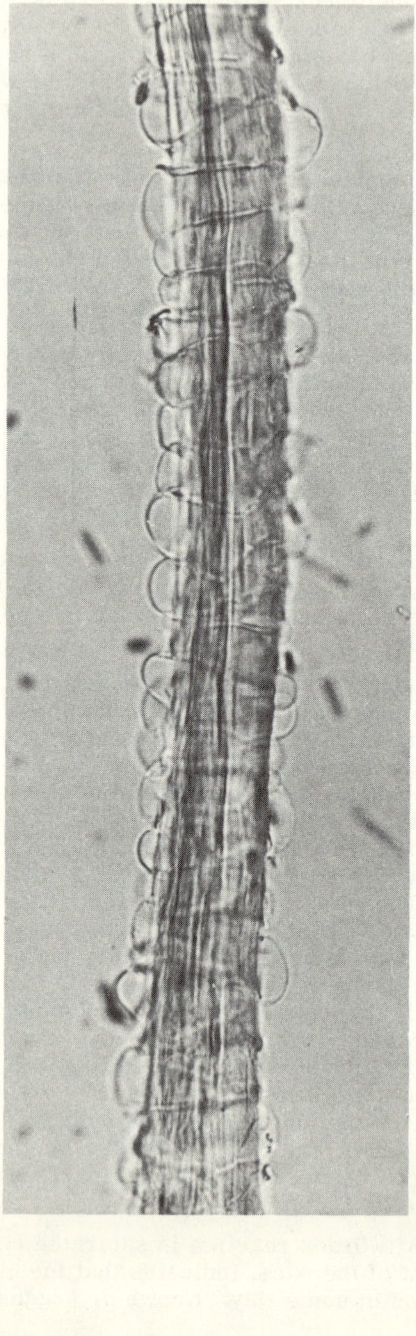

FIG. 6. Allwörden reaction on a merino wool fiber.

are: that the epicuticle has been removed or severely damaged; that the disulfide crosslinks have been stabilized by conversion to lanthionine [73]; that the disulfide bonds have been broken and blocked so that they cannot form sulfonic acid groups [73]; that the protein content of the cell has previously been degraded to fragments with low molecular weight that have already diffused out of the cell, or do so too quickly for swelling to occur [8]; that the epicuticle has been reinforced so that it is less extensible and suppresses swelling [78].

Sacs are also produced on wool fibers by bromine water. This is often called the bromine Allwörden reaction, but, following Leeder [10], it is better called the Herbig reaction after its discoverer [79], particularly as it is now known that the sacs differ considerably from those produced by chlorine. The Herbig sacs form more slowly and more variably than the Allwörden sacs, and they are bounded by a thicker membrane that appears to consist of epicuticle with various amounts of material from the other layers of the cuticle adhering to it [10,76,80-82]. There is some evidence that bromine attacks the endocuticle and the intercellular cement rather than, or as well as, the exocuticle [10].

E. Physical Properties of the Cuticle

The physical properties of wool fibers depend markedly on their content of absorbed water; this is true of the cuticle as well as of the whole fiber. Water acts as a plasticizer; it reduces the mutual attraction between acidic and basic groups and breaks many hydrogen bonds in the wool structure, thereby making the keratin softer. However, the disulfide bonds maintain their integrity, setting a limit to this softening and preventing the wool from dissolving. The cortex, which contains fibrils of highly ordered ("crystalline") material in an amorphous matrix, becomes anisotropic as water is absorbed, but this reaction is believed not to occur in the cuticle, which is, as far as can be determined, amorphous.

Wet keratin is highly deformable and can also recover its original dimensions completely after large deformations, up to 70% tensile strain. It is however, imperfectly elastic, since some energy is lost during any cycle of deformation and recovery. This hysteresis becomes more marked as the water content decreases. The tensile and torsional elasticity of the whole fiber has been much studied, but relatively little attention has been paid to the elastic behavior of the scales, mainly because of the difficulty of making quantitative experiments on objects so small and irregularly shaped. Micromanipulation has shown [8,83] that, as would be expected, the cuticle of normal, unmodified wool fibers can recover from quite large deformations in both the wet and the dry state; recovery occurs in times of the same order as, or less than, the period of deformation, and, as far as can be judged from qualitative experiments, is complete. It is to be expected that there is some hysteretic loss of energy in the course of a cycle of deformation, but this has not been demonstrated for the cuticle cells. The epicuticle adheres strongly to the exocuticle and is also highly deformable and recoverable (as is shown, in addition, by the extent to which it can stretch and recover during the Allwörden reaction). It can be damaged and torn off in small fragments by repeated rubbing, particularly under tension [8], but most of it remains intact after normal commercial processing.

The cuticle becomes transversely cracked when the whole fiber is stretched in the dry condition, and even in the wet condition when there are particular points of concentration of stress [84]. A photomicrograph of such cracking during extension of a dry human hair has been published by Lehmann [85]. The cuticle cracks longitudinally when the fiber is swollen about 30% diametrically by such reagents as formic acid [86], but not when it is swollen up to its normal limit of 16% diametrically by water.

III. FRICTIONAL PROPERTIES OF WET WOOL FIBERS, AND FELTING

A. Felting

It is a consequence of the scaly structure of the cuticle that a wool fiber is rougher when it is rubbed from tip to root (against scale) than when it is rubbed from root to tip (with scale). The difference can be felt quite easily if a wool fiber, or a hair, is drawn slowly between finger and thumb in each direction. This "frictional difference" is the basic cause of felting and is therefore of practical importance.

Felting is a process of compaction and entanglement of fibers that occurs when wool is agitated in water. (Normally, it does not occur to any appreciable extent unless some free water is present. The reasons for this are not fully understood; a number of factors appear to be involved.) Felting is used to produce industrial felts and nonwoven fabrics from loose wool. It is also used, in the finishing processes called "fulling" and "milling," to give a characteristic soft, uniform finish to certain kinds of wool fabric (those made on the woolen, as distinct from the worsted, system). When fabrics are felted, they contract in area. Such shrinkage is undesirable once the fabric has been finished and made into garments, so if wool garments are to be washed during use, felting must be prevented by an antifelting treatment. These treatments are commonly called "shrinkproofing," although shrinkage can arise from other causes besides felting (and can therefore occur with fabrics made from other kinds of fibers). Shrinkage due to felting has characteristic features which make it easy to recognize. In the first place, the other types of shrinkage usually reach a limit in one or two washes, but felting shrinkage continues to occur every time the fabric is washed, although its rate eventually decreases. With most other types of shrinkage, individual yarns in the fabric remain clearly visible; when felting occurs they are eventually blurred by the movement of individual fibers out of the yarns. On the microscopic scale, felting is characterized by the continual unidirectional movement of individual fibers towards their root ends. This occurs because each fiber in a fabric or other assemblage is held in position solely by friction against its neighbors. During washing or fulling treatments impulsive forces are applied to the fabric from many directions at random; those that drive a fiber towards its root end have to overcome a smaller frictional force than those directed to drive it toward its tip. The fibers therefore slip toward their root ends more often than toward their tips, and appear to migrate continually in one direction. In tightly constructed fabrics whole fibers do not usually move; parts move, and other parts are restrained.

A detailed discussion of how this migration causes shrinkage would not be appropriate here. There appear to be many different mechanisms, corresponding to different types of fiber assemblage [87-95]. A feature common to most of them is that any fibers put into tension by this migration pull their neighbors closer together, shrinking the fabric; but any that are put into compression will buckle and therefore be ineffective in expanding the fabric.

B. Definition and Measurement of Coefficients of Friction

The coefficient of friction f between a wool fiber and a plane surface, in either the with-scale (f_w) or the against-scale (f_a) direction, is defined as the ratio of the frictional force F opposing sliding in that direction during or at the moment of sliding, to the total normal force L pressing the surfaces together. The assumption is implicit that these forces are proportional to each other. When this is not the case, it is, strictly speaking, incorrect to speak of a coefficient of friction, although this often is done and the "coefficient" is then said to be dependent on the normal load.

Friction is commonly divided into two classes, static friction and kinetic friction. The coefficient of static friction is defined as the above ratio with F measured at the moment when sliding is about to begin; usually F at this moment has the maximum value attainable in a particular set of conditions. The coefficient of kinetic friction is defined as the above ratio, with F measured during sliding; it may depend on the sliding speed, and is usually less than the coefficient of static friction, except at high speeds. Conceptually, kinetic friction is the simpler of the two; it is relatively precisely defined and can be dealt with mathematically by energetics. Static friction is essentially an imprecise concept, because there is always some changing deformation and sometimes some actual slip before sliding can be perceived, and because, if there were really no movement before sliding begins, as is assumed in all simple treatments, no energy changes would be involved. However, for brevity it is usually necessary to limit the discussion to one or the other frictional coefficient; whenever possible static friction will be chosen because it is more familiar to nonspecialists in the field and has been measured more often in the case of wool fibers. When certain mechanisms are discussed, however, it will be necessary sometimes to refer to kinetic friction. The discussion that follows, although dealing only with static friction, can readily be extended to kinetic friction.

The coefficient of static friction, f, between two rigid, macroscopic bodies is measured by placing one on the other, pressing them together with a known force L, and measuring the tangential force (parallel to their common surface) needed to cause sliding. This force is equal and opposite to the force of static friction F. Because the bodies are rigid the load L is distributed uniformly, on the macroscopic level, over the bearing area, so the definition $f = F/L$ is directly applicable. A wool fiber, however, is a flexible, extensible filament, in which the tension and normal load can vary along its length; the definition therefore applies directly only to elementary areas of contact. It is customary to ignore the extensibility of the fiber and to use its flexibility in the measurement procedure. A number of methods of measurement have been used [96a]; by far the commonest is the capstan method. In the simplest form of this method, the wool fiber is hung over a rod and loaded with equal weights at both ends. The weight at one end is then gradually increased until the fiber begins to slip over the rod. At this instant the tension in the fiber at either end is equal to the weight hanging at that end, while the tension and the normal load vary monotonically along that part of the fiber in contact with the rod. If it can be assumed that there is a true coefficient of friction independent of the load, i.e., that the frictional force is proportional to the load, then it can be shown [96a] that $T_\theta/T_0 = \exp f\theta$, where T_0 is the tension at the lightly loaded end and T_θ is the tension at an angular distance θ around the rod. In the simple experimental arrangement described the tension is known for $\theta = \pi$. Hence f can be calculated. If f is not independent of L, however, much more complicated relations must be used [96a]. It is therefore advisable to choose an experimental arrangement in which f is constant [97].

The coefficients of interest in connection with felting are those pertaining to wool fibers in water. Because immersion of the simple capstan system in water would introduce large buoyancy effects, the surfaces were simply kept wet during measurement in much of the early work. This led to a nonlinear relation between F and L, apparently because surface tension increased the normal load above the nominal value. To avoid this the assembly has sometimes been inverted [97,98], or the angle of wrap around the rod has been changed [99], so that the system can be fully immersed. Under these conditions a constant coefficient, independent of load, can be obtained [97].

The choice of a suitable rubbing surface for frictional measurements is not always easy. Ideally, for work in connection with felting, it would be desirable to use a second fiber. The relatively simple capstan method could not then be used, and one would have to choose whether to use transverse, longitudinal, or oblique sliding; different angles might be more appropriate to different types of fiber assembly. Fiber-on-fiber experiments at any angle are difficult and have only occasionally been used [100-103]. When the capstan method is used, the rod is usually cut from a bulk solid; horn has been found very suitable, in spite of its irreproducibility, because it, like wool, is keratinous and therefore has elastic constants of the same order of magnitude as those of a wool fiber under any particular conditions. It also has a convenient physical form; when polished, it presents a smooth but cracked surface, so that a fiber sliding over it in the with-scale direction "sees" it as smooth, whereas one sliding in the against-scale direction "sees" it as rough, creating the same situation that exists when a fiber slides over another oppositely oriented fiber. However, for examination of particular aspects of the frictional behavior, it may be necessary to use other materials. Sometimes wool fibers or yarn are wrapped around the rod, so that sliding occurs on a wool surface; the variability between repeated measurements is usually much greater with this arrangement, so it is not often used.

In Table 1 are listed some typical values of the coefficients of friction of wool fibers against various surfaces in aqueous media. These data should not be used for cross comparisons between independent experiments, in order, for example, to determine the effect of different media or different surfaces, because the values obtained are sensitive to details of the experimental conditions.

C. Relation of the Scales to the Frictional Difference

It seems obvious that if a wool fiber is covered with unidirectional scales it will be rougher in one direction than in the other. There has, however, been considerable argument as to how the scales cause the frictional difference, or even whether they cause it. There are in fact a number of mechanisms, different ones coming into play in different circumstances, and in particular on different types of mating surface. It is necessary to understand these mechanisms if measurements made with fibers sliding over different surfaces such as horn are to be extrapolated to the case of interest in felting, fiber-on-fiber sliding. Usually the actual magnitudes of the coefficients in the latter case are not required; the typical question for which an answer is needed is: "If a shrinkproofing treatment changes the with-scale friction and the frictional difference in certain ways for a wool fiber

TABLE 1

Typical Values of the Coefficients of Friction of Wool Fibers in Various Aqueous Media (Wool Not Shrinkproofed)

Rubbing surface	Solutes in water	f_w	f_a	$\sigma = f_a - f_w$	Ref.
Horn					
probably cow's	Soap and soda, pH 10.7	0.04-0.24	0.25-0.42	0.13-0.28	[104]
	None	0.50	0.85	0.35	[105]
	Borate (0.05 M); anionic detergent	0.25	0.39	0.14	[106]
	Buffer, pH 9; nonionic detergent	0.23	0.52	0.29	[107]
	Borax (0.05 M), pH≈9; anionic detergent	0.21-0.25	0.37-0.42	0.16-0.17	[108]
	Buffer, pH 9; anionic detergent	0.17	0.31	0.14	[109]
	Soap	0.12	0.25	0.13	[110]
	None	0.17	0.44	0.28	[111]
ram's	Buffer, pH 4	0.3	0.6	0.3	[112]
	Buffer, pH 7	0.5	0.9	0.4	[102]
	Soap	0.11-0.18	0.30-0.48	0.17-0.30	[9]
	Anionic washing powder	0.15-0.34	0.54-0.61	0.27-0.47	[83]
	Nonionic detergent	0.18-0.39	0.43-0.71	0.23-0.32	[83]
unspecified	None	0.36	0.51	0.15	[113]
Wool fiber, oppositely oriented	Buffer, pH 2	0.19-0.24	0.34-0.42	0.15-0.18	[100]

	Buffer, pH 7	0.19–0.22	0.35–0.36	0.14–0.16	[100]
	Buffer, pH 9	0.15–0.21	0.32–0.34	0.13–0.17	[100]
	None	0.17	0.64	0.47	[101]
	None	0.15	0.34	0.19	[114]
Wool fibers, oppositely oriented	Borax, 0.05 M, pH≈9; anionic detergent	0.18	0.30	0.12	[108]
	None	0.23	0.55	0.32	[115]
Sawtoothed diffraction grating, oppositely oriented	Soap	0.23	0.62	0.39	[116]
Polytetrafluoroethylene	Nonionic detergent	0.13	0.25	0.12	[117]
Rubber	Sodium oleate	0.22	0.27	0.05	[118]
Ebonite	None	0.62	0.72	0.10	[105]
Wool yarn	Soap	0.12	0.23	0.11	[110]
	None	0.23	0.45	0.23	[111]
Glass, plate	5% soda ash	0.20–0.33	0.27–0.38	0.02–0.07	[119]
rod	None	0.16–0.31	0.25–0.36	0.05–0.11	[111]
filament	None	0.17	0.28	0.12	[120]
Casein	None	0.20	0.67	0.47	[111]
Felt	None	0.11–0.18	0.37–0.51	0.22–0.33	[121]

sliding over a certain surface, such as horn, will it have the same qualitative effect when a wool fiber slides over another wool fiber?" This question cannot be answered without some understanding of the processes involved in sliding.

The mechanism responsible for with-scale friction will be considered first, and the discussion will be limited to kinetic friction. If we neglect for the present the effect of the small angle by which the flat scale face is inclined to the axis, the mechanism is that common to friction between most fairly flat, weakly adhering, elastomeric surfaces. The inert epicuticle of the wool fiber adheres to the mating surface under the action of van der Waals' forces only. Since these are weak, sliding, when it occurs, does so at the geometrical interface without damage to either surface. The frictional force however, does not arise solely at the geometrical interface, but principally from deformation, with losses of energy, of thin subsurface layers behind one or both of the surfaces. Under the low normal loads that are usually applied to wool fibers during felting and during measurement of friction, the thickness of these layers is in the range 10^2-10^3 Å in with-scale sliding. (In against-scale sliding these layers may sometimes be as thick as 10^4 Å, because then in extreme cases the whole cuticle cell may be deformed.) Some (probably most) of the deformation is due simply to compression, perpendicular to the direction of sliding, under the pressure of the normal load between the surfaces, and some is due to shear of the subsurface layers parallel to the direction of sliding. Both of these deformations travel with the relative motion of the sliding surfaces, so that energy is lost during sliding. In general these two types of deformation interact and cannot be considered additive.

When a fiber slides against scale, the friction is still due to adhesion and deformation, but the asymmetry of the scales may cause the deformation processes to differ considerably from those involved in with-scale sliding. This asymmetry is usually the major source of the frictional difference, but other factors may also contribute. The following mechanisms have been proposed to explain how the scales cause the frictional difference.

1. The Ratchet Mechanism

When the ratchet mechanism is operative, the scales of the wool fiber engage like the teeth of a ratchet with asperities on the mating surface (Fig. 7a). These asperities may be the scales of a second, oppositely oriented, wool fiber, or they may be steep asperities or the sides of cracks in some other material. Because the scales and asperities are irregular, the ratchet will be ill-fitting; its fitting is, however, better than geometrical considerations might suggest because the scales can deform a little to fit the asperities. Strictly speaking, a "ratchet" should totally prevent movement in the against-scale

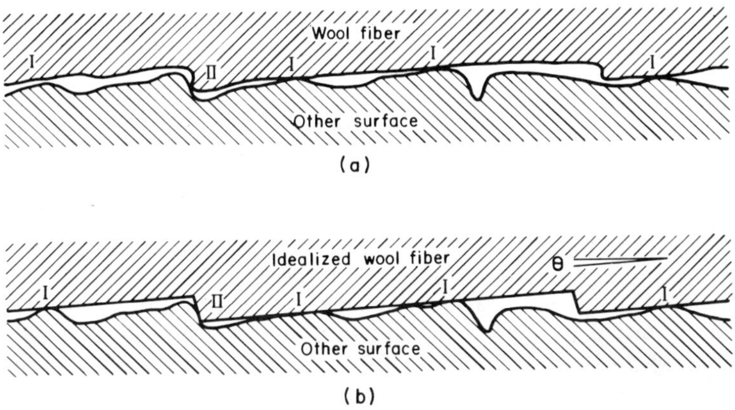

FIG. 7. Contact between a wool fiber and a flat surface with asperities. I. Contacts on the flat faces of the scales; II. contacts on steep faces. The lower diagram is an idealized version used for calculations. Reproduced from ref. 122, Fig. 4, courtesy of the International Wool Secretariat.

direction (unless forces great enough to break the teeth of the ratchet are applied), but this limitation is not intended by the use of the term "ratchet mechanism;" during measurement procedures, the fibers are forced to slide in the against-scale direction, the scales deforming rather than breaking, and even during felting some fibers slip in this direction occasionally.

That such engagement of the scales actually occurs has been established by microscopy and micromanipulation for wool fibers sliding along other, oppositely oriented, fibers and across diffraction gratings with saw-toothed profiles [8,9,83,116], and it has been shown indirectly that the scales also engage with the sides of the cracks in polished horn [8,9].

On the basis of the idealized case in Fig. 7b, it can be shown [123] that, approximately

$$f_w = \tan(\lambda + \theta) \tag{1}$$

and

$$f_a = (1 - m)\tan(\lambda - \theta) + \Sigma g/L \tag{2}$$

where (1 - m) is the fraction of elementary contacts made on flat faces of the scales (I in Fig. 7), m is the fraction made on steep faces

(II in Fig. 7), λ is the angle of friction, i.e., $\tan^{-1} f^*$ where f^* is the "true" or hypothetical coefficient of friction between the surfaces if the scales were quite flat, θ is the mean inclination of the flatter scale faces to the axis of the fiber (as in Fig. 7), Σg is the total force required to deform the scales that make steep-face contacts (or their mating asperities) sufficiently to permit against-scale sliding, and L is the total normal load pressing the fiber against the other surface. The frictional difference $f_a - f_w$ is therefore rather less than $\Sigma g/L$; physically this is because with-scale sliding is "uphill" at contacts of type I, whereas against-scale sliding is "downhill." However, it can be shown [122] that, qualitatively, changes occurring in $\Sigma g/L$ as a result of chemical or other treatments in general produce changes of the same type in the frictional difference (which is the measurable quantity).

Further references to theoretical discussions of the ratchet mechanism can be found in ref. 122. In general, the ratchet mechanism causes a higher value of the frictional difference than any other mechanism.

2. The Ploughing Mechanism

The frictional difference has sometimes been found to be quite large in circumstances in which it was clearly impossible for the ratchet mechanism to be operative, e.g., in the sliding of a wool fiber over a smooth rod of polytetrafluoroethylene (PTFE). The frictional difference in this case has been shown [117] to be due to ploughing of the scales through the PTFE in against-scale sliding; the scales actually cut out little chips of PTFE. In this case it has been deduced that, approximately,

$$f_w = f^*$$

and

$$f_a = f^* + kh^3 P_0/L,$$

where f^* and L have their previous meaning, h is the scale height, P_0 is the yield pressure of the PTFE, and k is a geometrical factor [122,124].

Ploughing is expected to occur in some circumstances during fiber-on-fiber sliding, but may then be expected to involve deformation rather than cutting [122].

3. Lincoln's Mechanism

Lincoln [125,126] considered, for the case of a wool fiber sliding over a surface of a bulk solid, the effect of the finite radius at the tips of the scales, which would lead to contacts of the type shown on the right in Fig. 8, that is, contacts not deep enough to touch the really steep part of the scale. At such contacts the scale can slide up the asperity, so that the ratchet mechanism is not operative. Lincoln's mechanism is operative on any surface having only rounded asperities. Equations of the same type as (1) and (2) still describe the coefficients of friction, but the term $\Sigma g/L$ is replaced by a smaller quantity. This would cause the frictional difference to become negative, were it not for the fact that a cylindrical fiber sliding over a flat surface makes a higher proportion of steep-face contacts than would be expected from the simple two-dimensional diagrams shown here.

It seems probable that this mechanism contributes to the frictional difference of a wool fiber sliding over any surface which has rounded asperities and which is too hard to be ploughed by the scales. However, Lincoln's mechanism seems unlikely to be of importance for sliding over another, oppositely oriented fiber or over any surface with steep asperities or cracks.

4. Flanagan's Mechanism

Flanagan [120] considered the particular case of a wool fiber sliding across a fine, transversely oriented glass filament, with a very short arc of contact. This geometry is relevant to fiber-on-fiber sliding in a mass of loose wool during the production of natural (nonwoven) felts. He calculated the frictional difference as a function of

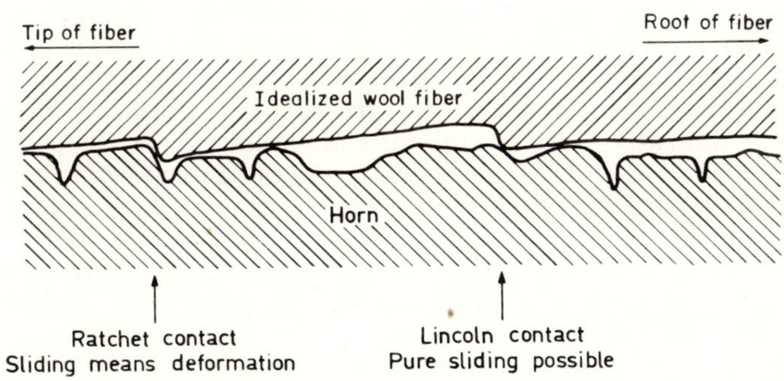

FIG. 8. Two types of steep-face contact between a wool fiber and horn. Reproduced from ref. 122, Fig. 5, courtesy of the International Wool Secretariat.

the diameter of the glass filament, and showed theoretically and experimentally that it fell to zero if the diameter exceeded a critical value, of the order of 100 μm in his experiments.

5. Grosberg's Mechanism

Grosberg [127] suggested that the frictional difference could arise from anisotropy of the scale tips. This theory has been discussed at length elsewhere [96b,122].

6. Rudall's Mechanism

A number of people [128-130a] have described a wooden model of a wool fiber made by Rudall. It had a serrated surface with a leaf of rubber glued to each tooth, and exhibited a frictional difference against glass. Various explanations of the effect have been suggested [128,129], but so little has been published about this model that it is difficult to assess its relevance to the sliding of wool fibers.

7. Martin's Mechanism

Martin [128] at one time suggested that the frictional difference was due, not to the geometrical shape or the mechanical properties of the scales, but to an asymmetry in the molecular fields at their surfaces. He postulated unequal fields in opposite senses parallel to the fiber axis, a situation which is probably thermodynamically impossible. A number of workers appear to have disproved this suggestion, see ref. 122.

In summary, of these six mechanisms which have been proposed to explain how the presence of scales on a wool fiber causes the frictional difference, the first two have been shown directly, by microscopy, to be operative in appropriate circumstances, and the next two are probably operative in certain other circumstances. It has been shown that in the case of the first two the elastic properties of the scales and/or of the surface layers of the mating material, expressed through Σg in the ratchet mechanism and P in the ploughing mechanism, are of major importance in determining the magnitude of the frictional difference. This conclusion will be referred to in the discussion of shrinkproofing.

IV. MODIFICATION OF THE WET FRICTIONAL PROPERTIES; SHRINKPROOFING

A. General Considerations

It is sometimes desirable to prevent felting shrinkage, e.g., when a finished garment is to be washed. Once, this result could only be achieved by keeping the force level low and the construction tight, so that fiber migration did not occur, or by dry cleaning in the almost total absence of water. Nowadays a number of shrinkproofing treatments, which either prevent or reduce felting, are available. These fall into several categories. It would be possible to prevent felting by removing the scales from wool fibers; treatments of this type are not used outside the laboratory because they are too damaging. Sometimes polymers are deposited on the fabric to "glue" the fibers together wherever they touch; in this procedure the only critical surface property is the critical surface tension of the fibers, which must be increased sufficiently to ensure that the polymer spreads over the fibers throughout the fabric [131]. The most used technique is to attack the cuticle of the fibers chemically. Such treatments are often called "degradative" because they attack, or degrade, the keratin in the cuticle; however, care is taken to confine the degradation to the cuticle, so that the fibers are substantially undamaged. Treatments of this type will be considered in some detail, as they depend essentially on modification of the elastic properties of the surface and near-surface layers. There are also treatments in which the fibers are coated with polymer, usually after prior degradative attack, without "glueing" them together. In such treatments the polymer appears to reduce the frictional difference by masking undertreated or untreated scales.

B. Shrinkproofing by Chemical Attack

The commonest agent used to degrade keratin for shrinkproofing is chlorine. It may be used either in solution in water (acid, neutral, or alkaline) or as a gas. When used in water, care has to be taken to confine the attack to the cuticle of the fiber either by exposing the wool for only a very short time, or by adding osmotically active solutes to the chlorine solution so that swelling of the wool is repressed and entry of the chlorine is slowed. (These precautions probably prevent the formation of Allwörden sacs, also). Other oxidizing agents have been used, e.g., permonosulfuric acid, potassium permanganate, and sulfuryl chloride [130b,132]. Alkalies have been used, but are not in commercial use at present [130c]. Treatment in a gaseous electrical discharge has also been found suitable when only low levels of shrinkproofing are required [133-135]. Reductive treatments have also been used but are generally less satisfactory [136].

These treatments do not all produce similar frictional changes in all circumstances. It has been shown, however, that many of them do have one common physical effect, which is that, after treatment, the scales are "softer" when wet than they were before treatment. This has been demonstrated, by micromanipulation of the fibers, for aqueous acid chlorination, gaseous chlorination, potassium permanganate treatment, and corona-discharge treatments [83]. The "softening" is exhibited not only as easier deformability but also as a partial loss of elasticity; i.e., the contents of the cuticle cells inside the plasma membrane (epicuticle) become less elastic and more viscous. The scale cells are consequently much more deformable at low rates of strain than at high rates; after mild treatments they behave in water like little bags of jelly, and after excessively severe treatments, like bags full of viscous liquid. The semipermeable plasma membrane, which is not perceptibly damaged by these treatments, prevents or hinders the escape of the contents unless they are degraded to a low molecular weight or the membrane is broken mechanically.

These physical changes result primarily from scission of the cystine disulfide bond in the exocuticle, with transformation to negatively charged sulfonic acid groups ($-SO_3^-$), which solubilize some of the keratin molecules so that the cuticle cells swell osmotically, and soften, in water. Acid chlorination, however, undoubtedly produces a higher level of shrinkproofing than treatment with other oxidizing agents, even when these agents also oxidize the disulfide bond as far as $-SO_3^-$. Alexander et al. [137] showed that acid chlorination was associated with oxidation of the tyrosine residues by chlorine, and Gordon [138] has shown that for at least one cystine-containing protein (insulin) chlorination facilitates hydrolysis of the tyrosine residue. Chlorination can be expected to have the same effect on keratin. Such hydrolysis greatly increases the number of soluble molecules in the cuticle cells, and hence the degree of osmotic swelling, and softening, of these cells in water [138a]. (If the hydrolysis is carried to the level at which all the molecular fragments are small enough to diffuse out through the epicuticle, the scales finally become flat rather than swollen. Shrinkproofing is obtained in this case also, but this level of treatment is not intentionally sought as it leaves the cortex of the fiber susceptible to mechanical damage.)

There has been controversy in the past [106,136,137,139,140] as to the necessity for hydrolysis of the protein chains in oxidative shrinkproofing. Careful examination of earlier results shows that only a relatively low level of shrinkproofing is achieved whenever there is no hydrolysis; high levels are achieved only when there is some hydrolysis. Now that the physics of shrinkproofing by these treatments is understood, the generalization can be made that some degree of shrinkproofing can be produced by the slight softening which results from oxidation of the disulfide bond alone, but for a high level of shrinkproofing more severe softening is necessary, and this can be produced if some hydrolysis of the protein chains accompanies the oxidation [138a].

This softening effect of degradative treatments is not exhibited by dry fibers unless treatment has been excessive, because, in the absence of water, hydrogen bonding and salt linkages reinforce the degraded protein structure. The softening exhibited by wet fibers is responsible, via frictional changes, for the shrinkproofing produced by these treatments. The frictional changes produced depend on the nature of the surface over which the fiber slides, and to some extent on the environmental conditions. It was at one time supposed, e.g., in ref. 105, that $f_a - f_w$ must necessarily be reduced if the scales are softened, but this is not the case. In particular, if the frictional difference is due to the ratchet mechanism, $f_a - f_w$ will not be reduced much until softening is quite severe. This has been established experimentally [122], and was in fact observed a number of times before the role of softening was understood; at that time it was often taken to indicate that there had been little attack on the scales. The finding is readily explicable if we consider the sequence of steps involved in the measurement of the coefficient of static friction in the against-scale direction [83,141]; the argument can be extended to kinetic friction. The fiber is first placed on the other surface, under load; a tangential force is subsequently applied and increased gradually until sliding begins. Most contacts are made on the flatter faces of the scales. If the fiber has not been subjected to chemical attack, the scales deform a little, elastically. This brings a few of them, say n_0, into a position to engage with the asperities of the other surface during against-scale sliding. Observations on two different types of surface have given values of 0.01 and somewhat greater than 0.1 for the ratio of n_0 to the total number of scales in the area of apparent contact, the second value being for contact with another fiber [8,116]. But if the scales have been softened and made viscoelastic, then during the few seconds or minutes that the tractive force is being increased before sliding begins, the scales, wherever they happen to be making flat-face contact, flow and flatten down against the other surface. This brings many more scale-tips n_1 into a position to make steep-face contact when sliding begins. After even a low degree of softening, n_1 may be equal to the total number of scales in the area of apparent contact. Softening therefore produces a large increase in the number of scales that engage and a decrease in the strength of each. It is difficult to predict where the balance will lie, particularly as the resistance of the scales to bending during sliding depends on their rate of strain, and therefore on the sliding speed. It has been found experimentally [9,83] that for sliding over horn the frictional difference is reduced only slightly, if at all, by mild chemical attack, but it is considerably reduced by more severe attack [83,122]. When the scales are very weak, $f_a - f_w$ must necessarily fall to a low value, as there is then little physical difference between against-scale and with-scale sliding.

When the frictional difference is due to ploughing, quite different results are obtained. A mild softening treatment, which has no perceptible effect on $f_a - f_w$ measured on horn, causes $f_a - f_w$ measured on PTFE to fall almost to zero. This has been shown to be caused by the scales (after treatment) becoming too soft to penetrate the PTFE, so that the ploughing component of friction is almost zero [117].

When the mating surface is another, oppositely oriented, fiber, the situation more nearly resembles sliding over horn than sliding over PTFE. Hence it can be deduced that severe shrinkproofing treatments owe at least part of their effectiveness to reduction of the frictional difference, but mild treatments must act in some other way.

This other way is by increasing the coefficient of with-scale friction f_w, which is the second principal effect produced by softening. The same degree of treatment that has little effect on $f_a - f_w$ measured on horn (though it has a large effect on $f_a - f_w$ measured on PTFE) has been shown to produce a considerable increase in f_w measured on horn [9,83]. It has, however, little effect on f_w measured on PTFE. These effects also are explicable if we consider the sequence of steps used to measure the coefficient against a surface such as horn. As before, during the period before sliding begins the normal load pressing the fiber against the other surface deforms the scales of an untreated fiber almost elastically, so that a definite area of contact is established almost immediately, but the scales of a treated fiber flow viscoelastically throughout this period, so that the area of real contact established between the two surfaces is very much greater for shrinkproofed than for untreated fibers. During sliding the thin volume surrounding this junction area is sheared until slip occurs at the interface, where there is only van der Waals bonding. Since the region that is strained in shear under the light loads applied to the wool fiber is very thin (10^2-10^3 Å), the rate of strain is very high; at the lowest sliding speed that can be perceived in the measurement of "static" friction ($\approx 5.10^{-3}$cm/sec in ref. 9) the rate of strain is of the order of 5,000 sec^{-1}. (This is much higher, also, than the rate of strain produced in bending the scales during against-scale sliding.) At this rate of strain the scales have almost the same elastic properties as in the untreated state, because simple entanglement of the long-chain molecules is enough to maintain their original moduli even if all crosslinks and hydrogen bonds are broken. It is therefore necessary to shear a much bigger area of unweakened material to cause sliding; the force required, which is equal in magnitude to the frictional force, is correspondingly increased. This result, which has been observed when the friction has been measured against horn [9,83], is to be expected when measurements are made against any surface as hard as or harder than the scales of an untreated fiber; it should therefore apply to fiber-on-fiber sliding in the felting situation, when both surfaces are softened to the same degree.

However, when f_w is measured against PTFE, which is softer than the scales of an untreated wool fiber and comparable in hardness to the scales of a fiber subjected to a mild shrinkproofing treatment, there is little increase in the area of real contact as a result of softening the scales, because the PTFE principally determines this area [117]. Hence the increase in f_w produced by the treatment is relatively small.

Examples of the effects of oxidizing treatments on f_w and $f_a - f_w$ are given in Table 2. In this table the shrinkage-reduction factor R is the ratio of the rate of shrinkage of untreated fabric to that of the shrinkproofed fabric. R may be taken as a measure of the severity of treatment, a value of unity corresponding to no treatment. A value of infinity would correspond to complete unshrinkability of fabrics made from the treated wool, but after such treatment the fabric is often found to expand on washing.

It may be deduced from these studies of frictional behavior that mild degradative treatments reduce felting shrinkage by increasing f_w so that fiber movement becomes more difficult, whereas more severe treatments act by increasing f_w and decreasing $f_a - f_w$, the latter effect reducing the likelihood of continual unidirectional migration of any fibers which do move. Both these mechanisms, however, are the result of one simple physical change, softening of the scales (in the wet state) by these treatments. The practical importance of the elastic properties of the surface layers of the wool fibers is thus evident.

V. FRICTION OF DRY WOOL FIBERS

Studies of the frictional properties of air-dry wool fibers have proceeded along very different lines from those of wet fibers. The frictional difference between the against-scale and with-scale directions has little practical importance when the fibers are dry. In some circumstances of practical interest it is in fact very small or zero; a typical example is the sliding of wool fibers over smooth metal pins as in carding or combing [142]. In other circumstances it may be appreciable, though smaller than for wet fibers and still unimportant; a typical example is interfiber friction during drafting and spinning. Partly because of the unimportance of the frictional difference, less attention has been devoted to understanding the basic mechanisms of friction in studies of dry fibers, and more has been devoted to empirical, special studies of the friction in particular situations occurring during processing. The aim of these studies has usually been to achieve greater understanding and control of fiber movement during processing. A number of other differences in the two fields of study have resulted. Dry-fiber friction is frequently measured at relatively high sliding velocities; wet-fiber friction usually at low or zero velocity. Dry-fiber friction is often measured with processing oil on the fibers; in wet-fiber studies the fibers are usually clean. Consequently theoretical studies of dry-fiber friction have often been concerned more with the behavior of the oil film than with that of the fiber surface itself. With dry fibers, more often than with wet ones, frictional forces may be measured without calculation of a coefficient

TABLE 2

Effects of Some Shrinkproofing Treatments on the Coefficients of Friction of Wool Fibers in Water with Wetting Agents[a]

Treatment	Rubbing surface	R^b	In 0.05% CF 10[c]			In 0.1% Vel[d]		
			f_w^e	f_a	σ	f_w	f_a	σ
None	Horn	1	0.23	0.47	0.23			
$KMnO_4$/salt	"	3.5	0.32	0.55	0.23			
None	"	1				0.16	0.57	0.41
1.1% dry Cl_2	"	10				0.53	0.80	0.27
1.7% dry Cl_2	"	30				0.57	0.69	0.13
2.4% dry Cl_2	"	Expands				0.47	0.68	0.21
None	"	1	0.39	0.71	0.32	0.34	0.61	0.27
Acid wet Cl_2	"	15-40	0.92	0.99	0.08	0.56	0.69	0.13
None	"		No control available					
Commercial dry Cl_2	"		0.97	1.12	0.14			
None	"	1				0.20	0.54	0.33
Corona with Cl_2	"	8				0.62	0.84	0.22

None	"	1	0.18	0.43	0.25
Corona without Cl_2	"	Not known	0.49	0.70	0.21
None	PTFE	1	0.13	0.25	0.12
$KMnO_4$/salt	"	3.5	0.15	0.18	0.03

[a]Reprinted from ref 122, p. 14, courtesy of the International Wool Secretariat.
[b]R is the shrinkage-reduction factor (see text).
[c]Triton CF 10 (Rohm and Haas, Phila., Pa.), nonionic.
[d]Vel (J. Kitchen and Sons, Sydney, Australia), anionic, with buffers.
[e]The values of f_w, f_a, and σ are means over several fibers. Small differences between σ and $f_a - f_w$ are due to rounding-off errors.

of friction, because in many of the experimental situations it is difficult to define the normal load on the fibers. In most cases in which a coefficient has been calculated the experimental situation was remote from practical processing conditions.

Some typical values of the coefficient of friction of air-dry wool fibers sliding over various materials are listed in Table 3. It should be noticed that a misleading impression could be gained by comparing this table as a whole with Table 1, because the largest source of variability in absolute value of the coefficients in both tables is clearly the experimenter! This results from variation in a number of factors, such as the degree of cleanliness or of deliberate contamination that was established and maintained during the experiments, the magnitude of the normal load between the surfaces, the likelihood of surface tension of thin contaminating films adding to the apparent load, and the length of the arc of contact of the wool fiber against the other surface. Whereas casual inspection of these two tables might suggest that for wool on horn the coefficient of with-scale friction could be higher for dry fibers than for wet, comparison of measurements made with the same experimental arrangements under both sets of conditions (e.g., in refs. 100, 105, 116, 144) shows the reverse to be true, not only for friction on horn but also on all of the materials examined except PTFE and felt.

It is widely believed that the low-speed, dry friction of the component fibers influences the "hand" or "feel" of a fabric or a handful of loose wool, especially its "smoothness" or "softness." respectively. There is, however, no convincing evidence that this is true for fabric, and there is evidence against it for loose wools of the types normally used for clothing [147,148], the "hand" of which is determined principally by the resistance of the sample of wool to compression. This resistance, in turn, depends on the mean fiber diameter and the crimp [147-149]. Only when the range of "wools" examined has been extended by the inclusion of very soft fibers, such as mohair, and/or very harsh fibers, such as karakul and goat hairs or very coarse wool, has a positive correlation been established between friction and harshness for untreated wools [147,148].

Wools shrinkproofed by degradative treatments are usually found to have a harsher "hand" than comparable untreated wools; the effect can be reduced or eliminated by application of a softening agent. The dry friction of such wools has not often been measured. It has been shown, however, that one such treatment (with alcoholic alkali) considerably increases the dry-fiber friction [102,150], and that the friction is decreased by the application of a softening agent [102].

Measurements of frictional forces in situations approximating processing conditions mostly fall into two categories, measurements of the force required to withdraw single fibers from an assemblage of fibers (sometimes embedded in pins) [151-154], and measurements of the force required to pull tufts of fibers over each other [152,155a,

TABLE 3

Typical Values of the Coefficients of Friction of Wool Fibers in Air, Usually at About 65% RH[a]

Rubbing surface	f_w	f_a	$\sigma = f_a - f_w$	Ref.
Horn, sheep's, polished	0.3	0.5	0.2	[143]
	0.20-0.34	0.50	0.30-0.16	[102]
	0.61	0.62	0.01	[105]
roughened	0.51	0.63	0.12	[105]
cow's, polished	0.49	0.53	0.04	[105]
roughened	0.40	0.52	0.12	[105]
Wool fiber, at 180°	0.10	0.13	0.03	[100]
	0.11	0.19	0.08	[144]
at 160°	0.15	0.73	0.58	[145]
at 90°	0.24	0.38	0.14	[102]
Saw-toothed diffraction grating	0.19	0.30	0.11	[116]
PTFE	0.13	0.34	0.21	[117]
Ebonite, polished	0.58	0.60	0.02	[105]
roughened	0.51	0.63	0.12	[105]
Glass, plate	0.18	0.21	0.03	[119]
rod	0.60	0.60	0.00	[105]
filament	0.13	0.18	0.05	[120]
Casein, polished	0.58	0.59	0.01	[105]
roughened	0.47	0.57	0.10	[105]
Felt	0.21	0.35	0.14	[121]
Wool cloth	0.23	0.34	0.11	[146]

[a]The values in this table are not directly comparable with those in Table 1 except where the same reference is given for both sets of measurements.

156-160]. Included in the second group are measurements of drafting force and of the force required to rupture a sliver. In both types of measurement the force is not purely frictional but contains a bulk-elastic component because the fiber has to be straightened or bent as it slides. A third, smaller group of measurements contains experiments on the sliding of single fibers over wires [142,155b]. These various types of experiments have shown that hydrodynamic forces are involved in sliding in the presence of oil. (This is not equivalent to saying that there is full "hydrodynamic lubrication.") The resistance to sliding is usually greater in the presence of oil than in its absence and increases with $v\eta$ where v is the speed of sliding and η is the viscosity of the oil. The function of the oil is to promote sliding rather than stick-slip motion, by increasing adhesion between the fibers, not to reduce the friction [153]. There is therefore an optimum amount of oil which should be added in most processing situations, and often an optimum viscosity may be found. Some authors (e.g., refs. 152,153,161) have calculated the forces between assemblies of fibers from measurements of single-fiber withdrawal forces and have obtained satisfactory agreement with measured values. More information on friction in processing can be found in ref. 96.

VI. SURFACE AREA AND ADSORPTION PROPERTIES

Although the scale structure of a wool fiber makes the fiber appear microscopically rough, this roughness does not greatly affect its surface area, as the flat surface of each scale is remarkably smooth unless the fiber has been damaged. This is shown by the fact that the surface area determined from adsorption measurements is only about 5-8 times that calculated from the fiber diameter on the assumption that the fiber is a smooth cylinder. A discrepancy of this magnitude can be accounted for by penetration of the adsorbate into submicroscopic cracks and fissures. Some values of the specific surface area (in two sets of units which are commonly used) are shown in Table 4.

Determination of the surface area by adsorption is not, however, straightforward. In the first place, the sorbate should not be absorbed by the wool; it is not possible to separate the adsorbed and absorbed moieties with any accuracy. For example, a "surface area" was at one time calculated from the water-sorption isotherm, and found to be 200 times that calculated from nitrogen sorption [163]. This was explained by an opening up of pores in the structure as a result of swelling on absorption of water. Sorption in this part of the isotherm, however, is now known to be due to absorption onto polar molecular sites [164].

Secondly, it is necessary to be able to estimate, from an adsorption isotherm that continues into the multilayer region, the hypothetical

TABLE 4

Determinations of Specific Surface Area of Wool Fibers

	Fiber diameter (μm)	Specific surface area	
		cm^{-1}	$m^2\ g^{-1}$
Calculation from diameter	24	1.7×10^3	0.13
Adsorption p-nitrophenol[a]	24	8×10^3	0.6
Adsorption N_2[b]	-	13×10^3	1

[a]Ref. 162.
[b]Ref. 163.

amount of sorbate which would be adsorbed when there was one, and only one, layer of molecules on the surface. Since multilayer adsorption begins before monolayer formation is complete, this estimation must be based on some theory. Fortunately, the estimate is not very sensitive to the theoretical basis chosen, because the forces involved in binding a monolayer are considerably greater than those involved in binding subsequent layers, so that monolayer formation is almost complete before multilayer formation is appreciable. It is usual nowadays to analyze the isotherm on the basis of some well-established theory; to obtain the values given in Table 4, the authors quoted used the theory of Brunauer et al. [165].

Thirdly, one sometimes wishes to separate the "external surface" from the total surface area. In the case of nitrogen adsorption, penetration into any cracks, depressions, and pores occurs rapidly, so that their surface area cannot be separated, by such measurements, from the geometrical surface area of the fiber. In the case of the p-nitrophenol experiments [162], however, an attempt was made to determine sorption solely on the "external" surface, by extrapolating the sorption isotherms to zero time.

The free energy of adsorption of nitrogen onto wool has been estimated at 20 erg cm^{-2} [163].

VII. WETTING PROPERTIES

A. Rain Repellency

Early work on the wetting properties of wool fibers dealt principally with wetting by water or aqueous solutions. The practical applications of the work were twofold: how to maintain high water repellency in fabrics for rain resistance, and how to ensure rapid wetting during processing or laundering. For the most part, work related to the second question was concerned more with the properties of detergents than with those of fabrics; this work will not, therefore, be discussed here. Studies of the first question were made mainly by Cassie and Baxter [155c,166,167], who extended the existing theory of the wetting of rough surfaces to porous or gridlike surfaces, which textile fabric surfaces approximate. Although their theory was general, they applied it specifically to wool fabrics, wool being the most water-repellent of the textile fibers in general use at that time.

Cassie and Baxter began by considering the contact angle θ formed by a liquid drop on a plane surface. They accepted the standard expression $\cos\theta = (\gamma_{SA} - \gamma_{LS})/\gamma_{LA}$, where γ_{SA}, γ_{LS}, and γ_{LA} are the surface tensions between solid and air, liquid and solid, and liquid and air, respectively. This expression has recently been criticized by Bikermann [168]. They also accepted the concept that the advancing and receding contact angles, θ_A and θ_R, differ even on a smooth, plane surface. It is now known (e.g., ref. 169) that this hysteresis may have several causes, such as an initial contamination of the surface, which is changed by contact with the water, and trapping of the liquid in depressions in the surface, which is never quite smooth [170-173]. They showed that on a porous or gridlike surface the apparent contact angles θ_D (advancing) and θ_W (receding) should often be much larger than θ_A and θ_R and confirmed this experimentally for a grating made of wires coated with paraffin wax. To take one example, θ_A and θ_R measured on a similarly coated plate, were 105° and 93°; the values of θ_D and θ_W predicted from this for a grating of a particular geometry were 145.5° and 138.5°, and the measured values were 143° and 138°. On textile yarns the agreement between theory and experiment was good for the advancing contact angle θ_D, considering the approximations that had to be made to apply the theory; e.g., for the individual wool fibers (in the not particularly clean condition in which they were used in the yarn) $\theta_A = 120°$; for a particular yarn structure the predicted value of θ_D was 148° and the measured value was 157°. Similar results have since been obtained by Hollies et al. [174], using solvent-extracted fibers with θ_A values from 65 to 97°. However, agreement was not good for the receding angle θ_W, which was much smaller than expected, mainly because films of water were formed between the fibers. In addition, the wetting characteristics of the individual fibers changed during prolonged contact with these films [155c,167]. It can now be seen that the reduction in θ_W due to film formation is essentially similar to the reduction of the "true" contact angle from θ_A to θ_R as a result of trapping of liquid in depressions on the surface. Film formation can be minimized by holding the fibers in the fabric in such a way that they cannot be drawn together by surface tension. Baxter and Cassie [155c,167] concluded that for maximum rain repellency a fabric, in addition to being made of water-repellent fibers, should have as large and uniform a separation of the component fibers as possible (short of the level at which raindrops could pass freely between the fibers), and that the fabric should be constructed to hinder the ends of the fibers from drawing together.

More recent work on wettability has been concentrated on this aspect of the importance of yarn and fabric construction, and in particular on the promotion of wetting by capillary action [174,175]; for example Hollies et al. [174] have shown that the yarn properties may be more important than the contact angles of the individual fibers in determining wettability.

The question of water-repellent finishes, obtained by chemically coating the fibers or the fabrics, will not be discussed here, as it is not peculiar to wool.

The accepted value for the advancing contact angle of water on a clean untreated wool fiber is about 90° [174,176]. Some measurements have also been made of the contact angle between a wool fiber and various organic liquids [177].

B. Effect of Chemical Treatments on Water Repellency

It has been known for a very long time, because the effect is so marked that it could not escape notice, that although clean, untreated wool is very difficult to wet with pure water, hours or even days being needed, wool that has been subjected to various chemical treatments, such as chlorination for shrinkproofing, wets rapidly, in minutes or even seconds. Correspondingly, the contact angles with water are reduced; some typical values have been given by Stewart and Whewell [176]. It has been accepted for a long time that this increased wettability is due to the production of polar groups on the surface of the wool fibers [178]. Thorsen [179] has found that exposure of wool fibers to a corona discharge increases their wettability by water, and has suggested the same explanation, as he found that the increases could be reversed by adsorption of cationic surfactants. Recently Millard [180] has demonstrated by X-ray electron spectrometry that disulfide sulfur in the surface of wool is oxidized to sulfonate or sulfate by corona discharge.

C. The Critical Surface Tension of Wool Fibers

Interest in the wetting properties of wool waned for some years, to revive in a new form about 1964 [131] because of its importance for various recently developed shrinkproofing treatments in which the wool fibers were coated with a thin film of polymer. It was argued that if a polymer is to be applied from a solution or an emulsion and subsequently to adhere firmly, it has to spread over the fibers; this means that it has to wet them at all stages from initial application through to some late stage in drying or curing, when it becomes too viscous to retract in the time available. Zisman's concept [181,182] of the critical surface tension (CST) of a solid appeared to be relevant to this problem, and a considerable amount of work has been done on measuring the CST of wool fibers. It should be noted that some measurements of the same quantity, under different names, had also been made much earlier (e.g., refs. 178,183,184).

The CST of a solid is the highest surface tension a liquid can have and still spread over that solid. It is measured by exposing the solid to a series of liquids of graded surface tension; the actual quantity observed may be any one of several, such as the contact angle, the surface conductivity, or, in the case of fibers, the proportion of fibers that sink or float when placed carefully on the liquid. The measurements should be made on individual fibers that do not touch each other; if yarns, fabrics, or other assemblies of fibers are used, and even if a few fibers are allowed to touch each other, the apparent contact angle will be too high because of the capillary effects referred to near the end of Section VII-A. This capillary effect has not been widely realized, and many measurements have been made on assemblies of fibers.

When this work began, the CST was regarded essentially as an empirical quantity. Since that time, a great deal of effort has gone into relating it to more fundamental quantities, and discussion still continues [185-189]. Kaelble [188] has shown that the CST is related to the surface free energy (or solid-surface tension) γ_s of the solid by

$$\mathrm{CST} = \gamma_s[(d_\ell d_s)^{1/2} + (p_\ell p_s)^{1/2}]^2$$

where d_ℓ and p_ℓ are the fractional proportions of the surface tension of the measuring liquid that arise from dispersion forces and polar forces respectively, and d_s and p_s are the corresponding proportions for γ_s. (Hence $d_\ell + p_\ell = d_s + p_s = 1$.) The polar components of the surface tensions can be zero, but the dispersion components are always finite. Since there is no interaction between the two types of force, there are two possible cases in which the only interaction between the liquid and the solid is due to dispersion forces, viz., when $p_s = 0$ and when $p_\ell = 0$. In the former case, that of a nonpolar solid, the CST = $d_\ell \gamma_s$, i.e., it depends on the nature of the measuring liquid. In the latter case (measurement with a nonpolar liquid), the CST =

$d_s \gamma_s$, and so is characteristic of the solid and is a measure of the dispersion-force component of its surface free energy. In the limiting case in which p_s and p_ℓ are both zero, the CST = γ_s, the whole of γ_s being caused by dispersion forces. In general, the CST is equal to γ_s when, and only when, the proportions of dispersion and polar components are the same in the liquid as in the solid—a condition difficult to fulfill.

Before these relations had been developed it appeared that the CST might be a useful empirical measure of the wettability of wool fibers. A liquid L should spread over the fibers if the CST of the wool, measured with a series of liquids homologous to L, is greater than the surface tension of L. This concept was applied by Feldtman and McPhee [131] to explain why it was necessary to attack wool fibers chemically before certain polymers could be made to spread through wool fabrics from a padded-on solution or emulsion and to adhere firmly after drying and curing. The condition for spreading and adhesion should be that the CST of wool fibers is greater than the surface tension of the polymer at all relevant stages (as defined above). Feldtman and McPhee approximated the surface tension of the polymer by the critical surface tension of a dried film of the polymer. Had the polymer and the measuring liquids been nonpolar, this would have been equal to the solid-surface tension of the solid polymer. However, the solid and the liquids were polar, so it is impossible to estimate the closeness of the approximation. Feldtman and McPhee were able to show that the CST of unmodified wool (measured on fabric) was less than the CST of the polymer, whereas that of the chemically pretreated wool was higher than that of the polymer. This appeared to explain the necessity for pretreatment. These arguments, developed for a padding process, would not necessarily apply to treatments in which a polymer is exhausted onto wool fibers.

The results of various measurements of the CST of wool and hair fibers, made with both hydrogen-bonding and non-hydrogen-bonding liquids, are summarized in Table 5. It can be seen that the values for the CST of normal, untreated wool, measured with aqueous liquids, fall into two distinct groups, about 45 dyn cm^{-1} and about 30 dyn cm^{-1}. The higher values were all obtained from measurements on assemblies of fibers; the lower values were obtained using single fibers, or on assemblies with allowance for capillary effects. Therefore, the lower values must be accepted as giving better values for the CST of untreated wool fibers. With non-hydrogen-bonding liquids values from 25 to 34 dyn cm^{-1} have been obtained. With whatever type of liquid the CST is measured, its value is increased by nearly all the chemical treatments that have been studied.

There is evidence from some of these studies [196,197] that the CST, measured with non-hydrogen-bonding liquids, depends on relative humidity, increasing as the RH decreases. Alter and Cook [196]

TABLE 5

Some Measured Values of the Critical Surface Tension (CST) of Wool and Human Hair Fibers[a]

Liquids	Method[b]	Untreated fibers CST (dyn cm^{-1})	Chemically treated fibers Treatment	Chemically treated fibers CST (dyn cm^{-1})	Ref.
Salt Solution	1	46	Alcoholic KOH	63-71	[178]
H_2O/butanol/NaCl	1	45	H_2SO_4	45	[184]
			SO_2Cl_2/solvent	53	
			Papain	53-64	
H_2O/butanol	2	45	Alkaline chlorine	50	[131]
			$KMnO_4$/salt	51	
			Alcoholic KOH	62-70	
			$KBrO_3$/salt	60-70	
H_2O/butanol	3a	45	Acid chlorine	72	[190]
			Gas chlorine	72	
H_2O/butanol	3b	$<$29-33	Chlorine	67	[191]
H_2O/butanol	4	30			[192]
H_2O/surfactants	5	30-33	$KMnO_4/Na_2SO_4$	Increased	[193]
			Alcoholic KOH	Increased	
H_2O/butanol	3b	27-30			[194]

		RH 0%	RH ~ 50%		RH 0%	RH ~ 50%	
Non-H-bonding	3b		27	HOCl		31	[195]
"	3b	34	26	Solvent	30	26	[196]
		28	25	Bleached	32-34	30	
"	3b	33	26	Bleached	34	30	[197]
		33	26	Plasma	33	33	
		33	26	Corona	33	33	
		33	28				
"	3b		30-31	Extracted		30-37	[194]
				Corona		>44	

[a]The measurements using aqueous solutions were all made on wool fibers. Those using non-hydrogen-bonding solvents were made variously on wool, mohair, and human hair.

[b]Methods: 1—rolling drop on sheet of fibers; 2—dipped yarn; 3a—sink/float with many fiber snippets touching; 3b—sink/float with individual fiber snippets; 4—calculated from observations of capillary wetting and measurements of contact angle; 5—contact angle on single fibers.

suggested that this was due to adsorption of a film of water as the RH increased. Pittman [197] has discussed the implications of such adsorption for practical problems in the application of polymers. The work of Rance [198], however, indirectly suggests that other explanations may be possible.

The CST of compressed, powdered wool, measured with non-hydrogen-bonding liquids, has been found to be 36 dyn cm^{-1} at 45% RH and 38 dyn cm^{-1} at 0% RH [197]. These values are higher than the corresponding values for whole wool or hair fibers unless they have been exhaustively extracted with mixtures of polar and nonpolar solvents [194]. The difference in values has frequently been assumed to be caused by the fact that the interior of the wool fiber is more reactive chemically than its epicuticular covering, but the higher value for powdered wool could be due simply to microscopic roughness of the surface of the compressed powder, or to the exposure of lipid-free protein.

VIII. SOILING AND SOIL REMOVAL

A. Soiling

Because the soiling of any textile is sensitive to its fiber type, fiber denier, construction, and finish, as well as to the environmental conditions and type of soil, it is difficult to separate out the effect of a single factor such as fiber type. This is reflected in disagreement between workers as to the ranking order of soilability of different fibers (see bibliographies in refs. 199,200). Most of the disagreement centers around apparel fabrics; there seems to be general agreement that for use in carpets wool is in practice one of the most soil-resistant fibers available. To some extent this high soil resistance is more apparent than real [201,202], although nonetheless valuable; a given proportion of soil is less apparent on a diffusely reflecting fiber such as wool than on a specularly reflecting or transparent one.

It is generally accepted that soil particles can be held on fibers by at least three mechanisms suggested by Compton and Hart [203]: by entrapment in the interstices between fibers or yarns (macro-occlusion), by entrapment in indentations, preexisting or newly formed, in the fiber surfaces (micro-occlusion), and by physical sorption. For sorption to occur, the fiber or the soil particles must be deformable at some stage during deposition, so that an interface of sufficient area is formed [203]. There is uncertainty as to the nature of the sorptive forces. Van der Waals forces must be involved, and it was at one time believed that electrostatic forces must be, too. It was expected that fibers that under normal atmospheric conditions acquired and held a high charge would consequently attract and hold soil particles. It has, however, been found impossible to demonstrate such a relation, and in fact its existence seems to have been disproved by results such as those of Schappel [201], who found that the soiling properties of blends of fibers corresponded to the mean of the properties of the component fibers, whereas the electrostatic properties of the blends corresponded to those of the components having the lowest charge.

The resistance of wool to soiling is increased by most shrink-proofing treatments of the chemical (degradative) type, such as chlorination [204]. Resin treatments may increase or decrease it [204]. These changes have not been satisfactorily explained.

The subject of soiling of textile fibers in general has recently been reviewed exhaustively [203a] and will also be dealt with in Part II of this book. Stain repellent finishes for wool have also been discussed in a recent publication [203b].

B. Soil Removal

There is disagreement also as to the ranking of different fiber types for ease of soil release in washing or dry-cleaning. One factor on which there appears to be agreement is that fibers with a polar surface are easier to wash clean than those with a nonpolar surface. (e.g., ref. 205). This generalization is in agreement with the generally accepted observation that wool made shrink resistant by chemical treatments such as chlorination is easier to clean by washing than untreated wool [204,206]; the former has a highly polar surface whereas the latter, if undamaged, has a nonpolar, hydrophobic surface. In a practical situation Benisek and Crawshaw [199] found that chlorinated wool had soil release properties very similar to those of nylon and somewhat better (especially for greasy soils) than those of Orlon, which in turn showed better soil release than the still less polar Crimplene.

The generally accepted explanation for the greater ease of washing fibers with polar surfaces is that they are more readily wetted by water, as has been directly demonstrated by measurement of the contact angle and work of adhesion between wool fibers and water before and after various treatments for shrink resistance [176]. There has, however, been some controversy recently as to exactly how this affects the detailed mechanism of soil removal. Warburton and Parkhill [206a,206b] have suggested that the most important factor is the mechanical force exerted by the superficial layers of polar material as they swell in water; Kissa [206c], on the other hand, considers that the rapidity of diffusion of water into the polar surface layers is of major importance, since it allows the water to detach the soil from the fiber side as well as from the perimeter. It has also been shown recently that, for ease of washing, the surface of the fibers should be oleophobic as well as hydrophilic [206d]. From time to time it has been suggested that the charges in the fiber surface have some direct effect. Normally, both the fibers and the particles of soil are negatively charged in water, so it might be expected that the repulsive force between them would increase as the charge on either or both increased. A number of investigators [207-209] have attempted to establish the existence of a correlation between ease of soil release during washing and magnitude of the ζ potential (see Section IX-B) in water. Harris [208] concluded that "although a direct correlation between ζ potential and detergency seems to exist, exact proof is lacking." More recently Jacobasch [209] has found such a correlation for particular types of fiber subjected to various modifications; he did not include wool in his study. Difficulty in establishing such a correlation is to be expected, since many other factors, in addition to the state of charge of the fiber surface, affect soil release.

Shrinkproofing treatments that coat wool fibers with a resin have been found to make the wool more difficult to clean by washing [204].

IX. SURFACE ELECTRICAL PROPERTIES

A. Static Charging

Wool fibers, like other fibers, can become charged by contact with dissimilar surfaces or even, in some circumstances, by contact with each other. The mechanisms involved will be discussed in Chapter 6. When clean and dry, wool is particularly likely to acquire a charge in this way, because it lies at the extreme positive end of the triboelectric series for textile fibers [210]. Normally, however, wool contains considerable quantities of water absorbed from the atmosphere and is contaminated with ionic substances. This lowers its resistivity; typical values are given in Table 6. The charge can therefore be dissipated more easily from wool than from synthetic fibers, so that static causes fewer difficulties in processing and less discomfort in wear with wool than with synthetics. Antistatic treatment of wool garments is unnecessary, but during processing, particularly at low relative humidities, it is usually necessary to apply films of antistatic agents to the surfaces of wool fibers. The general mode of action of these substances will be discussed in Chapter 6. Medley [212] has studied their particular mode of action in the processing of wool, and has found that the charge is normally dissipated by conduction through small droplets that bridge gaps such as those between the rollers and the wool in worsted spinning. This conclusion has received support from Löbel [213] and from Henry [214].

The level of static charging reached can also be reduced by prior exposure of the fibers to a corona discharge [215]. This treatment reduces the probability that charge will be generated, because it shifts the position of the wool toward the center of the triboelectric series.

TABLE 6

Electrical Resistivity of Some Textile Fibers at 30°C[a]

	Resistivity (Ω cm^2/cm)		
Fibers	50% RH	65% RH	85% RH
Nylon	200	30	2
Wool	5	1	0.1
Cotton	3×10^{-2}	5×10^{-3}	3×10^{-4}

[a]Based on ref. 211.

B. Zeta Potential

When a solid is immersed in a liquid an electrical double layer is commonly produced in the liquid at the interface. If the liquid is then forced to flow past the solid, this double layer is sheared at some plane. The potential difference between this plane and the bulk of the liquid is called the ζ potential. It can be measured in a number of ways, the commonest being by calculation from measurements of the streaming potential, which is the potential difference set up in the direction of flow when the liquid is forced through a porous plug of the solid. The liquid of interest is usually water or an aqueous solution. The electrical double layer at the surface results from the development of a charge in the solid at the surface, which attracts ions of opposite sign from the water; these in turn attract ions of the original sign to form a cloud between them and the bulk liquid. The source of the initial charge in the solid varies from system to system; in the case of keratin fibers immersed in aqueous solutions it is largely if not entirely due to ionization of chemical groups such as $-COOH$ and $-NH_2$ on or close to the surface [216]; consequently, changes in the ζ potential can be used to follow changes in ionization of the surface. The exact position of the shear plane in the electrical double layer is not known, and it is not necessary to know it for this purpose.

The usual procedure in using ζ potential for study of changes in the surface of wool or hair fibers is to plot the ζ potential or the streaming potential against the pH of an aqueous solution streaming through a plug of fibers, the ionic strength of the solution being kept constant. The pH at which either potential difference is zero is determined by interpolation. This point is called the zero point of charge (ZPC); it has in the past been called the isoelectric point of the surface.

A number of workers have measured the ZPC of clean, untreated wool or hair, obtaining values from 2.3 [217] to 4.5 [218]. The value most commonly found is 3.4 [219-221] or close to that value [216,222]. Cook and Smith [216], using human hair, found that prolonged rinsing (for about one week) in distilled water increased the ZPC from 3.3 to 4.5; Algie [217], using wool, has found that prolonged immersion in an aqueous solution, without any rinsing action, causes a similar increase.

The value of the ZPC is sensitive to changes that are confined to the surface of the fibers and that are therefore difficult to study by other methods. Cook and Smith [216] have demonstrated reduction of the ZPC of human hair to about 1.8 by cosmetic bleaching; this is ascribed to the formation of strongly acid sulfonic acid groups ($-SO_3^-$) on the surface of the hair as a result of oxidation of disulfide bonds. They have also demonstrated elevation of the ZPC of hair to 10.4 as a result of adsorption of a cationic polymer; this is the ZPC of a complete, thin layer of the polymer. Similarly, Algie [217] found that chlorination (as used for shrinkproofing) reduced the ZPC to below 2; this also is due to oxidation of disulfide bonds to sulfonic acid groups. Subsequent coating with a (different) cationic polymer increased the ZPC to about 5. Stigter [222] showed that deposition of a polyamide on the surface of wool fibers by interfacial polymerization (Wurlan process) increased the ZPC to a little greater than 5. Cook and Smith calculated from their values of streaming potential the excess of ionized groups of one sign over those of the other sign, per unit area of surface, at a particular pH; they obtained a value about one-quarter that to be expected from bulk titration measurements on hair.

The ζ potential of fibers and of soil particles may be relevant to the removal of soil from fabric (see Section VIII-B).

X. THE SURFACE AS A BARRIER

It is widely believed that there is some kind of a barrier to diffusion of dyes at the surface of wool fibers. This is based principally on the observation, frequently made under both practical and laboratory conditions (e.g., refs. 222-227), that dyeing is more rapid and more uniform when the surface of the wool fibers has been damaged. The damage can be either mechanical or chemical, imposed, for example, by degradative shrinkproofing treatments with chlorine or alcoholic caustic potash, or by extraction with organic solvents. Support for the concept of a barrier has come from studies of the kinetics of dyeing [228]. The evidence presented is that the graph of depth of penetration of dye against the square root of time is initially concave upward instead of linear, as would be expected if the material were homogeneous. However, it is claimed that this effect could also be due to pores or to convection [229]. Evidence has also been presented for the presence of a barrier to the diffusion of acids [227,230]. Studies of the diffusion of alcohols into wool [231,232] have shown no evidence of a barrier to the diffusion of these molecules. It has been reported [231] that the rate of absorption of alcohols is increased by prior extraction with hot ethanol, but this does not necessarily indicate the presence of a barrier at the surface, since recent work [28] suggests that such extraction modifies the whole cell-membrane network throughout the wool fiber. Bradbury et al. [28] suggest that it is this whole network, rather than any surface layer, which acts as a "barrier" to the diffusion of alcohols. At present, therefore, it can be accepted that there is probably some kind of a barrier, in the surface layers of wool fibers, to the diffusion of fairly large molecules such as those of dyes, but there is no generally accepted evidence of a barrier to the diffusion of smaller molecules.

The nature and thickness of the barrier to large molecules are not known. Lindberg and his colleagues [39,230] suggested that it was the epicuticle, 100 Å thick, and that the mechanical removal or disruption of the epicuticle was responsible for the observed facilitation of dyeing. Later, Lindberg [178,226,227] came to the conclusion that disruption of the epicuticle was not necessary, and that some relatively minor chemical modification, such as a change of polarity, was sufficient. More recent work has confirmed that disruption of the epicuticle is not necessary [8,28], and it has been suggested [8] that it is the whole cuticle (or some large part of it such as the exocuticle or the a layer) that acts as the barrier to diffusion of dyes. The exocuticle and in particular the a layer are known to be the principal sites of attack by degradative shrinkproofing treatments, and mechanical damage severe enough to be observed always extends into these layers.

ACKNOWLEDGMENTS

I am indebted to a number of my colleagues in the Commonwealth Scientific and Industrial Research Organization for discussion of various sections of this chapter, in particular Dr. I. C. Watt, Dr. B. E. Fleischfresser, and Mr. J. E. Algie for discussion of Sections VI, VII-C, and IX-B respectively. I am also grateful to Dr. M. G. Dobb and the Textile Physics Laboratory, University of Leeds, for permission to reproduce a hitherto unpublished electron micrograph (Fig. 3).

REFERENCES

1. J. D. Leeder and J. H. Bradbury, Textile Res. J., 41, 215-218 (1971).
2. R. D. B. Fraser and F. G. Lennox, Textile J. Australia, 37, 120-123 (1962).
3. E. G. Bendit and M. Feughelman, "Keratin," in Encycl. Polymer Sci. Tech. (N. M. Bikales, ed.), Vol. 8, Wiley, New York, 1968, pp. 1-44.
4. H. M. Appleyard and C. M. Greville, Nature, 166, 1031 (1950).
5. P. Kassenbeck, Bull. Inst. Textile France, No. 76, 1958, pp. 7-19.
6. M. Leveau, N. Cebe, and A. Parisot, Bull. Inst. Textile France, No. 42, 1953, pp. 7-10.
7. R. D. B. Fraser and G. E. Rogers, Biochim. Biophys. Acta, 16, 307-316 (1955).
8. K. R. Makinson, Textile Res. J., 37, 831-842 (1968).
9. K. R. Makinson, Textile Res. J., 38, 710-721 (1969).
10. J. D. Leeder, M.Sc. Thesis, Australian National University, Canberra, Australia, 1969.
11. M. G. Dobb, F. R. Johnston, J. A. Nott, L. Oster, J. Sikorski, and W. S. Simpson, J. Textile Inst. (Trans.), 52, T153-T170 (1961).
12. J. H. Bradbury and G. E. Rogers, Textile Res. J., 33, 452-458 (1963).
13. J. H. Bradbury and G. V. Chapman, Australian J. Biol. Sci., 17, 960-972 (1964).
14. J. H. Bradbury and J. D. Leeder, Australian J. Biol. Sci., 23, 843-854 (1970).
15. J. D. Leeder and J. H. Bradbury, Textile Res. J., 41, 563-568 (1971).
16. M. G. Dobb, R. Murray, and J. Sikorski, J. Microscopy 96, Pt. 3, 285-299 (1972).
17. J. Lindberg, B. Philip, and N. Gralén, Nature, 162, 458-459 (1948).
18. J. Lindberg, Textile Res. J., 19, 43-45 (1949).
19. N. L. R. King and J. H. Bradbury, Australian J. Biol. Sci., 21, 375-384 (1968).
20. W. McMurtrie, Wool and Other Animal Fibers, U.S. Government Printing Office, Washington, D.C., 1886.
21. K. von Allwörden, Zeits. angewandte Chem., 29, I, 77-78 (1916). (Abstracted in Chem. Abs., Jan.-June 1916, pp. 1436-1437.)
22. C. Kronacher and G. Saxinger, Zeits. f. Tierzüchtung u. Züchtungsbiol., 4:3, 201-211 (1925) and 5:3, 373-378 (1926); quoted by C. Kronacher and G. Lodemann, Technik der Haar und Wolleuntersuchung, Urban and Schwarzenberg, Berlin, 1930, pp. 147-8.
23. C. H. Müller, Zeits. Zellforsch. Mikroskop. Anat., A29, 1-43 (1939).

24. C. W. Hock, R. C. Ramsay, and M. Harris, J. Res. Natnl. Bur. Stds., A27, 181-190 (1941).
25. G. J. Schuringa and L. Algera, Biochim. Biophys. Acta, 6, 325-333 (1950).
26. D. F. O'Reilly, J. C. Whitwell, R. O. Steele, and J. H. Wakelin, Textile Res. J., 22, 441-447 (1952).
27. J. D. Leeder and J. H. Bradbury, Nature, 218, 694-695 (1968).
28. J. H. Bradbury, J. D. Leeder, and I. C. Watt, Proceedings 4th International Wool Textile Res. Conf., Berkeley, Calif., 1970, published in Appl. Polymer Symp., No. 18, 1971, pp. 227-236.
29. P. Alexander, J. Soc. Dyers Colourists, 66, 349-357 (1950).
30. R. L. Elliott, J. Soc. Dyers Colourists, 66, 470 (1950).
31. R. L. Elliott and B. Manogue, J. Soc. Dyers Colourists, 68, 12-14 (1952).
32. G. Lagermalm and N. Gralén, Acta Chem. Scand., 5, 1209-1210 (1951).
33. G. J. Schuringa, C. J. Schooneveldt, and A. J. Ultée Jr., Biochim. Biophys. Acta, 9, 506-514 (1952).
34. H. Zahn, Melliand Textilber., 32, 419-420 (1951).
35. H. Zahn, Textil-Rundschau, 7, 305-319 (1952).
36. R. L. Golden, J. C. Whitwell, and E. H. Mercer, Textile Res. J., 25, 334-342 (1955).
37. N. L. R. King and J. H. Bradbury, Australian J. Biol. Sci., 21, 375-384 (1968).
38. P. F. Lofts and E. V. Truter, J. Textile Inst., 60, 46-56 (1969).
39. J. Lindberg, E. H. Mercer, B. Philip, and N. Gralén, Textile Res. J., 19, 673-678 (1949).
40. H. Davson and J. F. Danielli, The Permeability of Natural Membranes, 2nd ed., Cambridge Univ. Press, London and New York, 1952.
41. J. D. Robertson, Prog. Biophys. Biophys. Chem., 10, 343-418 (1960).
42. S. J. Singer and G. L. Nicolson, Science, 175, 720-731 (18 Feb. 1972).
43. Monitor, New Sci. and Sci. J., 51, 613 (1971).
44. Monitor, New Sci. and Sci. J., 51, 439 (1971).
45. P. P. da Silva and D. Branton, J. Cell Biol., 45, 598-605 (1970).
46. R. L. Steere, J. Biophys. Biochem. Cytol., 3, 45-60 (1957).
47. H. Moor, K. Mühlethaler, H. Waldner, and A. Frey-Wyssling, J. Biophys. Biochem. Cytol., 10, 1-13 (1961).
48. J. A. Lucy, J. Theoret. Biol., 7, 360-373 (1964).
49. F. S. Sjöstrand, J. Ultrastruct. Res., 9, 340-361 (1963).
50. G. Vanderkooi and D. E. Green, Proc. Ntnl. Acad. Sci. (U.S.A.), 66, 615-621 (1970).
51. E. L. Benedetti and P. Emmelot, in The Membranes (A. J. Dalton and F. Haguenau, eds.), Academic Press, New York, 1968, pp. 33-120.
52. A. J. Dalton and F. Haguenau (eds.), The Membranes, Academic Press, New York, 1968.

53. K. Malhotra, Prog. Biophys. Molec. Biol., 20, 69-131 (1968-9; pub. 1970).
54. "Research News," in Science, 175, 1348-1350 (1972).
55. S. J. Singer, in Structure and Function of Biological Membranes (L. I. Rothfield, ed.), Academic Press, New York, 1971, pp. 145-222.
56. N. M. Green, "Proc. Biochem Soc.," in Biochem. J., 122, 37P-38P (1971).
57. R. A. De Deurwaerder, M. G. Dobb, and B. J. Sweetman, Nature, 203, 48-49 (1964).
58. H. Zahn and M. Biela, Textil-Praxis, 23, 103-106 (1968).
59. J. H. Bradbury, G. V. Chapman, and N. L. R. King, Proceedings 3rd International Wool Textile Research Conference, Paris, 1965, Vol. I, 359-364 (1965).
60. J. H. Bradbury, G. V. Chapman, and N. L. R. King, Australian J. Biol. Sci., 18, 353-364 (1965).
61. G. E. Rogers, in The Epidermis (E. W. Montagna and W. L. Lobitz Jr., eds.), Academic Press, New York, 1964, pp. 179-236.
62. D. E. Peters and J. H. Bradbury, Australian J. Biol. Sci., 25, 1225-1234 (1972).
63. M. G. Dobb, F. R. Johnston, J. A. Nott, L. Oster, J. Sikorski, and W. S. Simpson, J. Textile Inst. (Trans.), 52, T153-T170 (1961).
64. G. Lagermalm, Textile Res. J., 24, 17-25 (1954).
65. G. E. Rogers, Ann. N.Y. Acad. Sci., 83, 378-399 (1959).
66. P. Kassenbeck, in Structure de la Laine, Inst. Textile de France, Paris, 1961, pp. 50-68.
67. M. G. Dobb and J. Sikorski, in Structure de la Laine, Inst. Textile de France, Paris, 1961, pp. 37-49.
68. J. Sikorski and W. S. Simpson, Nature, 182, 1235-1236 (1958).
69. J. A. Swift, J. Roy. Microscop. Soc., 88, 449-460 (1968).
70. J. H. Bradbury, G. V. Chapman, N. L. R. King, and J. M. O'Shea, Australian J. Biol. Sci., 23, 637-643 (1970).
71. J. H. Bradbury and K. F. Ley, Australian J. Biol. Sci., 25, 1225-1234 and 1235-1247 (1972).
72. S. J. Leach, G. E. Rogers, and B. K. Filshie, Arch. Biochem. Biophys., 105, 270-287 (1964).
73. J. H. Bradbury and J. D. Leeder, Australian J. Biol. Sci., 25, 133-138 (1972).
74. C. W. Hock, R. C. Ramsay, and M. Harris, Am. Dyestuff Rep., 30, 449-456 and 469-470 (1941).
75. K. Stirm and H. Colle, Melliand Textilber., 16, 585-586, 667-668, 795-798 (1935).
76. G. J. Schuringa, T. Konings, and A. J. Ultée Jr., Textile Res. J., 23, 645-646 (1953).
77. J. B. Speakman, B. Nilssen, and G. H. Elliott, Nature, 142, 1035 (1938).
78. K. R. Makinson, Proceedings International Wool Textile Research Conference, Australia, 1955, Section E, pp. 460-471.
79. W. Herbig, Zeits. angewandte Chem., 32, 120 (1919).

80. M. Leveau, J. Langlois, and A. Parisot, Bull. Inst. Textile France, No. 34, 1952, pp. 9-14.
81. A. Parisot and M. Leveau, Textile Res. J., 23, 440-441 (1953).
82. R. D. B. Fraser and G. E. Rogers, Biochim. Biophys. Acta, 16, 307-316 (1955).
83. K. R. Makinson, Proceedings 4th International Wool Textile Research Conference, Berkeley, Calif., 1970, published in Appl. Polymer Symp., No. 18, 1971, 1083-1096.
84. K. R. Makinson, J. Textile Inst., 61, 151-153 (1970).
85. E. Lehmann, Melliand Textilber., 22, 453-456 (1941).
86. J. H. Bradbury and E. G. Chapman, Textile Res. J., 33, 666-667 (1963).
87. G. Berg, Melliand Textilber., 25, 145-150 (1944).
88. O. K. Dobozy, Textile Res. J., 28, 717-719 (1958).
89. K. R. Makinson, Textile Res. J., 29, 431-439 (1959).
90. K. R. Makinson, Textile Res. J., 29, 439-443 (1959).
91. K. R. Makinson, Textile Res. J., 30, 598-606 (1960).
92. S. Ikeda, J. Soc. Textile and Cellulose Ind. Japan, 16, 1014-1019 (1960).
93. G. F. Flanagan, Textile Res. J., 31, 576-577 (1961).
94. K. R. Makinson, Textile Res. J., 32, 364-374 (1962).
95. K. R. Makinson, in Wool Sci. Rev., No. 24, 1964, pp. 34-48.
96. H. G. Howell, K. W. Mieszkis, and D. Tabor, Friction in Textiles, Butterworths, London, 1959; (a) pp. 47-51 and 82-84; (b) pp. 99-103.
97. K. R. Makinson, J. Textile Inst., 61, 465-471 (1970).
98. A. J. P. Martin and R. Mittelman, J. Textile Inst., Trans., 37, T269-T280 (1946).
99. J. B. Sherman, E. Balasubramaniam, and K. J. Whiteley, J. Textile Inst., 59, 1-9 (1968).
100. J. Lindberg and N. Gralén, Textile Res. J., 18, 287-301 (1948).
101. A. K. van der Vegt and G. J. Schuringa, Textile Res. J., 26, 9-16 (1956).
102. E. H. Mercer and K. R. Makinson, J. Textile Inst. (Trans.), 38, T227-T240 (1947).
103. H. L. Röder, J. Textile Inst. (Trans.), 44, T247-T265 (1953).
104. M. Lipson and P. Howard, J. Soc. Dyers Colourists, 62, 29-32 (1946).
105. G. King, J. Textile Inst. (Trans.), 41, T135-T144 (1950).
106. J. R. McPhee, Textile Res. J., 30, 349-357 (1960).
107. J. R. McPhee, Textile Res. J., 31, 770-778 (1961).
108. J. H. Bradbury, Textile Res. J., 31, 735-743 (1961).
109. M. W. Andrews, A. S. Inglis, F. E. Rothery, and V. A. Williams, Textile Res. J., 33, 705-711 (1963).
110. E. F. Denby and M. W. Andrews, Textile Res. J., 35, 913-922 (1965).
111. M. A. Chaudri and K. J. Whiteley, J. Textile Inst., 60, 37-45 (1969).
112. E. H. Mercer, Nature, 155, 573-574 (1945).

113. B. K. Lohani, L. Valentine, and C. S. Whewell, J. Textile Inst. (Trans.), 49, T265-T278 (1958).
114. J. Lindberg, Textile Res. J., 18, 470-474 (1948).
115. Ch. Sakai, J. Soc. Fiber Sci. Tech. Japan, 21, 603-611 (1965).
116. K. R. Makinson, Textile Res. J., 37, 763-771 (1967).
117. K. R. Makinson, Textile Res. J., 39, 1072-1074 (1969).
118. K. R. Makinson, unpublished measurements (1966).
119. L. Bohm, J. Soc. Dyers Colourists, 61, 278-283 (1945).
120. G. F. Flanagan, Textile Res. J., 36, 55-65 (1966).
121. D. Frishman, A. L. Smith, and N. Harris, Textile Res. J., 18, 475-480 (1948).
122. K. R. Makinson, in Wool Sci. Rev., No. 42, 1972, pp. 2-16.
123. K. R. Makinson, Trans. Faraday Soc., 44, 279-282 (1948).
124. F. P. Bowden and D. Tabor, The Friction and Lubrication of Solids, Clarendon Press, Oxford, 1950, Chap. V.
125. B. Lincoln, J. Textile Inst. (Trans.), 45, T92-T107 (1954).
126. B. Lincoln, in Wool Sci. Rev., No. 18, 1960, pp. 38-50.
127. P. Grosberg, J. Textile Inst. (Trans.), 46, T233-T246 (1955).
128. A. J. P. Martin, J. Soc. Dyers Colourists, 60, 325-327 (1944).
129. J. Menkart and J. B. Speakman, Nature, 156, 143 (1945).
130. R. W. Moncrieff, Wool Shrinkage and Its Prevention, National Trade Press, London, 1953; (a) pp. 81-82; (b) Chapter 21; (c) Chapter 22.
131. H. D. Feldtman and J. R. McPhee, Textile Res. J., 34, 634-642 (1964).
132. A. D. Sule, Wool and Woollens of India, 3, 1-38 (1967).
133. W. J. Thorsen, Textile Res. J., 38, 644-650 (1968).
134. W. J. Thorsen and R. C. Landwehr, Textile Res. J., 40, 688-695 (1970).
135. P. Kassenbeck, Bull. Inst. Textile France, No. 18, 7-33 (1964).
136. A. J. Farnworth, J. Soc. Dyers Colourist, 77, 483-488 (1961).
137. P. Alexander, D. Carter, and C. Earland, J. Soc. Dyers Colourists, 67, 23-27 (1951).
138. P. G. Gordon, private communication, 1971.
138a. K. R. Makinson, Textile Res. J., 44, 856-857 (1974).
139. C. Earland, J. Textile Inst.,(Trans.), 48, T132 (1957).
140. C. Earland, Textile Res. J., 27, 587-588 (1957).
141. K. R. Makinson, Wear, 16, 287-292 (1970).
142. D. E. Henshaw, J. Textile Inst. (Trans.), 52, T537-T553 (1961).
143. E. H. Mercer, Nature, 155, 573-574 (1945).
144. J. Lindberg, Textile Res. J., 23, 225-236 (1953).
145. N. Gralén and B. Olofsson, Textile Res. J., 17, 488-496 (1947).
146. J. B. Speakman and E. Stott, J. Textile Inst. (Trans.), 22, T339-T348 (1931).
147. S. M. A. Shah and K. J. Whiteley, J. Textile Inst., 62, 361-374 (1971).
148. S. Smuts and R. I. Slinger, SAWTRI Tech. Rep., No. 163, S. African Wool Textile Research Institute, Port Elizabeth, South Africa (1972).

149. M. A. Ali, K. J. Whiteley, and M. A. Chaudri, J. Textile Inst., 62, 375-381 (1971).
150. C. M. Van Wyk, Onderstepoort J. Vet. Sci. Animal Ind., 21, 99-226 (1946), in particular pp. 180-181.
151. L. J. Postle and J. Ingham (with Appendix by D. R. Cox), J. Textile Inst. (Trans.), 43, T77-T90 (1952).
152. S. L. Anderson, D. R. Cox, and L. D. Hardy, J. Textile Inst. (Trans.), 43, T362-T379 (1952).
153. D. S. Taylor, J. Textile Inst. (Proc.), 46, P59-P83 (1955).
154. E. Dyson and A. E.-A. El-Bayoumi, Proceedings 4th International Wool Textile Research Conference, Berkeley, Calif., 1970, published in Appl. Polymer Symp., No. 18, Pt. II, 1255-1267 (1971).
155. Wool Industries Research Association, Wool Research 1918-1948, Wool Industries Research Association, Headingley, Leeds, England, 1949; (a) Vol. 6, Drawing and Spinning; (b) Vol. 4, Carding; (c) Vol. 2, Physical Properties, Chapter 5.
156. G. Brook and M. Hannah, J. Textile Inst. (Proc.), 46, P23-P39 (1955).
157. P. J. Kruger, J. Textile Inst. (Trans.), 58, T158-T168 (1967).
158. R. E. Belin, J. Textile Inst. (Trans.), 58, T169-T174 (1967).
159. A. Barella and A. Segura, Proceedings of 4th International Wool Textile Research Conference, Berkeley, Calif., 1970, published in Appl. Polymer Symp., No. 18, Pt. II, 1289-1301 (1971).
160. A. Kul and P. A. Smith, Proceedings of 4th International Wool Textile Research Conference, Berkeley, Calif., 1970, published in Appl. Polymer Symp., No. 18, Pt. II, 1467-1476 (1971).
161. M. Hannah, J. Textile Inst. (Trans.), 41, T57-T123 (1950).
162. C. H. Giles, A. P. D'Silva, and A. S. Trivedi, J. Appl. Chem., 20, 37-41 (1970).
163. J. W. Rowen and R. L. Blaine, Ind. Eng. Chem., 39, 1659-1663 (1947).
164. I. C. Watt and J. D. Leeder, J. Textile Inst., 59, 353-364 (1968).
165. S. Brunauer, Adsorption of Gases and Vapours, Vol. I, Physical Adsorption, Oxford University Press, Oxford, England, 1945, Chapter IX.
166. A. B. D. Cassie and S. Baxter, Trans. Faraday Soc., 40, 546-551 (1944).
167. S. Baxter and A. B. D. Cassie, J. Textile Inst., 36, T67-T90 (1945).
168. J. J. Bikerman, Koll. Zeits. u. Zeits. F. Polymere, 218, 52-56 (1967).
169. A. W. Adamson, Physical Chemistry of Surfaces, 2nd edn., Wiley-Interscience, 1967, p. 360.
170. H. W. Fox and W. A. Zisman, J. Colloid Sci., 5, 514-531 (1950).
171. H. W. Fox and W. A. Zisman, J. Colloid Sci., 7, 109-121 (1952).
172. H. W. Fox and W. A. Zisman, J. Colloid Sci., 7, 428-442 (1952).
173. W. A. Zisman, in Adhesion and Cohesion (P. Weiss, ed.), Elsevier, Amsterdam/London/New York, 1962, pp. 176-208.

174. N. R. S. Hollies, M. M. Kaessinger, and H. Bogaty, Textile Res. J., 26, 829-835 (1956).
175. A. M. Schwartz and C. A. Rader, J. Soc. Cosmet. Chemists, 13, 245-250 (1962).
176. J. C. Stewart and C. S. Whewell, Textile Res. J., 30, 912-918 (1960).
177. F. Minor, A. M. Schwartz, E. A. Wulkow, and L. C. Buckles, Textile Res. J., 29, 940-949 (1959).
178. J. Lindberg, Textile Res. J., 23, 585-588 (1953).
179. W. J. Thorsen, Proceedings of 4th International Wool Textile Research Conference, Berkeley, Calif., 1970, published in Appl. Polymer Symp., No. 18, Part II, 1171-1175 (1971).
180. M. M. Millard, Anal. Chem., 44, 828-829 (1972).
181. A. H. Ellison, H. W. Fox, and W. A. Zisman, J. Phys. Chem., 57, 622-627 (1953).
182. E. G. Shafrin and W. A. Zisman, J. Phys. Chem., 64, 519-524 (1960).
183. H. A. Schuyten, J. W. Weaver, and J. D. Reid, Am. Dyestuff Reptr., 38, 364-366 (1949).
184. L. Wikstroem, E. Soervik, M. Cednaes, and B. Olofsson, Proceedings of 1st International Wool Textile Research Conference, Australia, 1955, Vol. C, 257-276.
185. F. M. Fowkes, Ind. Eng. Chem., 56, 40-52 (1964).
186. E. A. Boucher, Nature, 215, 1054 and 1071 (1967).
187. J. R. Dann, J. Coll. Interface Sci., 32, 302-320 and 321-331 (1970).
188. D. H. Kaelble, Physical Chemistry of Adhesion, Wiley-Interscience, New York, 1971, p. 154.
189. W. J. Murphy, M. W. Roberts, and J. R. H. Ross, J. Chem. Soc. (Faraday Trans.), I, 68, 1190-1199 (1972).
190. C. A. Anderson, R. E. Belin, H. D. Feldtman, and M. T. Goldsmith, Textilveredlung, 5, 846-854 (1970).
191. E. Weideman, SAWTRI Technical Report, No. 178, South African Wool and Textile Research Institute of the CSIR (1970).
192. Y. Nakamura, K. Kamada, and T. Kondo, J. Soc. Fiber Sci. Technol. Japan, 26, 263-271 (1970).
193. Y. Nemoto and Y. Terada, Nagoya-Shi Kogyo Kenkyasho Kenkyu Hokoku, No. 45, 1-5 (1971); per Chem. Abs., 76, No. 35115 (1972).
194. B. Fleischfresser, private communication, 1973, to be published.
195. J. P. Mutchler, J. Menkart, and A. M. Schwartz, Advan. Chem. Ser., No. 86, 7-14 (1967).
196. H. Alter and H. Cook, J. Coll. Interface Sci., 29, 439-443 (1969).
197. Pittman, Proceedings of 4th International Wool Textile Research Conference, Berkeley, Calif., 1970, published in Appl. Polymer Symp., No. 18, Part I, 593-599 (1971).
198. P. W. Rance, J. Coll. Interface Sci., 41, 588-593 (1972).
199. L. Benisek and G. H. Crawshaw, Proceedings of 4th International Wool Textile Research Conference, Berkeley, Calif., 1970;

published in Appl. Polymer Symp., No. 18, Part II, 1199-1210 (1971).
200. L. Benisek, Textile Res. J., 42, 490-496 (1972).
201. J. W. Schappel, Textile Res. J., 26, 211-226 (1956).
202. G. C. East and J. P. Ferguson, J. Textile Inst., 64, 273-278 (1973).
203. J. Compton and W. J. Hart, Textile Res. J., 24, 263-264 (1954).
203a. G. M. Venkatesh, N. E. Dweltz, G. L. Madan, and R. H. Alurkar, Textile Res. J., 44, 352-362 (1974).
203b. J. Lewis, Wool Sci. Rev., No. 48, 42-58 (March, 1974).
204. J. L. Cave and O. A. Swanepoel, Textilveredlung, 4, 299-304 (1969).
205. R. P. Harker, J. Textile Inst. (Trans.), 50, T189-T222 (1959).
206. A. C. Goodings, Am. Dyestuff Reptr., 40, P55-P60 (1951).
206a. C. E. Warburton, Jr. and F. J. Parkhill, Textile Chemist and Colorist, 5, 113-117 (1973).
206b. C. E. Warburton, Jr. and F. J. Parkhill, Textile Chemist and Colorist, 5, 233-234 (1973).
206c. E. Kissa, Textile Chemist and Colorist, 5, 232-233 (1973).
206d. S. Smith, B. Johannessen, and P. O. Sherman, Textile Chemist and Colorist, 5, 39-44 (1973).
207. E. K. Goette, J. Colloid Sci., 4, 459-484 (1949).
208. J. C. Harris, Textile Res. J., 28, 912-928 (1958).
209. H. J. Jacobasch, Textilveredlung, 5, 385-391 (1970).
210. S. P. Hersh and D. J. Montgomery, Textile Res. J., 25, 279-295 (1955).
211. S. P. Hersh and D. J. Montgomery, Textile Res. J., 22, 805-818 (1952).
212. J. A. Medley, J. Textile Inst. (Proc.), 48, P112-P120 (1957).
213. W. Löbel, J. Textile Inst., 59, 457-459 (1968).
214. P. S. H. Henry, J. Textile Inst., 59, 459-460 (1968).
215. R. C. Landwehr, Textile Res. J., 39, 792-793 (1969).
216. H. E. Cook and I. T. Smith, Proceedings of 4th International Wool Textile Research Conference, Berkeley, Calif., 1970, published in Appl. Polymer Symp., No. 18, 663-672 (1971).
217. J. E. Algie, private communication, 1972.
218. A. M. Sookne and M. Harris, Am. Dyestuff Reptr., 28, P593-P596 (1939).
219. J. S. Stanley, J. Phys. Chem., 58, 533-536 (1954).
220. S. M. Neale and R. H. Peters, Faraday Soc. (Trans.), 42, 478-487 (1946).
221. M. Harris, Am. Dyestuff Reptr., 21, 399-403 (1932).
222. D. Stigter, J. Colloid Sci., 19, 268-278 (1964).
223. E. G. H. Carter and R. Consden, J. Textile Inst. (Trans.), 37, T227-T236 (1947).
224. G. L. Royer, H. E. Millson, and C. A. Amick, J. Soc. Dyers Colourists, 63, 214-224 (1947).
225. H. E. Millson and L. H. Turl, Am. Dyestuff Reptr., 39, P647-P656 (1950).

226. J. Lindberg, Textile Res. J., 23, 67-76 (1953).
227. J. Lindberg, Textile Res. J., 23, 573-584 (1953).
228. J. A. Medley and M. W. Andrews, Textile Res. J., 29, 398-403 (1959).
229. B. H. Butcher and E. L. Cussler, J. Soc. Dyers Colourists, 88, 398-400 (1972).
230. J. Lindberg, Textile Res. J., 20, 381-384 (1950).
231. J. D. Leeder and M. Lipson, J. Appl. Polymer Sci., 7, 2053-2066 (1963).
232. I. C. Watt, J. Appl. Polymer Sci., 8, 1737-1750 (1964).

Chapter 4

SURFACE GEOMETRY OF SYNTHETIC FIBERS

Francis L. Scardino

Textile Department
Philadelphia College of Textiles and Science
Philadelphia, Pennsylvania

I. INTRODUCTION

The surface geometry of synthetic fibers is relatively simple in comparison with the surface geometry of natural fibers. This simplicity is a reflection of the uniformity and of the quality control in the production of synthetic filaments vs. the seemingly unlimited variations found in the properties of naturally occurring fibers. For example, cotton has an ever changing cross-sectional shape and a variation in cross-sectional area (fineness) along the length of a given fiber and between fibers. Moreover, wool has a circular cross-sectional shape of varying diameter and a complex scaly surface. The scales vary in size and in depth in any given wool fiber and between fibers. By comparison, synthetic fibers tend to have little variation in cross-sectional shape and in cross-sectional area along their length, and the microscopic surface roughness normally present is fairly uniform in size and distribution. Finally, on a macroscopic level, other major differences can be found in fiber length and crimp geometry. Cotton and wool tend to have a tremendous array of fiber lengths, whereas synthetic fibers are continuous or when cut, relatively uniform in length. While natural fibers tend to have tremendous variations in crimp geometry and frequency, synthetic fibers (when crimped) tend to have fairly uniform crimp properties.

The purpose of this chapter is to describe the surface geometry of synthetic fibers and the effect it has on processing such fibers.

II. COMPONENTS OF FIBER SURFACE GEOMETRY

It is quite convenient to subdivide the surface geometry of synthetic fibers into three distinct categories. These categories are fiber surface roughness, cross-sectional shape, and crimp. Crimp, like texturing, is not added to all synthetic fibers produced and is certainly a special consideration. However, when crimp is added to a synthetic fiber, both the surface geometry and the processing behavior of the fiber are substantially affected. It is in this context that crimp is included as a component of fiber surface geometry.

Similarly, length and fineness (diameter or linear density) could be considered elements of fiber surface geometry: length because of surface continuity and fineness because of its effect on the radius of curvature of the fiber surface. However, length and fineness are considered by most technologists to be fiber geometric parameters rather than parameters of fiber surface geometry. Crimp on the other hand can fit into both categories.

Practically any process subsequent to the production of synthetic filaments significantly, if not substantially, affects fiber surface geometry. Among these processes are texturing, bulk additives to polymer, surface modifications, and several other processes of a mechanical, chemical, or thermal nature. Only the raw, or basic, components of fiber surface geometry will be covered in this chapter.

A. Fiber Surface Roughness

It has long been a common practice among fiber producers to incorporate something other than polymeric materials in the manufacture of synthetic fibers. An example of this is the use of a delusterant such as titanium dioxide (TiO_2); the amount of delusterant used varies according to the desired effect. For example, based on the weight of the fiber, the inclusion of 2% delusterant would produce a "dull" fiber while half that amount would produce a "semidull" effect. It might be pointed out that, under normal circumstances, the incorporation of such small proportions of delusterant does not substantially affect the mechanical properties of the synthetic fibers.

Agglomerates of delusterant particles located near the surface of synthetic fibers cause asperities as shown in Fig. 1. Such protuberances, or asperities, are very typical of the size of the microscopic roughness found on the surface of synthetic fibers. These asperities are about 0.5 μm in diameter and about 0.25 μm in height. Within limits, the frequency of asperities and, therefore, the surface roughness increases with an increase of delusterant. For example, in Fig. 2, polyethylene terephthalate (PET) fibers containing 0.1 and 2.0% TiO_2 delusterant are shown to have comparatively smooth and rough surfaces, respectively. The PET fiber containing 2.0% TiO_2 has twenty times by volume the amount of delusterant as the PET fiber containing 0.1% TiO_2. This difference in the concentration of agglomerates is brought out in the fiber cross sections shown in Fig. 3. It is pointed out that the size distribution of the TiO_2 agglomerates is approximately the same in the two samples. The PET fiber containing 2.0% TiO_2 has seven to eight times the number of asperities on the fiber surface as does the PET sample containing 0.1% TiO_2. (By squaring the cube root of the 20:1 ratio, one should expect a ratio of 7.5:1, approximately.) This ratio has been confirmed by actual surface scanning of the smooth and rough fibers [1].

B. Fiber Cross-Sectional Shape

Most synthetic fiber is produced with a circular cross-sectional shape by extrusion through round holes. However, synthetic fiber can be made with a variety of cross-sectional shapes, many of which are of commercial importance. These noncircular cross-sectional shapes often have surface geometric profiles with protuberances or projections which are an order of magnitude greater than those of asperities (several micrometers versus a fraction of a micrometer, for example.) Furthermore, slight deviations from noncircularity of cross-sectional shape can lead to substantial differences in fiber processing behavior and in the performance characteristics of textile structures. Fiber cross-sectional shape affects the cohesion and bulkiness (volumetric

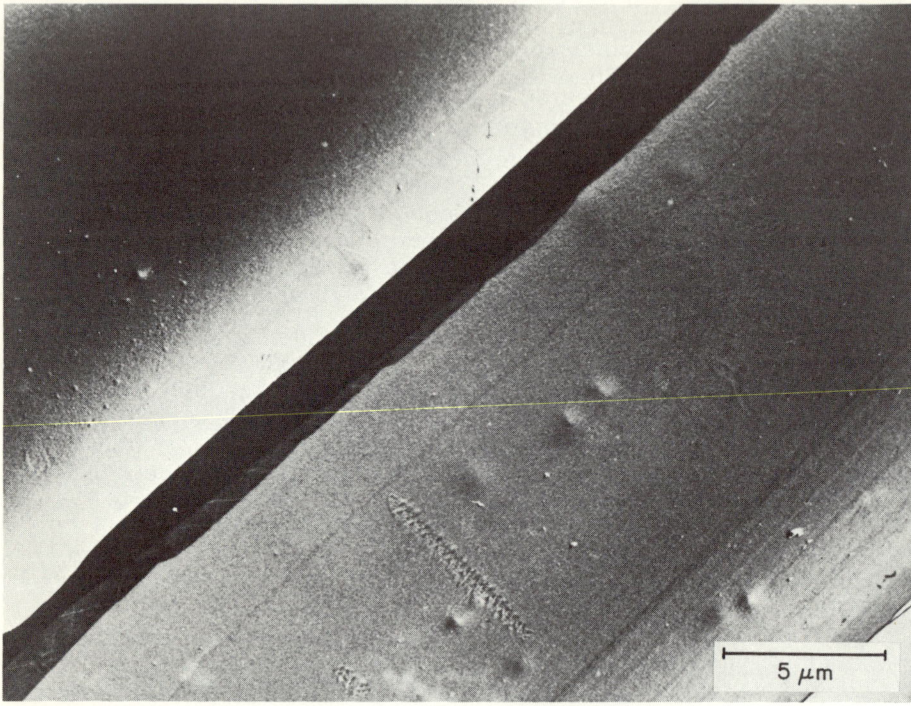

FIG. 1. Electron photomicrographs showing asperities caused by agglomerates of TiO_2 delusterant near the surface of a fiber. Reprinted from ref. 1, p. 875, courtesy of Textile Research Journal.

packing density) of fiber assemblies. Also, the relative luster and stiffness (bending modulus) of a fiber are affected.

A number of idealized diagrams of cross-sectional shapes of equal cross-sectional area are shown in Fig. 4. It is pointed out that, while these idealized shapes are representative of the variety possible, commercially produced synthetic fibers would not ordinarily be so perfectly shaped. Figure 4 is not to be confused with actual fiber cross-sectional shapes.

It is well known that textile fibers are quite flaccid and, once released from tension, are never found in a perfectly linear configuration, similar to that of a rigid rod for example. Rather, fibers tend to twist, spiral, or curl about their own axis. In fibers with a smooth, circular cross-sectional shape, as represented by Fig. 4a, the effect of this tendency to twist or spiral is minimal. In fibers with nonsmooth, noncircular cross-sectional shapes, however, a spiraling

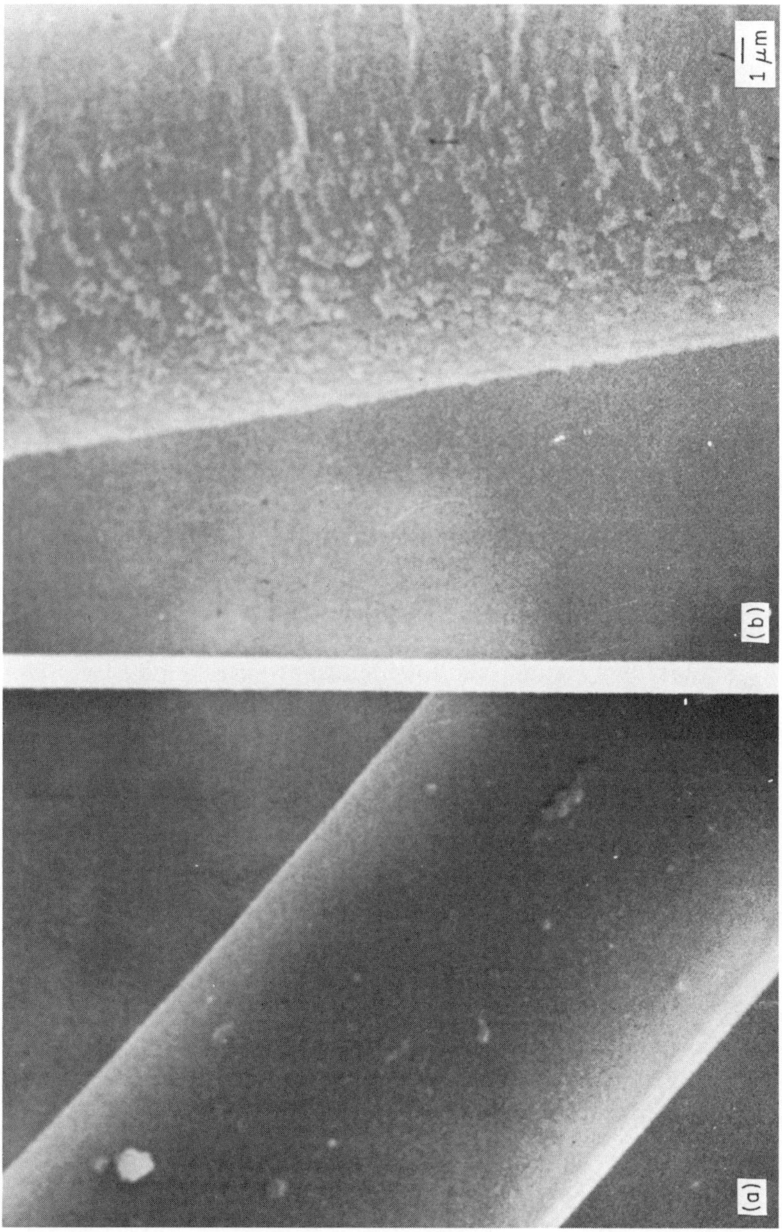

FIG. 2. Electron photomicrographs showing the surfaces of a smooth (a) (0.1% TiO_2 delusterant) and of a rough (b) (2.0% TiO_2 delusterant) polyester model fiber. Courtesy of Textile Research Institute.

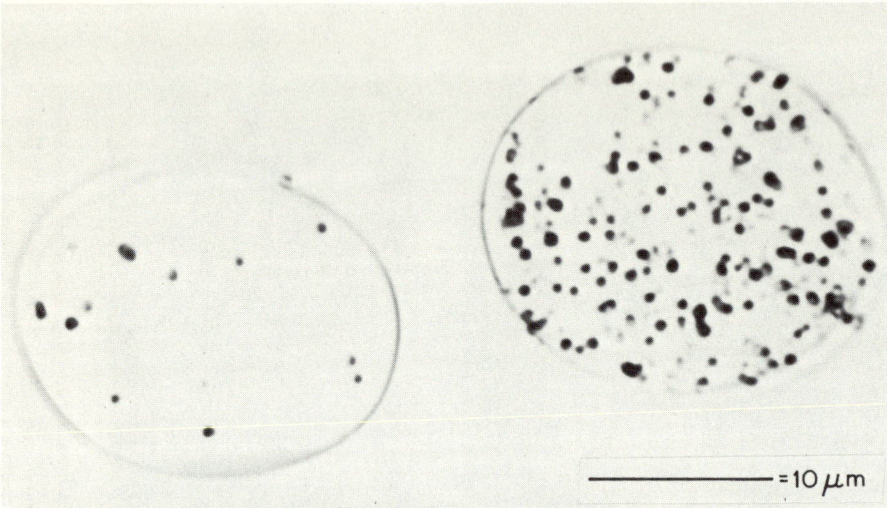

FIG. 3. Phase contrast photomicrographs of 1-μm thick sections of smooth (left) and rough (right) model fibers showing TiO_2 agglomerates. Reprinted from ref. 1, p. 876, courtesy of Textile Research Journal.

edge is generated (similar to that of a screw) thereby restricting the potential for interfiber contact and for high fiber packing density. The less circular the cross-sectional shape of the fiber, the larger the restriction.

An indication of the restrictive tendency is shown by the circles around each cross-sectional shape in Fig. 4. The potential area of contact with contiguous fibers over a very short segment of fiber length can be approximated by the arcs of fiber surface perimeter that are congruent with the circle circumscribing the idealized cross-sectional shape. Thus, if a fiber with a smooth, circular cross-sectional shape (Fig. 4a) had a potential for interfiber contact of 100%, the potentials for the other shapes shown in Fig. 4 would vary, approximately, as follows: 4b (rough, circular) 30%, 4c (tetrakelion) 27%, 4d (trilobal) 16%, 4e (elliptical) 11%, and 4f (triangular) 2%. Once again, Fig. 4b suggests the dramatic effect of asperities on the surface of a fiber with a circular cross section. It is pointed out, however, that when asperities are superimposed onto a noncircular cross section, the effect of the asperities is masked by the dominant noncircular profile.

C. Fiber Crimp

When synthetic filament is to be cut into staple form (short fiber of definite length) for the manufacture of spun yarn or for the produc-

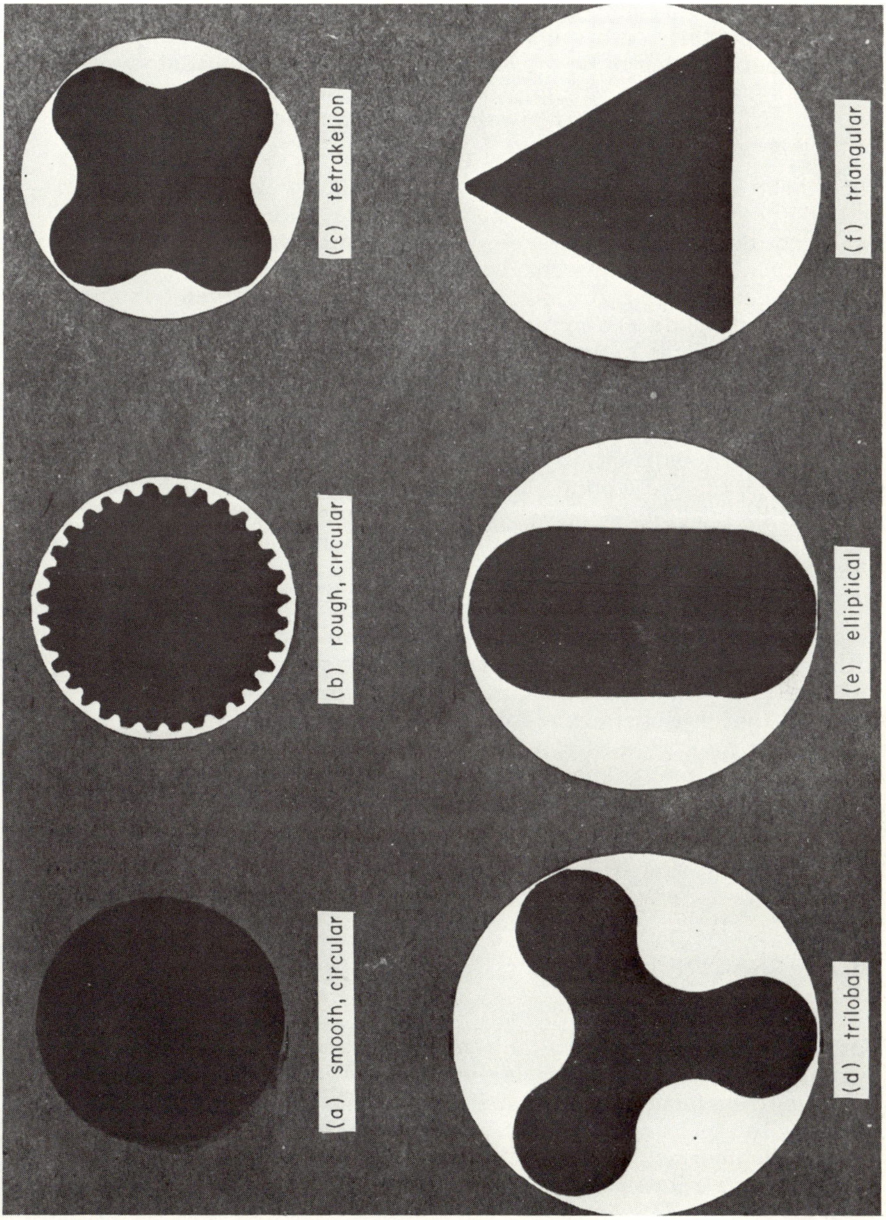

FIG. 4. Idealized diagrams of various fiber cross-sectional shapes, of equal cross-sectional areas, inscribed within circles generated by rotating the shapes about their centers. Courtesy of Textile Research Institute.

tion of nonwoven fabrics, the filaments are usually crimped before being cut. This is usually accomplished by imparting mechanical distortion in a compacting or gear box process. Crimp is necessary for the efficient manipulation of fiber in the early stages of mechanical processing, especially on opening and carding equipment. Crimp allows the fiber bundles to be more easily grasped and separated into individual fibers. Also, a crimp geometry is essential for the cohesion or strength of in-process fiber assemblies (web and sliver) and for the uniformity of fiber processing.

The planar sawtooth crimp geometry common in synthetic staple is easily detected by visual inspection, especially in unopened stock. What may not be apparent, however, are the prominent nodules and related surface deformations resulting from a typical mechanical crimping process. These projections can be easily observed by straightening (unbending) a crimped fiber under low tension. Under such low tension, these crimp nodules, or deformations, tend to project out of the line of the fiber axis for a distance equivalent, at least, to the fiber diameter. Under greater tension (sufficient for removal of crimp), the fiber becomes quite linear and the projections of the crimp nodules are substantially less prominent. Under extreme fiber tension (beyond the yield point), these nodules become obscure. The frequency of the nodules depends on fiber crimp frequency. For example, in a synthetic fiber with 12 crimps per inch, one would expect to find a crimp nodule every millimeter along the fiber length, approximately.

In summary, in the surface geometry of synthetic fibers, the spatial prominence of crimp is an order of magnitude greater than that of fiber cross-sectional shape and several orders of magnitude greater than that of fiber surface roughness. Thus, it is to be expected that the role of these components of fiber surface geometry during textile processing should be substantially different. Moreover, it should be expected that the prominence of the role of each component might vary according to the stage of processing or to the structure of the fiber assembly.

III. CHARACTERIZATION OF FIBER SURFACE GEOMETRY

Various techniques are available for the characterization of each of the components of fiber surface geometry. Microscopy can be used to detect fiber surface roughness and cross-sectional shape. Measurements under low-power magnification can be made to characterize fiber crimp geometry [2], and cyclic force-extension tests can provide many parameters relating to the uncrimping of fibers. However, it has been difficult to develop a technique which, by the integration of these individual components of fiber surface geometry, can be used to establish a complete and a meaningful fiber surface profile.

A. Microscopic Techniques

For many years, the ordinary light microscope was the only equipment available for the study of fiber surface geometry. Although the application of electron microscopy techniques provided far greater detail than regular microscopy, the interpretation and limitation of surface replicas left much to be desired. The development of the scanning electron microscope with manipulative stage for fiber specimen repositioning has made other microscopy techniques quite limiting and impractical, if not obsolete, for the study of fiber surface detail. This is easily substantiated by the scanning electron photomicrographs shown in Chapter 7.

In addition to "getting a picture" of the surface roughness or detail of a fiber, however, it is also desirable to determine the frequency, or amount, of roughness relative to other fibers or to other surfaces; the greater the surface area scanned, the more practical the data. Many scanning techniques that were developed for the characterization of surface roughness in films and bulk materials are simply not suitable for scanning the surface of a single fiber. This is due mainly to the unique structural form of textile fibers.

B. Fiber Surface Analyzer

Recognizing the need, a device was developed a number of years ago at Textile Research Institute to characterize the geometric roughness of fiber surfaces [3]. The device, referred to as the fiber surface analyzer, was designed to count the number of asperities in a straight path along the surface of a fiber for a 2-mm segment. Two parameters were used to characterize surface roughness: namely, asperity count and average asperity height, the latter being computed from the observed accumulative asperity height. The fiber surface analyzer was used to characterize the surfaces of a variety of fibers. The analyzer was also quite sensitive to gross or bulk fiber surface modifications on natural fibers (e.g., mercerization of cotton, shrinkproofing of wool). One main limitation of this analyzer, as with all physical probing or physical scanning devices, is in the interaction between scanning element and the fiber specimen surface. Fine surface detail cannot be completely "sensed" during scanning of a specimen.

Despite the clarity and depth of fiber surface detail possible with the scanning electron microscope and the practical roughness count possible with the fiber surface analyzer, one very important and relevant parameter remains quite elusive: the "real area of contact" between a fiber and a contiguous surface or among contiguous fibers in an assemblage [4]. It is pointed out, once again, that the actual surface geometry of a fiber is determined by a combination of the following: (1) fiber surface roughness, (2) fiber cross-sectional shape, (3) fiber crimp or texture, (4) fiber spatial configuration (twisting, buckling, bending, etc.), and (5) bulk additives or finishes, if any. Obviously, no simple technique or "black box" approach has been developed for indexing the total fiber surface profile.

C. Cohesion and Friction Measurements

Indirectly, however, some progress has been made toward getting a "feeling" for real area of contact among contiguous fibers, and between fiber and contiguous surface. For example, under carefully controlled conditions, dynamic cohesion tests on slivers and rovings [1,5] can be used to characterize the drafting behavior of a variety of samples and thereby detect any differences due primarily to fiber surface geometry. Also, Hertel et al. have used a shearing-type cohesion measurement on carefully prepared multilayered card web structures to investigate the effect of fiber surface geometry and surface modifications [6,7]. Still another approach is to measure, under proper conditions, fiber or yarn friction. It is fairly well known that frictional force or response is related to area of contact between fiber specimen and contiguous frictional surface [8]. Several technologists have applied this concept with good practical results.

All of these techniques detect a frictional response that is based on relatively large areas of fiber surface contact. As a measure of the true area of fiber surface contact, therefore, the cohesion and frictional data can be quite meaningful. It is pointed out, however, that many variables and interactions are involved in such measurements. Either by sample preparation or by the application of proper statistical techniques, great care must be taken in removing all differences other than fiber surface geometry in the samples characterized.

In summary, much work remains to be done in characterization of fiber surface geometry in the direction of predicting the relative area of contact among fibers or between fibers and a contiguous surface during processing.

IV. ROLE OF FIBER SURFACE GEOMETRY IN TEXTILE PROCESSING

A. Role of Fiber Surface Roughness

Following the development of the fiber surface analyzer, a series of mechanical processing experiments were performed at Textile Research Institute to establish the role of fiber surface roughness in textile processing and products [1,5]. Using the smooth and rough model fibers shown in Section II-A, samples were specially prepared, crimped, and cut into lengths suitable for processing on the cotton, woolen, and worsted yarn manufacturing systems. No processing finishes were used (other than that of the producer) since these would be expected to alter the fiber surface characteristics and hence mask any differences that might exist in processability and assembly properties. Otherwise, normal processing procedures and conditions were used. Based on these studies, certain general conclusions can be reached, which are given in the following summary.

1. Fiber Separation

Geometrically smoother fibers (those having fewer asperities on the surface) were found to be more difficult to separate from one another during opening and carding. The rougher fiber samples appeared to open more easily and more thoroughly than the smooth samples in the same number of operations. Also, more uniform card webs were produced from the rougher samples.

2. Cohesion in Loose Fiber Assemblies

Geometrically smoother fibers lead to greater cohesion in fiber assemblies, as indicated by static and dynamic cohesion tests on slivers and rovings. Several tables of data show a clear and consistent tendency. Also, the differences in cohesion are rather substantial and statistically quite significant. From a practical point of view, it is suggested that the greater cohesion exhibited by smooth fibers can, itself, be used as a control mechanism during drafting of fiber assemblies, under the proper conditions. However, loose staple fiber assemblies prepared from rough fibers are easier to draft, especially in massive form.

3. Blending of Smooth and Rough Fibers

In blends of smooth and rough model fibers in slivers, the cohesion was found to increase consistently with an increasing percentage of smooth fibers, as shown in Fig. 5. Also shown in Fig. 5 is that the higher the percentage of smooth fiber in the blend, the less the actual draft during processing at given machine settings, as reflected by the linear densities of the fiber assemblies before and after several pin draftings.

In slivers composed of blends of the polyester models with cotton fiber (Fig. 6) and with wool fiber (Fig. 7), the cohesion was found to increase steadily with an increasing proportion of smoother fiber. These data suggest that fiber surface roughness may be quite useful as an index for blend compatibility of various fibers, or as a basis for predicting the processing behavior of blends. This conclusion was substantiated even further in a subsequent experiment testing the effect of fiber surface roughness on fiber migrational behavior in the processing of blends [9].

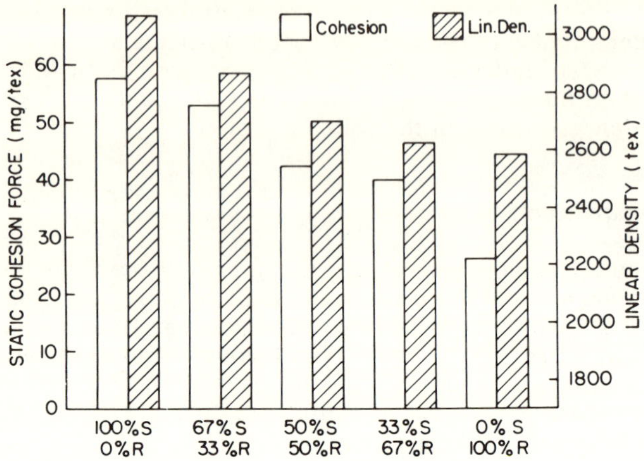

FIG. 5. Cohesion and linear density of sliver samples blended on the worsted system. S signifies smooth and R rough fibers. Reprinted from ref. 1, p. 878, courtesy of Textile Research Journal.

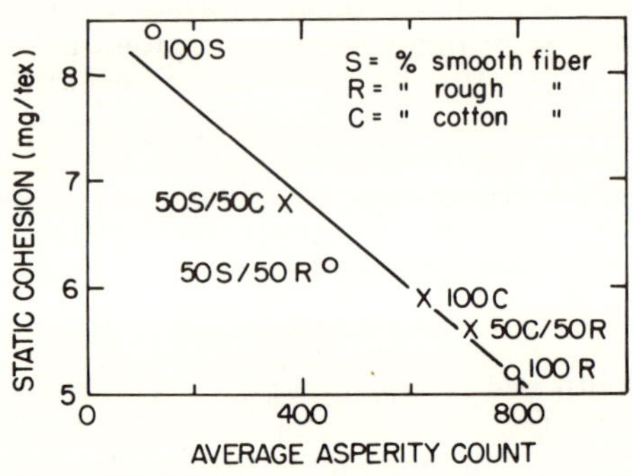

FIG. 6. Cohesion of slivers composed of model polyester and cotton fibers, including blends, as a function of average asperity count. Data: (O) on model fibers; (×) on cotton and model fiber blends. Reprinted from ref. 5, p. 986, courtesy of Textile Research Journal.

4. Staple-Yarn Twisted Structures

The degree of twist normally used in staple yarns was found to restrict the relative movement of fibers enough to mask the direct effects of fiber surface roughness in the tensile loading of yarns. However, the influence of roughness on the behavior of assemblies during drafting, in the early stages of processing, was reflected in the evenness, final linear densities, and average tensile properties of the yarns. The tendency for twist to mask differences in roving cohesion due to

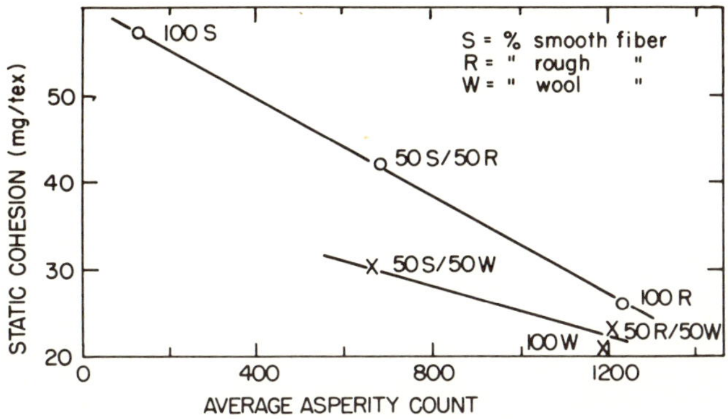

FIG. 7. Cohesion of slivers composed of model polyester and wool fibers, including blends, as a function of average asperity count. Data: (O) on model fibers; (×) on wool and model fiber blends. Reprinted from ref. 5, p. 986, courtesy of Textile Research Journal.

fiber surface roughness is illustrated in Fig. 8. The ratios of the breaking tenacity of the smooth model rovings to that of the rough model rovings (S/R) are plotted as functions of twist factor. It can be readily seen that the tenacity ratio decreases with an increase in twist to a point where there is no significant difference in tenacity. At this point, the roving samples are well beyond conventional draftability and have essentially become coarse staple yarns because fiber slippage under tensile stress is minimal. Two practical conclusions that relate to productivity can be drawn from these data. First, fibers having smooth surfaces require less roving twist for satisfactory packaging, handling, and performance in spinning. Secondly, fibers having smooth surfaces require less twist to reach the optimal twist level than fibers with rougher surfaces.

5. Filament Yarn Friction

Smooth model fibers in continuous filament form were found to lead to higher frictional forces in yarn-to-yarn and in yarn-to-metal tests than rough model fibers. The samples used were 70-denier, 34-filament polyester yarns with no twist. Results of these friction tests are given in Table 1. In all cases, the smooth samples show a greater final tension and the differences are considered highly significant according to the techniques employed in the test. The greater final tension represents higher frictional force.

6. Summary

In summary, the effect of fiber surface roughness in textile processing is that fibers with geometrically smoother surfaces (fewer

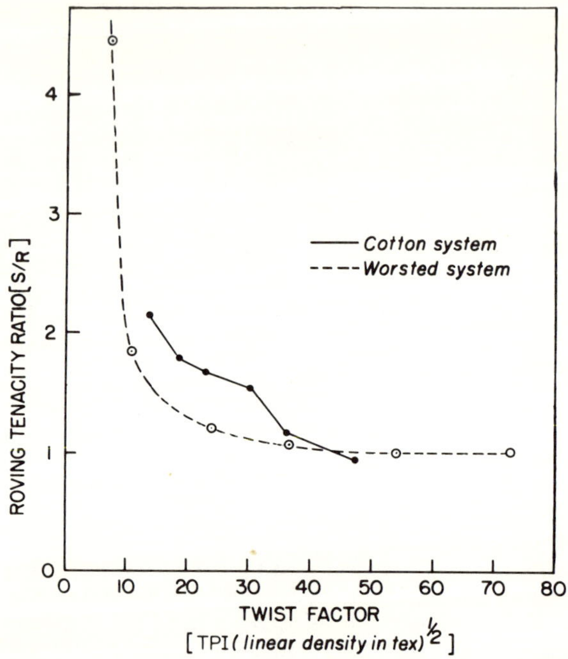

FIG. 8. Effect of increased twist on the smooth-to-rough tenacity (cohesion) ratio of roving samples prepared on the worsted and cotton systems. Reprinted from ref. 5, p. 984, courtesy of Textile Research Journal.

asperities) lead to greater cohesion and friction than fibers with geometrically rough surfaces. The explanation for these findings is that the geometrically smoother fibers offer greater area of contact between neighboring fibers or contiguous surfaces and, hence, greater opportunity for the action of cohesive forces, perhaps at the molecular level. It seems possible that some of the effects formerly attributed to "lubrication" by processing finishes may, in fact, result from the manner in which these finishes affect fiber surface geometry.

However, it must be restated that all of the above-mentioned effects and conclusions are based on different levels of fiber surface roughness that are not masked, in whole or in part, by processing finishes. It must also be kept in mind that under normal circumstances various amounts of lubricating and cohesive finishes are applied for improved processability. These finishes certainly mask to some degree the differences in fiber surface roughness, as brought out in the first chapter.

TABLE 1

Effect of Fiber Surface Roughness on Yarn Friction[a]

Continuous-filament yarn sample tested	Initial tension (g)	Final tension (g)	
		Smooth	Rough
Yarn-to-metal test			
With finish	25	60	40
Finish removed	25	80	52
Yarn-to-yarn test			
With finish	15	41	31
Finish removed	15	46	35

[a]Reprinted from ref. 5, p. 987, courtesy of Textile Research Institute.

B. Role of Fiber Cross-Sectional Shape

In an extension of the mechanical processing studies at Textile Research Institute, attention was given to the influence of fiber geometry on the mechanical properties of assemblies during processing [10]. Specifically, the major objective was to determine the effect and relative importance of the following fiber variables: surface roughness, cross-sectional shape, linear density, crimp, and staple length. Each variable was tested with two samples, differing as widely as feasible in that property. Thirty-two fiber samples were specially prepared and processed on the worsted yarn manufacturing system, once again, with no finish other than the producer's. In general, the important conclusion was that all of the variables are statistically significant factors controlling the cohesion and the related properties of loose staple-fiber assemblies but that length (first) and crimp (second) were the most highly significant.

Relative to the effect of fiber cross-sectional shape, it was found that fibers of circular cross section produce greater cohesion in webs and slivers than do trilobal fibers. For example, this finding is substantiated quite clearly by the data given in Table 2. In general, the drafting tenacities of the samples composed of fibers having circular cross sections (the 4 series) are shown to be substantially greater than the drafting tenacities of the counterpart samples composed of fibers having trilobal cross sections (the 8 series). Also, shearing cohesion tests were conducted by Hertel et al. [6,7] on card-web battings prepared from similar fiber samples with circular and

TABLE 2

Dynamic Cohesion of Sliver Samples[a]

Fiber sample	Drafting tenacity, mg/tex					
	Machine direction[b]			Reverse direction[b]		
	Card sliver	1st PD[c] sliver	2nd PD sliver	Card sliver	1st PD sliver	2nd PD sliver
Circular cross sections						
4A 3-in. with crimp	44	41	40	45	44	42
4B 6-in. with crimp	48	69	61	50	69	62
4C 3-in. no crimp	20	22	9	21	10	10
4D 6-in. no crimp	24	19	17	24	20	17
Trilobal cross sections						
8A 3-in. with crimp	36	30	20	34	35	24
8B 6-in. with crimp	33	18	20	36	26	21
8C 3-in. no crimp	20	4	1	16	4	1
8D 6-in. no crimp	13	10	5	16	11	5

[a]Reprinted from ref. [10], p. 566, courtesy of Textile Research Institute.
[b]Machine direction indicates that the sliver sample moved through the drafting analyzer in the same direction as it moved from the card, through first pin drafting, etc., in processing.
[c]PD = pin drafting.

trilobal cross sections. The battings composed of fibers with circular cross sections were found to lead to greater cohesion during shear deformation of the assemblies.

The explanation for this tendency is that a circular cross section provides a greater area of contact between adjacent fibers than does a trilobal cross section; because, with twist in the fibers, the trilobal will be in contact only where adjacent ridges cross one another, whereas on the circular fibers, there are no such twisting ridges. This situation is somewhat analogous to the differences in friction between a twistless multifilament yarn and the same yarn with twist. It is well known that the twisted multifilament yarn makes much less contact with a contiguous frictional surface because of the spiraling filament ridges in the yarn structure caused by the helical twist geometry.

In still another analogous example [11], card slivers from six different cottons were slack-mercerized in order to modify the surface geometry of the fibers contained. The treatment was found to reduce surface roughness and to produce a more circular fiber cross-sectional shape. Static and dynamic tests indicated that the smoother, more circular cross sections brought about a substantial increase in the cohesion of the card slivers.

In summary, the trend common to all these examples is that fibers with smoother, more circular cross sections lead to greater cohesion and friction, in general.

C. Role of Fiber Crimp

From the same study mentioned at the beginning of the previous section (IV-B), crimp emerged as one of the dominant factors in staple-fiber processing. The necessity of crimp is so well established that it is only under some rather unusual circumstances that crimpless staple fibers are processed. As would be expected, the findings of the processing experiment give substantial proof of the importance of crimp in the early stages of processing. Conclusions of the study relating to crimp are given in the following summary.

1. Fiber Separation

In the opening and carding processes, the crimp geometry in synthetic fibers is very pronounced. Fibers with crimp were found to open much easier and much more thoroughly than uncrimped fibers. For example, many unopened bundles of fiber were observed in card webs composed of the uncrimped sample. This undesirable condition was very seldom noticed in webs composed of crimped fibers. Crimp also leads to a more uniform distribution of fibers on the working surfaces of the opening and carding equipment.

2. Card-Web Structures

Fiber crimp was found to lead to greater cohesion and resilience in card web structures. The cohesion properties of various card web samples are given in Table 3. It can be observed that the samples with crimp (the first two in each group) have substantially greater tenacity than the samples without crimp (the last two in each group). Also, as an important component of an interaction, crimp in the 6-in. samples (the second sample in each group) leads consistently to the highest extension to maximum force in the webs. The cyclic extension properties of a variety of card-web samples are given in Table 4. Without exception, the samples with crimp (the A' s and B' s) can be seen to have significantly higher resilience in each cycle than the samples without crimp (the C' s and D' s). It is pointed out that the higher resilience implies greater elastic recovery. (In a perfectly elastic body, the resilience is unity, while in a completely inelastic structure, it is zero.)

Obviously, the regular crimp wave provides a mechanism for short-range entanglement of the ratchet type. It is thought that the greater cohesion and elasticity found in card webs are due not only to the elasticity that their geometry confers on the individual crimped fibers, but also to the effectiveness with which, because of this geometry, the elasticity can be transmitted through fiber-to-fiber interaction to the whole assembly.

3. Cohesion in Sliver

Static and dynamic tests on sliver samples indicated that crimp leads to greater cohesion. Referring to Table 2 once again, it can be seen that samples with crimp (the A' s and the B' s) have greater drafting tenacities than the samples without crimp (the C' s and the D' s). From drafting-force measurements under dynamic conditions similar to those employed in this experiment, Plonsker and Backer [12] also concluded that crimp leads to higher drafting force.

4. Twisted Structures

In twisted rovings and yarns made from staple fibers, the direct role of crimp on the cohesion or breaking tenacity of the twisted structures is masked. Once again, this is due to the fact that, in a yarn with optimum twist, fibers have no substantial relative motion during deformation of the assembly. Also, under normal conditions, fibers removed from roving and yarn tend to lose many of the crimp characteristics that they possessed before being processed.

However, the contributions of crimp during the earlier stages of processing, especially toward the uniformity and bulkiness of fiber assemblies, are reflected in the properties of the resulting yarns. Particularly evident are the higher breaking tenacities, on the average, which are related to the more uniform yarns resulting from fibers with crimp. Under normal conditions, other qualities such as fewer defects (slubs or slugs) and somewhat more fullness are associated with yarns made from fibers with suitable crimp.

5. Needled Felts

In a study of cohesion in needled felts [13,14], crimp was found to be a very influential factor. Fibers having no crimp have greater cohesion in felts than fibers with crimp. For example, results of force-extension behavior of various needled felts prepared from crimped and uncrimped nylon fiber are shown in Table 5. From these results, it can be seen that in all cases, fibers having no crimp produce the greater maximum force, in both the machine and the transverse directions. However, the felts composed of crimped fibers show higher extensions to maximum force than do those composed of noncrimped fibers. It can be clearly seen that in all of the samples the fibers are oriented mainly in the machine direction. Also, it is pointed out that the needling in felts is somewhat analogous to twist in rovings, in the tendency to restrict relative fiber motion during deformation of the assemblies. All of these conclusions are valid whether based on observed maximum force or on normalized values.

Obviously, the crimp geometry in staple fiber brings about a spatial hindrance that works against fiber packing during the needling process. In the case of crimped fibers, the hindrance leads to greater thickness of the battings, less effectiveness of the needling action, and thicker felts, in general, than found with uncrimped fibers. This is the basis for the lower cohesion and greater extension to maximum force in felts prepared from crimped fibers.

6. Summary

In the early stages of staple-fiber processing, crimp is most essential for processability and plays a dominant role over surface roughness and cross-sectional shape. This relatively important role of crimp is also observed in the mechanical properties of loose staple-fiber assemblies. In the latter stages of processing when fibers are much more linearized and in more intimate contact with one another, the role of surface roughness and cross-sectional shape becomes quite prominent and the role of crimp subsides. In tightly twisted or entangled textile structures, the effects of fiber surface geometry are masked. However, the fiber-fiber and fiber-machine interactions that occur during processing, exclusively through fiber surface contact, are reflected in the qualities and performance characteristics of the textile structures.

TABLE 3

Cohesion Properties of Card-Web Samples[a]

Fiber				
Sample	Description[b]	Maximum cohesive tenacity (mg/tex)	Textile modulus (mg/tex)	Extension to maximum force (%)
1A	S-C-3-c-ℓ	1.20	6.24	38
1B	S-C-3-c-L	1.26	4.88	64
1C	S-C-3-nc-ℓ	0.97	6.68	25
1D	S-C-3-nc-L	0.87	5.25	57
8A	R-T-6-c-ℓ	2.45	6.11	57
8B	R-T-6-c-L	1.60	3.84	65
8C	R-T-6-nc-ℓ	0.66	1.91	50
8D	R-T-6-nc-L	0.44	1.04	49
4A	R-C-6-c-ℓ	3.49	9.08	60
4B	R-C-6-c-L	2.72	4.94	116
4C	R-C-6-nc-ℓ	0.94	4.73	33
4D	R-C-6-nc-L	0.86	3.98	35
7A	S-T-6-c-ℓ	1.25	6.62	33
7B	S-T-6-c-L	1.27	5.56	43
7C	S-T-6-nc-ℓ	0.91	5.36	28
7D	S-T-6-nc-L	0.64	2.34	46
2A	R-C-3-c-ℓ	4.17	11.88	61
2B	R-C-3-c-L	6.02	21.76	75

2C	R-C-3-nc-ℓ	0.84	4.86	32
2D	R-C-3-nc-L	1.63	6.19	71
5A	S-T-3-c-ℓ	1.07	5.97	35
5B	S-T-3-c-L	1.59	6.53	58
5C	S-T-3-nc-ℓ	0.77	4.08	34
5D	S-T-3-nc-L	1.05	4.85	42
3A	S-C-6-c-ℓ	2.86	11.51	38
3B	S-C-6-c-L	2.37	9.31	75
3C	S-C-6-nc-ℓ	1.60	12.39	22
3D	S-C-6-nc-L	0.95	5.57	40
6A	R-T-3-c-ℓ	3.00	7.07	72
6B	R-T-3-c-L	3.67	10.14	79
6C	R-T-3-nc-ℓ	0.71	2.44	47
6D	R-T-3-nc-L	0.97	3.75	43

[a]Reprinted from ref. 10, p. 562, courtesy of Textile Research Institute.

[b]Explanation: S = smooth surface, R = rough surface, C = circular cross section, T = trilobal cross section, 3 = 3 den (nominal), 6 = 6 den (nominal), c = crimped, nc = no crimp, ℓ = 3 in., L = 6 in.

TABLE 4

Cyclic Extension Properties of Card Webs[a]

Sample	Force[b] (g)				Resilience			
	1st cycle	2nd cycle	3rd cycle	4th cycle	1st cycle	2nd cycle	3rd cycle	4th cycle
1A	2.0	1.5	1.4	1.2	0.28	0.36	0.47	0.43
1C	1.8	1.3	1.1	1.0	0.23	0.26	0.33	0.36
2A	3.6	3.0	2.7	2.5	0.24	0.51	0.57	0.56
2B	6.1	5.2	4.6	4.4	0.13	0.44	0.45	0.56
2C	1.0	0.8	0.7	0.6	0.09	0.35	0.35	0.35
2D	1.7	1.4	1.2	1.2	0.11	0.30	0.37	0.40
3A	5.5	4.2	3.8	3.4	0.18	0.46	0.56	0.55
3B	5.8	4.5	4.2	3.8	0.22	0.37	0.43	0.43
3C	2.6	2.0	1.7	1.6	0.16	0.28	0.33	0.44
3D	1.8	1.4	1.2	1.2	0.12	0.25	0.29	0.30
4A	1.8	1.3	1.2	1.2	0.22	0.56	0.61	0.61
4B	1.5	1.3	1.2	1.2	0.24	0.40	0.51	0.53
4C	1.5	1.2	1.0	1.0	0.09	0.33	0.33	0.36
4D	1.6	1.3	1.2	1.1	0.11	0.32	0.33	0.35
5A	1.4	1.0	0.8	0.8	0.16	0.45	0.54	0.59
5B	2.0	1.6	1.4	1.4	0.15	0.34	0.40	0.39
5C	0.8	0.6	0.6	0.6	0.11	0.27	0.31	0.33
5D	2.2	1.8	1.6	1.5	0.14	0.24	0.25	0.27

6A	2.2	1.7	1.6	1.4	0.22	0.48	0.56	0.56
6B	3.0	2.6	2.4	2.2	0.15	0.41	0.45	0.56
6C	1.0	0.6	0.6	0.5	0.10	0.23	0.25	0.26
6D	1.4	1.0	0.9	0.9	0.11	0.27	0.28	0.35
7A	1.6	1.1	1.0	0.9	0.13	0.45	0.62	0.69
7B	1.8	1.4	1.3	1.2	0.16	0.38	0.40	0.45
7C	0.8	0.6	0.4	0.4	0.09	0.25	0.27	0.39
8A	2.3	1.8	1.6	1.5	0.27	0.61	0.62	0.77
8B	1.7	1.3	1.2	1.1	0.20	0.44	0.53	0.61
8C	0.6	0.5	0.5	0.4	0.17	0.38	0.39	0.49

[a]Reprinted from ref. 10, p. 563, courtesy of Textile Research Institute.
[b]At 30% extension.

TABLE 5

Cohesion Properties of Nylon Felts[a]

Fiber sample	Needling frequency (ppsi)	Measured area density (oz/sq yd)	Maximum force (lb)	Extension to maximum force (%)
		Machine direction		
Crimped	277×2[b]	16.0	175	134
Noncrimped	277×2	13.5	530	57
Crimped	454×2	15.6	285	128
Noncrimped	454×2	17.1	582	55
Crimped	554×2	18.4	417	127
Noncrimped	554×2	16.7	684	68
		Transverse direction		
Crimped	277×2	16.1	38	202
Noncrimped	277×2	-	-	-
Crimped	454×2	15.2	56	220
Noncrimped	454×2	17.5	176	120
Crimped	554×2	16.6	118	185
Noncrimped	554×2	16.2	276	104

[a]Reprinted from ref. 13, p. 9, courtesy of the Laboratories of the U.S. Army at Natick.

[b]The sample was needled at the rate of 277 penetrations per square inch (ppsi) on each side in successive processes.

V. EFFECT OF TEXTILE PROCESSING ON FIBER SURFACE GEOMETRY

Several studies have shown that processing has a significant, if not substantial effect on fiber surface geometry. The finishing of synthetic fabrics can bring about the most substantial changes according to the desired results. Once again, the following remarks refer to the effects of mechanical processing on fiber surface geometry.

A. Effect on Fiber Surface Roughness

Using the smooth and rough model fibers, it was found that asperity count decreases substantially in both samples with the successive stages of processing [15]. This indicates that processing tends to reduce fiber surface roughness at the microscopic level. The reduction of fiber surface roughness in the samples used is evidently the result of the removal or deformation of agglomerates of TiO_2 near the fiber surfaces. It is quite common for several of these surface agglomerates to fall out during processing, especially during the drafting operations.

B. Effect on Fiber Cross-Sectional Shape

It is also well known that processing tends to introduce some sporadic modifications in the cross-sectional shape and the surface contour of synthetic fibers. These modifications are usually in the form of surface deformations from embossing by rolls, fluted surfaces, and other fibers. However, there exists also much fiber surface damage brought about by beaters, card clothing, faller pins, etc. All of these modifications tend to be macroscopic or considerably larger in effect than those associated with the removal or the depression of surface asperities.

C. Effect on Fiber Crimp

It has been long established that the crimp geometry and crimp properties of synthetic staple fibers decay with the ensuing stages of processing. Whereas the crimp geometry is quite pronounced before and during the early stages of processing, an erosion of the crimp is quite noticeable in fibers removed from rovings and yarns. These tendencies are supported both by cyclic force-extension tests on fiber and by visual estimates of crimp in fiber [16].

Conversely, it has been found that fibers without crimp do indeed acquire crimp-like characteristics as a result of processing [10].

However, these characteristics are not as uniform and are not as great in effect as those found in fibers that have been purposely crimped before processing. The crimp acquired by crimpless stock is a result of a complex deformation from bending, folding, buckling, and twisting of the fibers during the early stages of processing. The deformation usually begins with the passing of unopened bundles and clumps of fiber between rolls and other points or lines of pressure. In this manner, the fiber bundles and the individual fibers as well are "crimped" before complete fiber separation and orientation.

It is concluded that processing tends to roughen up the surface geometry of synthetic fibers on a macroscopic scale. Within the fiber surface deformations brought about by processing, however, there can be a smoothening effect on a microscopic or submicroscopic level.

ACKNOWLEDGMENT

Most of the material used in this chapter was extracted from publications of Textile Research Institute. The author deeply appreciates the generosity and encouragement of the Institute staff and supporters.

REFERENCES

1. F. L. Scardino and W. J. Lyons, Textile Res. J., 37, 874 (1967).
2. R. H. Brand and R. E. Scruby, Textile Res. J., 43, 544 (1973).
3. S. C. Sheier and W. J. Lyons, Textile Res. J., 34, 410 (1964).
4. J. H. Dillon, Third Edward R. Schwarz Memorial Lecture, 1969 ASME Textile Engineering Conference, p. 36. Available through American Society of Mechanical Engineers, United Engineering Center, New York.
5. F. L. Scardino and W. J. Lyons, Textile Res. J., 37, 982 (1967).
6. C. C. Cheng and K. E. Duckett, Textile Res. J., 42, 580 (1972).
7. K. L. Hertel and R. Lawson, Textile Res. J., 40, 472 (1970).
8. H. G. Howell, K. W. Mieszkis, and D. Tabor, Friction in Textiles, Butterworths, London, 1959.
9. F. L. Scardino and W. J. Lyons, Textile Res. J., 40, 573 (1970).
10. F. L. Scardino and W. J. Lyons, Textile Res. J., 40, 559 (1970).
11. F. L. Scardino, Textile Bulletin, April, 1967, p. 1.
12. H. R. Plonsker and S. Backer, Textile Res. J., 37, 673 (1967).
13. W. J. Lyons, F. L. Scardino, and B. C. Goswami, Technical Report 70-13-CE, August 1969, Clothing and Personal Life Support Equipment Laboratory, U.S. Army Natick Laboratories, Natick, Mass., p. 9.

14. F. Ko and F. L. Scardino, unpublished report, Philadelphia College of Textiles and Science, Philadelphia, Pa., 1970.
15. F. L. Scardino and W. J. Lyons, Textile Res. J., 37, 1005 (1967).
16. W. J. Lyons and J. H. Dusenbury, Final Report, Textile Research Institute, Princeton, N.J., December, 1959.

Chapter 5

ABRASION OF TEXTILE SURFACES

Ruth L. Galbraith

School of Home Economics
Auburn University
Auburn, Alabama

I. INTRODUCTION

Abrasion is the wearing away of a substance caused by rubbing against another substance. Abrasive damage to textiles during actual use may be caused by rubbing against many different types of external abradants, e.g., another fabric; a hard, smooth surface such as a desk top; or a hard, rough surface such as concrete or brick. In addition, internal abrasion can occur during flexing or bending of a fabric, which causes yarns to rub against other yarns within the fabric or fibers to rub against other fibers within a yarn [1,2]. Thus the type and magnitude of abrasive stresses that a fabric must withstand during use will vary with the type of garment or article, the activities of the wearer or user, and perhaps, the amount of ease allowed by the fit of a garment to the body.

Many researchers hoped that abrasion tests could be developed which would predict the durability or serviceability of a fabric during use. Such hopes have never been realized because (1) no one abrasion instrument has been devised which will either simulate or correlate with all the various types of abrasive stresses, and (2) actual wear usually involves mechanical stresses other than rubbing plus the action of various chemical agents on the fabric during laundering, etc. However, salvage studies [3,4] of worn textiles have indicated that abrasion is a major cause of failure of these textiles. Stoll [4] stated that the wear on army field uniforms was "30 per cent plane abrasion, 20 per cent edge and protection abrasion, 20 per cent tear, and 10 per cent other mechanical action." In addition to contributing to the failure of a fabric, abrasion may also contribute to changes in fabric appearances such as fuzzing, pilling, or frosting and to changes in fabric performance properties long before fabric rupture occurs. The consumer is often concerned about these less drastic changes in fabric structure that occur with progressive wear as well as with the final failure of the fabric.

II. MECHANISMS OF ABRASION

Backer [1] proposed three mechanisms to explain the mechanical breakdown of fabrics during abrasion by external forces. These were frictional wear, cutting, and plucking or snagging. He theorized that the relative magnitude of each in any given abrasive action would depend on the nature of the abradant, the behavior of the fibers within the fabric structure, and the conditions of rubbing.

Direct frictional wear occurs when the abrading surface is relatively smooth [1]. It depends only on the normal load present at the interface of the two materials and is usually considered the least severe form of abrasion. Of the three theories (surface roughness, molecular adhesion, and ploughing) developed to explain frictional resistance to relative motion between two bodies in contact, the latter two have gained the greater acceptance. The adhesion theory postulates that the surface atoms of two bodies being rubbed together form bonds or adhesions when they are moved into each other's electrostatic field of attraction. These adhesions must be ruptured by the addition of outside forces if the two surfaces are to be able to move across each other. The energy released at rupture causes vibration of the surface atoms and is dissipated as heat. Cotton fibers, given small amounts of frictional abrasion, often show a smoothing of surface rugosites as the first evidence of abrasion damage [5].

The ploughing theory involves movement of a hard material along the surface of a softer material. The damage may affect the surface smoothness of the softer material by causing bruising or mashing

[3,5] and may also affect its ability to withstand subsequent mechanical stresses during use.

Indirect frictional wear [1] may be caused by the transmission of frictional forces along the lengths of the abraded surface fibers. These forces may cause fiber slippage within the yarn structure if the interfiber frictional forces within the yarn are low or contribute to later tensile or bending rupture of the fibers caused by dynamic fatigue stresses if the interfiber cohesive forces within the yarn are high [1,2].

Surface cutting of fibers may occur when the projections of the abrading surface are sharp and also small relative to the surface of the fiber [1]. Microscopic studies of abraded fibers by Clegg [3] and Rollins et al. [5] indicate that nonwrinkle-resistant cotton fibers often exhibit surface gouges and fibrillation which could occur from a cutting action. In addition to this direct damage, complex stress patterns leading to fiber tensioning, bending, or slippage may be developed [1].

Plucking or fiber snagging can occur when the abradant protuberances are large relative to the fiber diameter and the pressure of the abradant on the fabric is high [1]. Plucking may result in fiber rupture on first contact, fiber slippage within the yarn, or vertical displacement. Loosely woven fabrics, fabrics with relatively long floating yarns, and loosely twisted yarns would be subject to greater plucking damage than tightly woven fabrics, those having relatively short yarn cross-over lengths, and tightly twisted yarns [2]. Textured or bulked filament synthetic yarns are also subject to fiber plucking or snagging. Yarn snagging has been cited [6] as a serious problem with polyester double-knit fabrics. Brown [7] was reported as stating that this snagging might be the result of using high-strength continuous filament yarns having low fiber entanglement in a fabric construction having a relatively high yarn mobility. Fuzzing, pilling, and snagging of knit fabrics can be decreased by increasing the stitch density and decreasing the length of yarn in each loop [6,8,9].

III. MEASUREMENT OF ABRASION RESISTANCE

A. Abrasion Instruments

In 1946, Dawson [10] listed and described more than 50 abrasion- and wear-testing instruments. Three other testers, the Schiefer [11], Stoll [4], and the Accelerotor [12], have been developed in the United States since that time. In addition, experimental work with sand abraders and sand-blast testers has been carried out [13]. The American Society for Testing and Materials test method D-1175 [14] gives instructions for abrasion testing using six of these instruments: Accelerotor, Schiefer, Stoll (inflated diaphragm and flex), Taber, and

Wyzenbeek. All six testers are designed to rub a fabric with an abradant, using varying levels of force in order to cause rapid wear or damage to the textile. The rate of abrasion caused by laboratory instruments is nearly always much greater than that caused by actual use and, thus, does not allow time for the textile structure to recover from one stress before another is applied. The six testers differ in the types of abradants used, the loads under which the abradants are applied, the levels of tensile stress applied to the fabric before the abrasion is started, the degree of fabric flexing and bending that occurs during abrasion, and the uniformity of stress application across the specimen surface. Because they differ greatly in the relative amounts of frictional, cutting, and plucking forces which they apply to fabrics, the results obtained with one instrument do not always correlate well with those obtained with another [13,15,16].

1. Flat Abrasion

The Taber, Wyzenbeek, Schiefer, and Stoll (inflated diaphragm) instruments all apply abrasive stresses to a fabric while it is held flat or relatively flat under tension. A Taber abrader applies multidirectional abrasion in an annular pattern using rubber-base or vitrified-base abrasive wheels having a $\frac{1}{2}$-in. thickness. A Wyzenbeek uses emery paper attached to the surface of an oscillating cylinder to apply a predominantly cutting type unidirectional abrasion to fabrics held against the cylinder.

The Schiefer instrument applies multidirectional abrasive stresses uniformly to the entire specimen surface. The abradant can be varied to give either frictional or cutting types of abrasion. When a Schiefer abrasion tester was equipped with a spring wire abradant so that it applied frictional abrasion [15], it was found to cause the least damage to cotton and nylon fabrics of the three instruments (Schiefer, Stoll inflated diaphragm, and Accelerotor) studied. Photomicrographs of fabric cross sections (Fig. 1) indicate that the Schiefer abrasion compressed and flattened the fabric structure while shaving off the tops of the yarn crimp crowns. Very little disarray of fibers within the yarns was produced and few damaged cotton fibers were found until the upper abrasion levels were reached. No significant nylon fiber damage was found.

The Stoll can be used to apply either unidirectional or multidirectional stresses to the center of a specimen. Fabric tension during a Stoll test is obtained by inflating the rubber diaphragm over which the specimen is mounted. This causes the emery abradant to apply a predominantly cutting stress concentrated in the center of the specimen. Figure 2 illustrates the severe fiber cutting found in a yarn removed from a cotton fabric given 90 cycles of Stoll multidirectional abrasion using No. 0 emery paper under a load of 1 lb and diaphragm pressure of 4 lbs/sq in. [15].

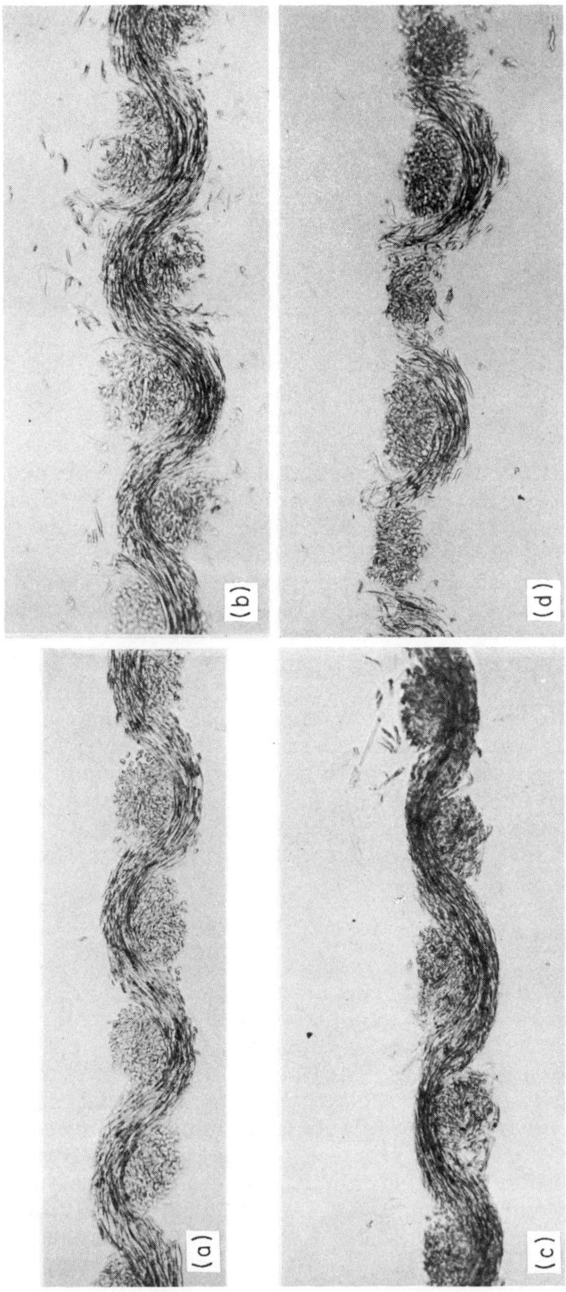

FIG. 1. Cross-sectional views of a cotton print cloth (a) before abrasion, and after (b) Accelerotor, (c) Schiefer, and (d) Stoll inflated diaphragm abrasion.

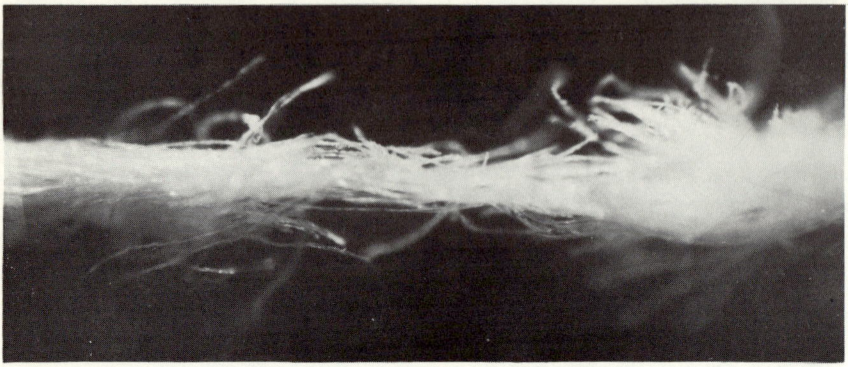

FIG. 2. Localized fiber cutting caused by Stoll inflated diaphragm abrasion of a cotton fabric.

As a group, flat abrasion instruments tend to be very sensitive to variations in the surface smoothness of fabrics, to yarn diameter, and to fabric thickness [15,17-19]. Any variation in yarn size (such as a filling rib yarn), a weave pattern that causes some yarns to float above the fabric plane, or an imbalance in yarn crimp between the warp and filling yarns will cause the raised yarns to be abraded first and, therefore, sustain the most damage [15,20,21]. Usually, thicker, heavier fabrics require a higher number of abrasive cycles to cause fabric rupture than do thin, lightweight fabrics. However, the surface appearance of these fabrics may be so damaged that consumer dissatisfaction would occur long before a hole appeared in the fabric. Flat abrasion testers tend to be less sensitive to inherent fiber abrasion resistance than are flex- or tumble-type abrasion testers [15,20]. End-use items for which flat abrasion testing might give good correlation with in-service performance are furniture upholstery and carpets.

2. Flex Abrasion

The Stoll flex abrader applies unidirectional abrasion to a fabric strip which is drawn across an abrasive bar with a reciprocating motion. The fabric is held under tension and is bent or flexed as it rubs against the abradant. Stoll [4] reported that flex abrasion most nearly correlated with combat course performance data on army fabrics because it measured the effects that variations in weave pattern have on relative abrasion resistance of fabrics. Flex abrasion results are affected by other yarn and fabric geometric factors such as yarn size, yarn twist, yarn crimp and crimp balance, as well as fabric thread count and weight [20,22,23]. Inherent fiber toughness also affects flex abrasion results [4,20]. Rollins et al. [5] reported that flex abrasion caused surface smoothing, erosion of the primary wall, and fiber fibrillation of cotton fibers if abraded wet and extensive bruising and mashing of the fibers if abraded dry. Crosslinked cotton fibers are more apt to show transverse or diagonal fracture failures, less surface fiber damage, and less fiber fibrillation than non-wrinkle-resistant cotton fibers.

3. Tumble Abrasion

An Accelerotor applies an abrasive force by using a whirling impeller to drive a fabric specimen in a zig-zag path around a cylinder which can be fitted with liners that vary in smoothness [12]. Thus, the relative amounts of frictional, cutting, and abrasive stresses applied can be varied by the researcher. It is the only abrasion tester that does not hold the fabric specimen under tension while it is being abraded; instead it allows the fabric to bend, flex, and try to move away from the impeller. An Accelerotor applies a mixture of flat and flex abrasion uniformly over both the front and back surfaces of a fabric specimen. Tumble abrasion resistance of a fabric is highly dependent on the inherent fiber toughness and less sensitive to differences in yarn crimp balance or weave patterns than is flat abrasion resistance.

When Accelerotor abrasion, using a smooth plastic liner, was applied as a frictional fatigue stress to a non-wrinkle-resistant cotton print cloth [24], it was found that much of the loss in fabric breaking strength seemed to be occurring as a result of fiber disarray within the yarn and subsequent loss of interfiber cohesion rather than from fiber damage and loss from the fabric. This type of abrasion tends to raise fiber ends above the fabric surface and, therefore, fabric thickness tends to first increase and then decrease with subsequent increments of abrasion [15,24]. As a result of the microscopic study of fabric, yarn, and fiber structures (Fig. 3), the following sequence of changes was observed:

1. The first steps in abrasion damage were fiber dislocation from the yarn with fibers remaining parallel to the yarn, followed successively by fiber breakage, and then more fiber dislocation.
2. As the outer fibers were broken off, the remaining fibers expanded in space until they met the restraining action of the yarn twist. Consequently, the yarn diameters did not greatly decrease but there was a loss of interfiber cohesion within the yarn.
3. As the amount of abrasion was increased, the yarns tended to rotate out of place within the fabric structure (Fig. 4).

Good correlation was found between the types of changes in yarn and fabric structures caused by Accelerotor abrasion and those caused by use and laundering of cotton sheets (Figs. 3 and 4). However, the Accelerotor did not cause quite as much fiber disarray as did the wear plus laundering of the sheet. Haycook [25] and Warfield and Fickle

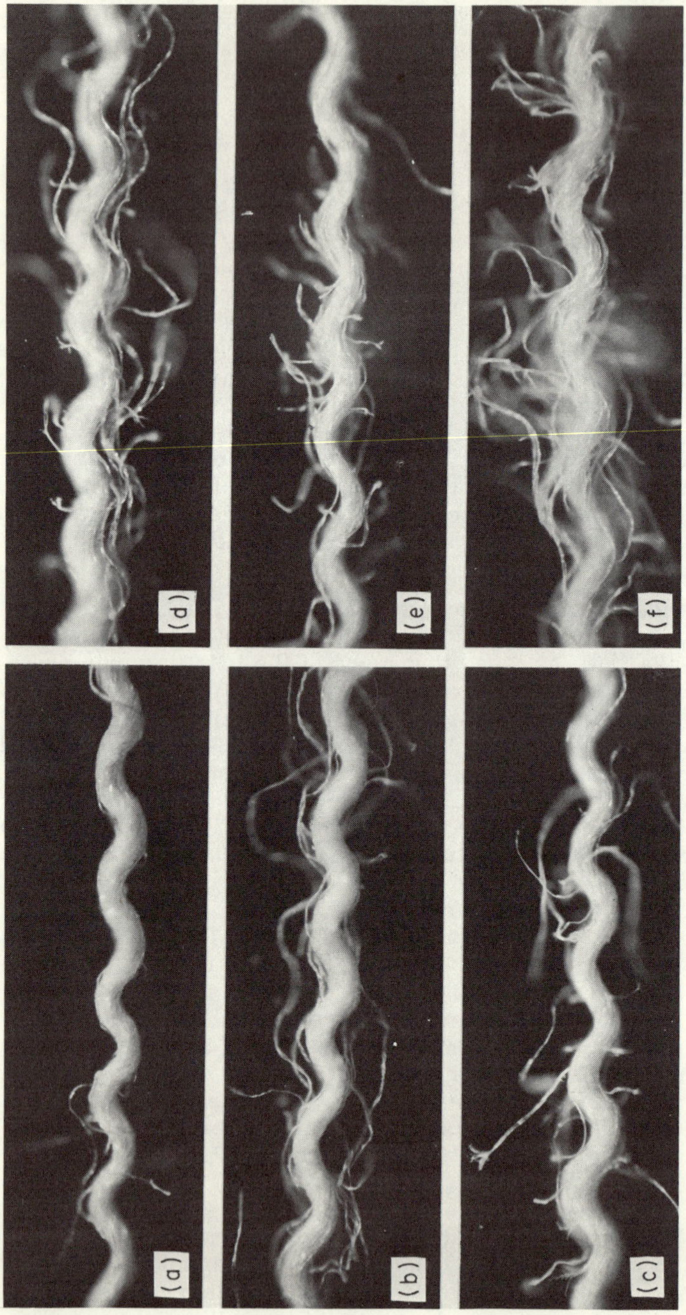

FIG. 3. Yarns removed from cotton sheeting fabrics (a) before abrasion; and after (b) 1, (c) 20, (d) 80, and (e) 480 min of Accelerotor abrasion using a plastic liner; and (f) from a worn and laundered sheet.

FIG. 4. Cotton sheeting fabrics (a) Accelerotor abraded 480 min using a plastic liner and (b) worn and laundered.

[26] also reported that Accelerotor abrasion caused changes in yarn structures similar to those caused by laundering and tumble drying or by wear plus laundering and tumble drying but with laundering and/or wear causing a greater degree of fiber disarray than Accelerotor tumbling (see Fig. 5). Young and Morris [27] reported that wear caused greater tearing out of fiber ends and fiber disarray from shirt fabrics than did laundering alone.

Accelerotor abrasion with a grit liner has been found [28,29] to cause much greater fiber cutting and fibrillation especially at the tops of the crimp waves than does frictional abrasion (see Fig. 6). Rollins et al. [5] reported that Accelerotor abrasion with an emery liner

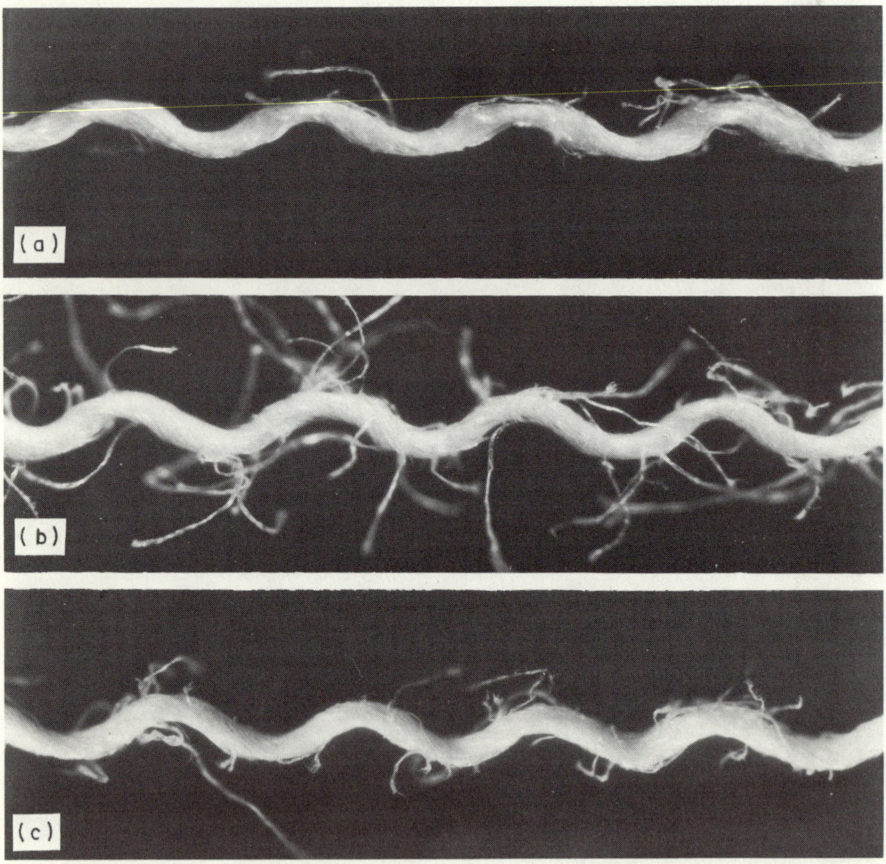

FIG. 5. Cotton yarns removed from (a) unabraded fabric, (b) from laundered and tumble-dried fabric, and (c) from Accelerotor-abraded (wet plus dry abrasion using a rubber liner) fabric.

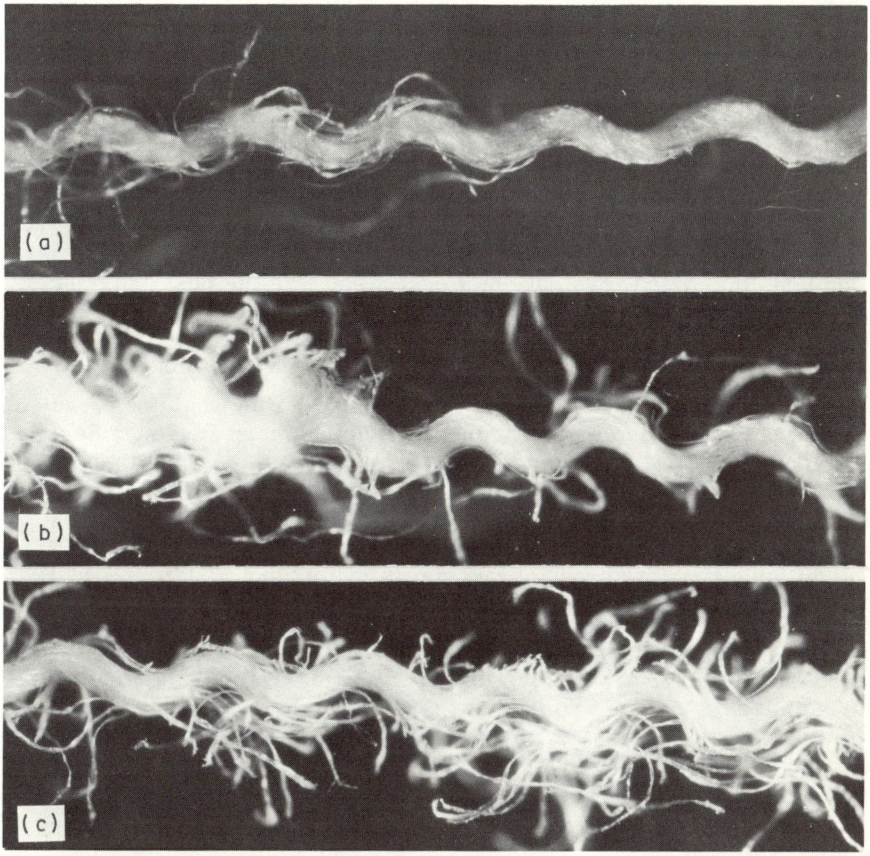

FIG. 6. Fiber disarray within cotton yarns removed from (a) unabraded fabric, (b) fabrics Accelerotor abraded using a rubber liner, and (c) fabrics Accelerotor abraded using an emery liner.

caused bending fractures in cotton fibers plus chipping and gouging of the fiber surface.

4. Edge Abrasion

Edge wear or abrasion occurs when a fabric is worn along a folded edge. It often occurs at the bottom of a hem or cuff, along a pleat or crease edge, along a facing edge, or in shirt collars at the points or where the collar turns over. Two methods [30,31] have been developed to test for this type of abrasion: (1) Accelerotor abrasion of a specimen folded and stitched to produce a folded edge; and (2) laundering plus tumble drying of mock trouser cuffs. Both Markezich [30] and the Gulf Coast American Association of Textile Chemists and Colorists Section [31] found good correlation between the results of these two tests, especially when the washing machines were comparatively new. If the washing machine tubs were chipped or damaged, then the abrasive damage to the cuff edge was greatly increased. The Accelerotor method was more rapid and more easily controlled than the laundered cuff method. Prato and Morris [32] also found good correlation between an Accelerotor method and a laundering method. They concluded that visual evaluation of edge wear was the best and easiest method for determining degree of damage. Other evaluation methods used were light transmission, weight loss, breaking strength, energy to break, and a chromaticity shift after staining with Simon's stain.

Although it might be expected that edge abrasion would be closely related to flex abrasion, researchers at the Southern Regional Research Laboratory, U.S. Department of Agriculture, Agricultural Research Service [22,23] have found this assumption to be false. Indeed, the greater degree of fabric bending present during edge abrasion caused the results of edge abrasion and flex abrasion tests to be almost complete opposites. Plain weave and oxford weave fabrics exhibited the best flex abrasion resistance but the poorest simulated cuff laundry resistance, while basket weave, mock leno, and warp rib fabrics had poor flex abrasion resistance but good edge wear resistance during laundry. Fabrics woven from finer yarns had better flex abrasion resistance but poorer edge wear resistance than those woven from coarser yarns. The presence of coarser yarns in the filling direction, especially in the twill weave fabrics, increased resistance to edge wear during laundering.

Goynes and Rollins [33] observed that edge wear of cotton fibers during laundering produced peeling of the outer layers as fibrillar sheaths followed by extensive breaking into fibrillar bundles. Edge wear during tumbling in an automatic dryer without previous wetting caused smoothing of the fiber surfaces followed by diagonal or transverse cracking of the fibers. Little or no fibrillation occurred when fabrics were tumbled dry.

B. Evaluation of Abrasion Damage

Once a specified amount of abrasion has been applied to a fabric, the problem becomes one of measuring the effects of that abrasion [2,34]. Various criteria have been suggested [3,5,15,34-36] of which the most often used are (1) visual evaluation of yarn breakage, formation of a hole, or change in surface appearance or color; (2) changes in mechanical properties of the fabric such as weight, thickness, stiffness, air permeability, or breaking strength and elongation; and (3) microscopic study (both optical and scanning electron) of changes in fabric, yarn, and fiber structures. Of these, the number of abrasion cycles needed to cause a fabric hole or rupture and loss in breaking strength have been used most often.

When the number of abrasion cycles needed to cause yarn or fabric rupture is used as the measure of abrasion resistance, a high degree of variability in the data is obtained, especially in interlaboratory testing [2]. Most abrasion testing is extremely sensitive to the tension placed on the specimen as it is mounted in the specimen holder, to the degree of roughness in the abradant, and to the pressure of the abradant on the specimen [1,2]. Abradant uniformity throughout a series of tests is a particularly difficult factor to control. If the abrasion is applied for a specified number of cycles and visual examination is used as the criterion for fabric deterioration, then difficulty in quantifying the degree of damage arises. Sometimes specimens are rated against photographic standards as in Markezich's work on edge wear [30] and in the frosting test method [37] in order to obtain a numerical rating value that can be analyzed statistically.

Changes in mechanical properties are often used to measure abrasion damage, but problems arise because successive application of equal increments of abrasion does not always cause physical property changes that are directly proportional to the amounts of abrasive stress applied. Fabric weight loss measures the amount of fiber or fiber fragments removed during abrasion but does not measure the degree of damage sustained by the fiber and yarn structures remaining in the fabric [2]. Both Ball [34] and Caldwell [38] reported that fabrics often gained weight during the early stages of abrasion. Ball attributed this weight gain to deposition of abradant particles within the fabric but Caldwell hypothesized that fiber napping during the abrasion process increased the amount of moisture absorbed by the fabric during subsequent reconditioning of the abraded specimens before weighing.

Changes in fabric thickness caused by abrasion are small, especially when the specimen is subjected to flat abrasion [2,15]. When tumble, flex, or laundry abrasion is applied, teasing out of fiber ends tends to increase fabric thickness during the initial abrasion stages, with decreases to the original fabric thickness occurring only after severe abrasion damage [24,34].

Fabric breaking strength decreases with abrasion, but not always in direct proportion to the abrasive stress applied nor to the weight loss from the fabric [15,24,34,39]. In many cases [15,24], the strength losses are much greater than would be predicted from fabric weight losses. This discrepancy could occur because of loss of interfiber cohesion within yarns, especially those made from staple fibers [15,24] or from serious damage to the tensile properties of fibers remaining in the fabric [1,2]. However, Hamburger [40] stated that strength appeared to offer the fewest objectionable features as a criterion for abrasive degradation although it was not a perfect measure of such damage.

Following the pattern of Clegg's earlier research [3], many of the recent studies of abrasion [5,15,24,25,27-29,33] have included optical and scanning electron microscope evaluations of fiber, yarn, and fabric structures both before and after abrasion. Types of fiber damage found have been bruising, mashing, and chipping of the fiber surface, and transverse or diagonal cracking across the fiber, sometimes with notched or forked ends [3,5,15,41,42]. Cotton fibers tend to fibrillate, especially when abraded wet and when not cross linked [5,33,43].

Studies of yarn structures [15,24-27,29] have indicated that fiber plucking or teasing is a major factor in abrasive damage to textiles and that abrasion tends to peel off successive layers of the yarn structure, especially when there is a high degree of cutting force acting on the yarn crimp crowns [24]. Fabric photographs, especially cross sections [15,25,26], have indicated that a spreading of fibers within the yarn structure occurs and that yarns rotate out of place within the fabric structure when abrasive damage is severe.

IV. FACTORS AFFECTING ABRASION RESISTANCE

The abrasion resistance of a fabric is dependent upon many factors including fiber properties, yarn structure and size, fabric weight and geometry, fabric finishes including lubricating and sizing agents, and the type of abradant [1,2,39]. For hydrophilic fibers, the amount of moisture present also affects abrasion resistance [2,5,25,29,33,43,44].

A. Fiber Properties

1. Mechanical Properties

Hamburger [15] stated that the inherent abrasion resistance of a fiber should correlate with the energy absorption of a mechanically conditioned fiber (one free of secondary creep) and the ability of the fiber to release this energy after stressing without the occurrence of failure. The area under the stress-strain curve is considered to be one of the best measures of the ability of a fiber to absorb energy, or fiber toughness. The larger the area under the mechanically conditioned stress-strain diagram, the higher the abrasion resistance. Hamburger listed five fiber properties required for high abrasion resistance: (1) low modulus of elasticity; (2) large immediate elastic deflection; (3) high ratio of primary to secondary creep; (4) high magnitude of primary creep; and (5) high rate of primary creep.

Because abrasion is usually a repeated stress caused by forces of comparatively low magnitude, the strength, elongation, and elastic recovery during and after repeated stress applications are important factors in abrasion resistance. McNally and McCord [2] theorized that the elongation and elastic properties were more important than strength, especially when fiber bending and flexing were included as part of the abrasive stress. These authors also indicated that flexural and shear properties of fibers were important factors in abrasion resistance.

There is no one study available that measures the inherent abrasion resistance of every fiber available today under the same conditions of fiber length, size, yarn and fabric geometry, and abrading conditions. Indeed it would be very difficult to obtain a hierarchy or ranking of fibers by relative abrasion resistance because of the inherent differences in length and diameter of the natural fibers. Hamburger [40] abraded filament yarns containing a minimum of twist in order to remove the confounding effects of yarn- and fabric-form factors. The length and diameters of man-made fibers can be controlled but it is not possible to control these properties in natural fibers such as cotton and wool. In addition a fiber ranking for flat or plane abrasion resistance would not be exactly the same as one for flex or tumble abrasion.

Instead of an abrasion series analogous to an electromotive series, fiber technologists have developed groupings of fibers by relative abrasion resistance as measured in many different studies and from observation of fiber behavior in actual wear situations. The fibers now considered to have very high abrasion resistances are nylon, followed by polyester with the olefin fibers just slightly less resistant than the first two types [2,39,40,45]. These fibers are characterized by having high strengths, high elongations, and good to excellent elastic recovery [45].

The acrylic and modacrylic fibers have less inherent abrasion resistance than do nylon and polyester fibers but more than do wool, cotton, or the rayon fibers [17,28,39,45]. These fibers might be classed as having "good" abrasion resistance. The acrylics and modacrylics tend to have slightly lower tenacities than do the nylon and polyester fibers but exhibit high elongation and good elastic recovery [45]. The elongation and elastic recovery properties of the synthetic fibers vary so much with fiber type within a generic class that it is sometimes difficult to develop rankings for these properties that can be applied to an entire generic group.

The fibers with moderate abrasion resistance include wool, cotton, and the high wet-modulus rayons [2,28,29,46]. Wool is a relatively weak fiber with high elongation and excellent elastic recovery, while cotton is a stronger fiber having low elongation and comparatively poor elastic recovery. However, despite the great disparity between the physical properties of these two fibers, they have been reported [46,47] to be similar in abrasion resistance. McNally and McCord [2] and Schiefer et al. [46] hypothesized that cotton's high shear strength contributed to an abrasion resistance higher than would have been predicted from its energy absorption after repeated stressing. Warfield

[29] reported that the high wet-modulus rayon fabrics in her study were similar to cotton in both wet and dry abrasion resistance and better than a conventional type viscose rayon.

Various sources [2,17,40,46,47] have found that viscose rayon and acetate fibers exhibit the lowest degree of abrasion resistance. These fibers are comparatively weak and have relatively high elongations at the breaking point, but do not exhibit good elastic recovery. Consequently, much of the energy absorbed during first extension is not recoverable and energy absorption during repeated stress cycles is relatively low. Susich [47] theorized that the unrecovered high stretch of a fiber would reduce the cross-sectional area of the fiber so that concentrated stresses could develop from small abrasive forces.

2. Fiber Shape

Fiber length, diameter, and cross-sectional shape have been reported [2] to influence abrasion resistance by affecting the interfiber cohesion within yarns and the magnitude of stresses that would develop within the fiber as a result of abrasive forces. When fibers are removed from yarns, the loss of interfiber cohesion can cause rapid deterioration of the yarn structure [15,24,48]. Longer fibers are usually more difficult than shorter ones to remove from a yarn and also make stronger yarns at equal twist and size [49]. Studies [17,47, 50] have indicated that filament yarns are usually more abrasion resistant than staple yarns containing the same fibers.

Although fine fibers produce stronger yarns, the use of moderately coarse fibers may improve abrasion resistance [17,51,52] because they require more force than do fine fibers to cause rupture. However, fibers that are too coarse may develop high strains on the outer portion of the bend if fiber flexing or bending is involved [51]. There has been little work reported on the effect of fiber cross-sectional shape on abrasion resistance, but McNally and McCord [2] stated that flat, elliptical, or hollow fibers might wear better than round ones.

B. Yarn and Fabric Structure

The yarn and fabric geometric factors that influence a fabric's abrasion resistance are those that affect the ability of the fabric to absorb energy repeatedly without failure or rupture [53]. Geometric factors in yarn and fabric construction affect the total energy absorption capability of a fabric, as well as the evenness with which abrasive stresses are distributed throughout all elements of the fabric. They also affect the ability of fibers and yarns to move away from the abradant but then regain or retain their original positions within the fabric. Backer and Tanenhaus [20] postulated that a large area of contact between fabric and abradant would allow a better distribution of the abrasive stresses, thus decreasing the localized load at any one fiber point. This would decrease frictional wear, surface cutting, fiber plucking and slippage, and tensile fatigue. Those geometric factors which have been found to be most important in determining fabric abrasion resistance are yarn twist, size, ply, and crimp and fabric thickness, thread count, and type of weave.

1. Yarn Twist

Yarn twist is the means used to form fibers into yarns. Consequently, the amount of twist inserted has a major effect on the degree of fiber binding within the yarn. As twist is increased, the interfiber frictional forces are increased and the tendency for fibers to be plucked from the yarn is reduced [2,20]. However, high amounts of twist may also stiffen the yarn so that it does not flatten or rotate under the abradant pressure. This could lead to stress concentration on relatively few high points of the yarn rather than a more even distribution of abrasive stress throughout the yarn. High twists also reduce the tensile energy absorption capability of yarns [49,54]. Several studies [17,20] have shown that fabric abrasion resistance rises as twist is increased to an optimum point and then decreases as twist is further increased beyond this optimum. The optimum level of twist necessary for a staple yarn is higher than that for a filament yarn. A more recent series of studies [22,23] indicated that varying the twist multipliers from 3.10 to 4.10 for 10/1 and 15/1 cotton yarns had no significant effect on either the flex abrasion resistance or edge-wear properties of the durable-press fabrics containing these yarns.

2. Yarn Size

The use of larger yarn sizes usually improves the resistance of fabrics to flat or plane abrasion [17,19,51,54]. This increased abrasion resistance is the result of the presence of a larger number of fibers so that more fibers must be ruptured before the yarn fails. There may also be better stress distribution over the larger number of fibers present in the surface layers of a large yarn than in the smaller number of fibers needed to form the surface layers of a fine yarn [2,20]. In weave patterns such as sateens or warp-faced twills where one set of yarns predominates on the fabric surface, it is advantageous to use larger yarns for the floating yarns but there is less value in increasing yarn size in the opposite direction.

McNally and McCord [2] theorized that the use of large yarn sizes might be less advantageous in fabrics subjected to large amounts of flex abrasion than in fabrics which would be abraded flat because of the high strains that might be induced in the fibers lying on the outside curvature of the bent yarn. A recent USDA study [23] indicated that the flex abrasion resistance of 7.5-9.0 oz per sq yd durable-press cotton suiting fabrics decreased as filling yarn size was increased from 15/1 to 7.5/1. However, the edge wear on simulated trouser cuffs during laundering was better, especially for twill-weave fabrics, when the larger filling yarns were used in spite of the fact that the degree of fabric bend in an edge-wear test is nearly double that in a flex-abrasion test. These results might indicate that the better yarn mobility within the fabric gained by employing fewer yarns of larger size more than offset any additional fiber strains caused by bending the larger yarns when the degree of fabric bending was severe.

3. Yarn Ply

In one study [50], it was found that the abrasion resistance of a fabric could be increased greatly by the use of a two-ply yarn rather than a singles yarn of equal count. Another study [15] indicated that when fiber abrasion and cutting was severe at the crown of the yarn crimp wave, the outside layer of fibers in singles yarns tended to peel away from the central core of the yarn (Fig. 7). When two-ply yarns of equivalent size were used, this peeling effect was restrained by the presence of the second ply. Instead fibers were cut off across the top of the crown and tufts of cut fiber ends were formed at each side of the crown.

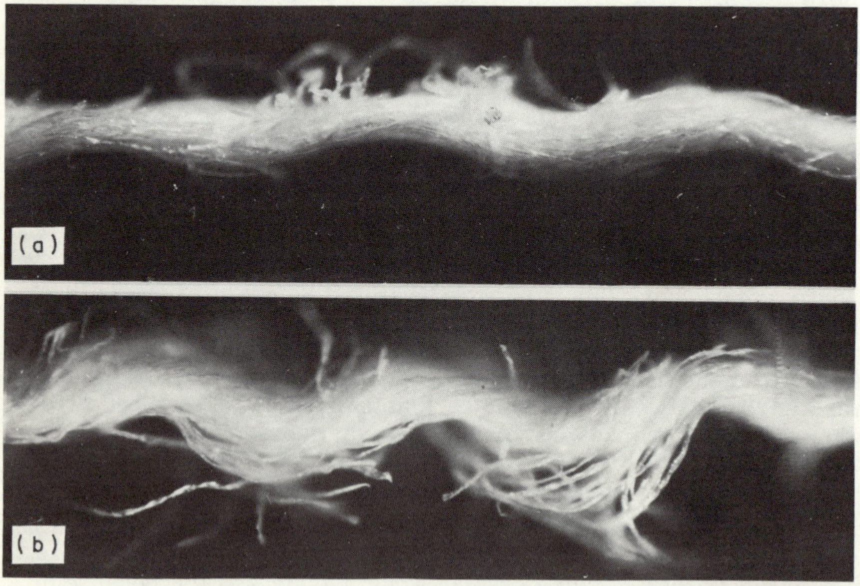

FIG. 7. Effect of Stoll, inflated diaphragm abrasion on (a) singles and (b) 2-ply yarns.

4. Yarn Crimp

The amount of crimp present in a yarn determines how high that yarn will rise above the fabric plane. Since abrasive degradation usually occurs first at the surface of a fabric, the projection of a yarn or set of yarns to the fabric surface increases abrasive damage to those yarns [20]. Thus, the balance between warp and filling yarn crimps often determines which set of yarns will be preferentially damaged and to what degree. In a plain-woven fabric, the highest degree of abrasion resistance would probably be obtained by having the same yarn size, yarn crimp, and thread count in both the warp and filling directions. This type of structure would give the most even stress distribution across the largest number of yarn crowns, thus decreasing the localized load at any one point. When the crimp balance shifts so that one set of yarns contains more crimp than the other, the set containing the most crimp rises further out of the fabric plane and sustains the first, and usually the greatest, amount of damage. The other set of yarns will be protected until the crimp crowns of the first set have been worn down enough for the abradant to come into contact with the protected yarns [15,20]. Any change in crimp balance between warp and filling yarns during laundering or wear may change both the degree of abrasion and the point of abrasive attack.

If the fabric wear pattern for a particular use indicates that tensile stress will be placed predominantly on one set of yarns (usually the warp), then designing fabrics with a high filling crimp has been found to prolong the wear life of the fabric [20]. This improvement in wear occurs because the filling yarns absorb abrasive stresses during wear so that the warp yarns are left relatively undamaged to absorb the tensile stresses.

5. Fabric Thickness

Fabric thickness is largely determined by the factors of yarn diameters and yarn crimp [20]. Consequently, just as increased yarn size usually increases fabric abrasion resistance, so does an increase in fabric thickness increase the number of rubbing cycles needed to cause failure. However, if the surface appearance becomes damaged to the point of consumer dissatisfaction with garment appearance before a yarn is ruptured or a hole is formed, then the additional abrasion resistance of the thicker fabric no longer contributes to a longer garment wear life. Changes in surface appearance are often more of a problem than is fabric rupture in pile fabrics such as corduroy, velvet, or velveteen.

6. Thread Count

Increasing the thread count of a fabric increases the number of yarn interlacings and, therefore, the number of yarn crowns over which the abrasive stress can be distributed. The larger number of yarn crossover points also increases the fiber and yarn cohesion within the fabric and decreases the probability that fibers will be plucked or slipped out of place within the structure [2,20]. Consequently, increasing the thread count up to an optimum point tends to increase abrasion resistance, especially to flat abrasion [19,20,55]. However, if the thread count becomes so high that it restricts yarn mobility, causes yarn jamming during use, or causes the yarn crowns to form high, rigid knuckles, then abrasion resistance decreases, especially if the abrasive stress also involves large flexing or bending stresses [2,20,23,56].

7. Weave Type

The influence of weave pattern on fabric abrasion resistance is highly dependent on the type of abradant used, the degree of fabric flexing or bending occurring during abrasion, and the direction of stress application [2,20,22,23]. If the abrasion is flat and multidirectional in nature, plain weaves provide the best fiber binding and, if yarn size and crimp are balanced, the most equal exposure of warp and filling yarns at the fabric surface. However, modified plain weaves, such as the oxford or basket, can provide greater yarn mobility than do taffeta weaves and, thus, allow the yarns to absorb stress by moving away from the abradant. Twill and sateen weaves can provide even greater yarn mobility or can provide adequate yarn mobility while reaping the advantages of weaving with higher thread counts than is possible with plain weaves.

The counterbalance to the advantages of greater yarn mobility in twill and sateen weaves is the increasing vulnerability of the longer floats as they form more and more of the fabric surface. Longer float lengths are also more vulnerable to plucking and snagging both fibers and entire yarns. However, the yarn floats in twill and sateen weaves can be used to protect the opposite set of yarns from abrasive damage. This becomes valuable when, e.g., the warp yarns must withstand the greatest degree of tensile stress during wear and can be protected against abrasive damage by the filling yarn floats in a sateen weave [20]. Backer and Tanenhaus [20] reported that in twill and sateen weaves, the maximum abrasion resistance of the float system was obtained when the abrasive stress was applied parallel to the direction of the floats. However, Ruppenicker et al. [23] reported that durable-press suiting fabrics woven with both twill and sateen weaves exhibited better flex abrasion resistance when the abrasive action was applied perpendicular to or across the float than when it was applied parallel to the floats. These authors reported that plain-weave fabrics gave better flex abrasion resistance than did fabrics having oxford, basket, twill, or sateen weaves, although the abrasion resistance order was reversed for simulated trouser cuff edge wear during laundering. Thus, high yarn mobility was an important factor in edge wear but was less important than good fiber binding and balanced distribution of abrasive stress over both warp and filling yarn systems in the flex abrasion tests.

C. Fabric Finishes

Fabric finishes added to improve such properties as hand, wrinkle resistance, or stain or soil resistance often alter the abrasion resistance of the fabrics to which they are added.

1. Lubricants

Lubricants such as the silicones or polyethylene compounds may improve flex abrasion resistance by increasing fiber and yarn mobility within the fabric structure [2, 57-59]. They have less effect on the flat abrasion resistance of a fabric [2], since this property is more dependent on good fiber binding than on fiber and yarn mobility. Contradictory results have been reported for the effects of the cationic softeners, often used in home laundry, on abrasion resistance. Murry [60] reported that these softeners increased the tear strength and elongation of resin-treated viscose fabrics but reduced the flat abrasion resistance slightly and had little effect on the flex abrasion resistance. The Midwest Section, American Association of Textile Chemists and Colorists [61] reported that the cationic softeners lowered the tumbling abrasion resistance of cotton print cloth and corduroy, acetate tricot, and nylon tricot fabrics but improved the abrasion resistance of polyester/cotton shirting, acrylic double knit, and polyester double-knit fabrics.

2. Coatings

Coating finishes often increase the flat abrasion resistance of a fabric by increasing the fiber binding and by covering the fabric face with some type of sizing or polymeric material but, in the past, have been less successful in improving flex abrasion resistance [2]. However, recent work with flexible polymer latices such as polyacrylates and polyurethanes has indicated that thin polymer depositions on fiber, yarn, and fabric surfaces can increase both the flex and edge-wear abrasion resistance of durable-press cotton fabrics [5,62-65]. These films are tough and flexible so they have less tendency to stiffen the fabric or to separate from the fabric than did earlier vinyl and rubber coatings. They appear to protect the fiber from surface damage because the coating rather than the fiber absorbs the abradant energy [5].

3. Wrinkle-Resistant Finishes

Because of the poor elastic properties of the cellulosic fibers, a large proportion of the apparel, bed linen, and drapery fabrics containing these fibers are reacted with crosslinking resins to improve fabric wrinkle resistance. Such finishes increase the elastic recovery but reduce both the tenacity and elongation of the cellulosic fibers (especially cotton) and, thus, drastically lower the ability of these fibers to absorb energy and resist abrasion [66,67]. The lowering of resistance to flex, tumbling, or edge abrasion is particularly severe and usually results in a shorter total wear life for an all cotton fabric unless the loss is compensated for by using heavier, stronger fabric constructions [2].

However, several studies [25,29,68] have indicated that, when abrasion conditions are mild (low abradant pressures and predominantly frictional abrasion), the resin finish may increase rather than decrease abrasion resistance. Gagliardi and Nuessle [68] theorized that the improved elastic recovery in the crosslinked fibers offset the losses in energy absorption when the abradant stresses were low or intermittent. Reeves [69] reported that resin finishing of cotton lessened the tendency for fiber ends to be teased to the fabric surface. This fact was confirmed by both Haycook [25] and Warfield [29]. In both studies, the resin-finished fabrics had greater weight losses than the nonresin or pure-finished fabrics but retained a greater percentage of their original breaking strengths after abrasion. These authors concluded that the better retention of fibers within the yarn structure of the resin-finished fabrics (see Fig. 8) contributed to greater interfiber friction and, therefore, higher fabric breaking strength even though fiber fracture and loss of fiber fragments was greater for these fabrics than for their pure-finished counterparts.

Rollins et al. [5,33] found that the type of abrasive damage altered when cotton fibers were reacted with crosslinking resins. Untreated cotton fibers, abraded dry, showed surface smoothing, bruising, and mashing. As abrasion damage became more severe, transverse cracks appeared and triangular wedges were pinched out of the fiber sides. When the fibers broke, fiber ends were mangled and frayed. Crosslinked fibers showed little surface distortion after abrasion but developed transverse cracks and fractures. When the fibers broke, there was little evidence of frayed, fibrillated, or bruised fiber ends. Thus, crosslinking caused more rapid fracture of fibers but helped the fiber resist fibrillation and the development of surface hairiness.

D. Fiber Blends

The development of fiber blending was given strong impetus by the consumer's desire for wrinkle-resistant clothing coupled with the great loss in abrasion resistance sustained by cotton when resin finished. Although many different fiber blends are used for various types of fabrics, the most used is the polyester/cotton blend. The use of such blends does increase the abrasion resistance of the blended fabrics over that of equivalent 100% cotton fabrics as indicated by the

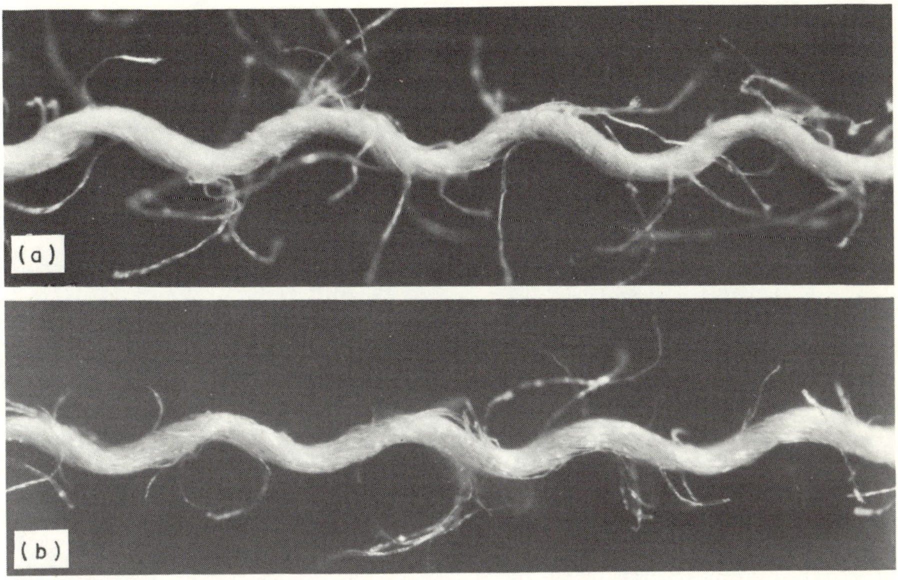

FIG. 8. Fiber disarray in yarns removed from (a) nonresin-finished and (b) resin-finished cotton fabrics given a treatment involving dry abrasion followed by laundering.

comparative weight-loss data graphed in Fig. 9. Adding a durable-press finish to the blended fabric increased the weight loss caused by abrasion but not as much as it would for an all-cotton fabric. However, blending a more abrasion-resistant fiber with a less abrasion-resistant one causes other problems such as frosting and the premature knee burst found in boys' jeans. Frosting occurs when two fibers differing in both color and abrasion resistance are combined in a fabric. This results in a color change in abraded areas, e.g., knees, seats, and elbows, when one fiber is preferentially removed from the fabric by abrasion.

When fiber blends were first developed, it was learned that blending stronger fibers with weaker ones did not always increase the tenacity of the blended yarn over that of an equivalent yarn made entirely of the weaker fiber. For example, blending nylon with cotton to increase abrasion resistance led to a significant loss in yarn strength [70] but blending nylon with wool increased yarn tenacity [70-72]. Hamburger [73] attributed this phenomenon to the elongation balance between the blended fibers. When two fibers differing in modulus of elasticity and total elongation (such as nylon and cotton) are blended, both fibers in the yarn elongate up to the breaking point of the less extensible fiber [72,74]. After the rupture of these fibers, the stress

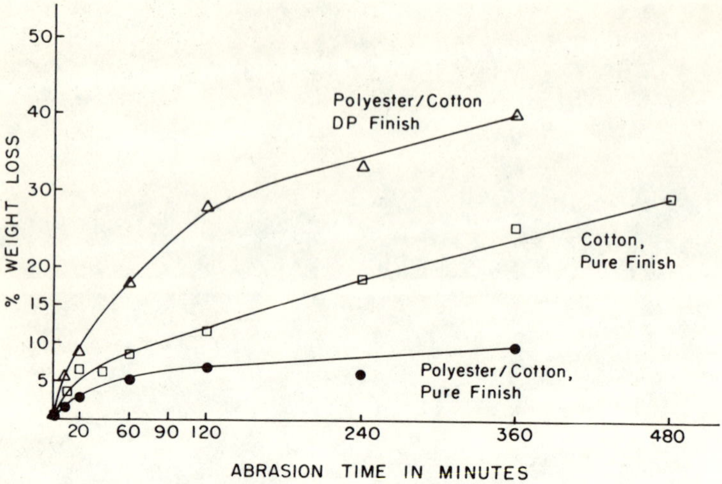

FIG. 9. Percent weight loss from cotton, polyester/cotton, and durable-press polyester/cotton fabrics caused by Accelerotor abrasion using a plastic liner.

is transferred to the more extensible fibers in the blend. This stress concentration on only part of the fibers in the yarn leads to a reduction in yarn tenacity.

Several authors [75,76] have reported the presence of a double-rupture phenomenon in both nylon/cotton and polyester/cotton blends. These authors indicated that the cotton fragments continued to contribute to the ultimate strength of the yarn and fabric, even though the fiber had been broken, by helping to retain the interfiber cohesion within the yarn. However, subsequent agitation may shake the fiber fragments from the yarn, leading to loss of intrayarn fiber binding and eventual loss of strength. Canter, Jones, and Weaver [75] postulated that it was such a sequence (tensile stress during wear followed by agitation during laundering) that led to the knee-burst phenomenon in boys' jeans in which the knee area would rupture even though the jeans showed little visual signs of wear.

Later work [77,78] indicated that tumble abrasion in an Accelerotor could cause sufficient cotton fiber fragmentation in durable-press polyester/cotton blends so that the fabric no longer exhibited a double-rupture phenomenon when subjected to tensile stress after abrasion. Elias [78] found that frictional tumble abrasion using a smooth-surfaced plastic liner had little effect on the fiber-length distributions of either the cotton or polyester fibers in a nonresin finished 65% polyester/35% cotton print cloth (Table 1). When a durable-press finish was added to the same fabric, abrasion again had little effect on the

TABLE 1

Effects of Accelerotor Abrasion Using Smooth-Surfaced Liner on Fiber Length Distributions in Polyester/Cotton Fabrics with and without a Durable-Press Finish [78]

Abrasion time (min)	Numbers of fibers of each length in 1 in. of yarn															
	Polyester								Cotton							
	Pure finish				Durable-press finish				Pure finish				Durable-press finish			
	1''/4	1''/2	3''/4	1''	1''/4	1''/2	3''/4	1''	1''/4	1''/2	3''/4	1''	1''/4	1''/2	3''/4	1''
0	5	13	15	23	5	10	14	21	14	13	12	5	23	22	8	2
1	7	12	16	22	7	13	14	23	18	18	10	5	40	22	7	1
10	6	12	14	22	7	15	14	26	13	16	10	5	42	15	3	8
20	8	15	14	24	8	15	15	20	19	16	11	5	29	8	7	0
60	5	15	15	19	8	14	14	22	18	20	11	6	52	4	8	0
120	5	14	15	20	6	12	12	21	10	17	9	4	37	2	0	0
240	6	13	15	24	8	15	16	23	14	20	10	5	20	1	0	0
360	5	11	14	20	6	12	13	17	9	11	12	6	0	0	0	0

fiber lengths of the polyester fiber but progressively fractured the cotton fiber. As the abrasion time was increased, the numbers of longer cotton fibers present in 1 in. of yarn became fewer and the number of short fragments ($\frac{1}{4}$ in. or less) became larger at first. Later, even the shortest fragments were progressively removed from the yarn structure. No cotton fibers of any length could be found in the durable-press fabrics after 360 min of abrasion time. In this same study, tensile tests on fibers remaining in the abraded yarns indicated that both the polyester and cotton fibers that were left sustained only minimal losses in strength or elongation.

The above data indicate that a different type of abrasion mechanism than the ones considered in previous studies of single-fiber nonresin-finished fabrics is responsible for fabric failure in this case. Much of the previous work indicated that abrasion damage occurred first on the outside of the yarn structure and was transmitted to the inner fibers only after the outer layers of fibers were peeled or rubbed off the yarn. In the blended durable press fabrics, the outer layers of polyester fibers can appear essentially undisturbed even though the cotton fibers throughout the entire fiber have been fractured into extremely small fragments and these fragments have been shaken out of the fabric during subsequent agitation.

The presence of the resin finish on blended fabrics does help retain fiber binding within the yarn just as it does for 100% cotton fabrics. Again, this retention of yarn structure reduces fabric fuzzing so that the fabric surface appears smooth and relatively undamaged. It also contributes to retention of a higher percentage of the original fabric breaking strength over a longer period of time even though the fabric weight loss is greater than that for the same fabric without a resin finish (Fig. 10).

E. Moisture Content

The amount of moisture present in a fabric has little effect on the abrasion resistance of hydrophobic fibers but does effect that of hydrophilic fibers. Most of the reported research in this area has been carried out using cellulosic fibers, especially cotton. Rollins and co-workers [33] and Chippindale [43] reported that wet abrasion of cotton fibers caused extensive fibrillation of the fibers whereas dry abrasion caused more surface erosion, bruising, and mashing. When resin-finished fibers were abraded wet, the fibrils tended to peel off in sheets or straps rather than as individual fibrils. Dry abrasion of crosslinked fibers caused transverse cracks and fractures.

Since water absorption increases both the strength and elongation of cotton fibers, McNally and McCord [2] theorized that cotton's abrasion resistance would be higher wet than dry. Caldwell [38] and Haycook [25] found this to be true when specimens were abraded in

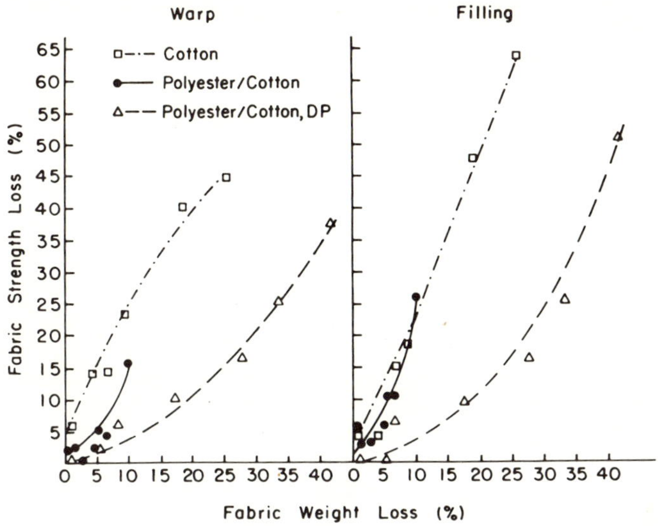

FIG. 10. Relationship between the fabric weight and strength losses caused by Accelerotor abrasion (plastic liner) of cotton, polyester/cotton, and durable-press polyester/cotton fabrics.

an Accelerotor, using either a grit liner with wet-out specimens or a rubber liner with 150 ml of water in the chamber. However, de Gruy et al. [79] found that wet-out cotton yarn specimens required fewer Stoll flex abrasion cycles to cause rupture than did dry yarns.

Warfield [29] found that wet abrasion of nonresin (pure) finished cotton using either a rubber or emery liner in an Accelerotor caused slightly lower weight losses but greater fabric breaking-strength losses than did dry abrasion (see Table 2). Resin-finished cotton exhibited essentially the same weight loss as the pure-finished cotton when abraded dry but higher weight losses when abraded wet. However, the losses in breaking strength were not as great for the resin-finished cotton as for the pure finished. When abrasion conditions were mild (rubber lined), the resin-finished specimens abraded wet had lower strength losses than those abraded dry but, when the cutting action of an emery liner was used, breaking strength losses were higher wet than dry. Haycook [25] and Reid [44] confirmed Warfield's findings that tumble abrasion with water in the chamber was a mild abrasion procedure that caused relatively low strength losses in resin-finished cottons.

Both Caldwell [38] and Warfield [20] found that rayon fabrics were much less abrasion resistant wet than dry. This would be expected because of the great loss in fiber strength when wet. In contrast, the

TABLE 2

Effect of Varying Accelerotor Abrasion Conditions on Weight and Strength Losses of Various Cellulosic Fabrics [29]

Fiber content, finish and property measured	Abrasion treatments				
	Conditioned, rubber liner	Conditioned, grit liner	Wet-out, rubber liner	Wet-out, grit liner	Water in chamber, rubber liner
Cotton, pure					
% Weight loss	1.7	10.7	0.1	8.9	1.8
% Strength loss					
per warp yarn	6	44	24	-[a]	15
per filling yarn	7	68	12	-[a]	11
Cotton, wrinkle resistant					
% Weight loss	2.0	9.6	2.8	15.7	4.3
% Strength loss					
per warp yarn	8	14	2	57	2
per filling yarn	17	54	0	71	+3
Rayon, pure					
% Weight loss	2.0	9.6	7.1	45.6	3.4
% Strength loss					
per warp yarn	2	40	65	-[a]	62
per filling yarn	+2	45	35	-[a]	24

HWM rayon, pure[b]					
% Weight loss	1.9	15.3	0.2	4.6	0.1
% Strength loss					
per warp yarn	5	41	30	67	6
per filling yarn	+2	33	5	32	2
HWM rayon, wrinkle resistant					
% Weight loss	7.2	23.8	0.7	12.9	0.2
% Strength loss					
per warp yarn	10	47	14	78	0
per filling yarn	+6	75	+2	51	+2

[a]Specimens exhibited so many holes that valid breaking strength data could not be obtained.
[b]HWM = high wet modulus (rayon).

high wet modulus rayon fabrics in the Warfield study had wet abrasion resistance properties more like those of the cotton fabrics than like the conventional viscose rayon fabric (Table 2). However, these fabrics lost less weight but more strength when abraded wet than when abraded dry. Resin finishing tended to increase both weight and strength losses during abrasion of the high wet modulus rayon fabrics.

REFERENCES

1. S. Backer, Textile Res. J., 21, 453-468 (1951).
2. J. P. McNally and F. A. McCord, Textile Res. J., 30, 715-751 (1960).
3. G. G. Clegg, J. Textile Inst., 40, T449-480 (1949).
4. R. G. Stoll, Textile Res. J., 19, 394-415 (1949).
5. M. L. Rollins, I. V. DeGruy, T. P. Hensarling, and J. H. Carra, Textile Res. J., 40, 903-916 (1970).
6. C. Reichman, Knitted Outerwear Times, 37, 56-58 (Oct. 28, 1968).
7. J. L. Whitfield, Daily News Record (October 29, 1969).
8. D. L. Munden, Knitted Outerwear Times, 37, 267-269, 480 (May 2, 1968).
9. M. M. Amis, M.S. Thesis, Univ. of Illinois, Urbana, Ill., 1970.
10. T. R. Dawson, J. Rubber Res., 15, 65 (1946).
11. H. F. Schiefer, L. E. Crean, and J. F. Krasny, Textile Res. J., 19, 259-269 (1949).
12. H. W. Stiegler, H. E. Glidden, G. J. Mandikos, and G. R. Thompson, Am. Dyestuff Reptr., 45, 685-700 (1956).
13. C. J. Pope and E. P. Parker, Textile Chemist and Colorist, 1, 345-353 (1969).
14. American Society for Testing and Materials, Committee D-13, Annual Book of ASTM Standards, Part 24, 1973, pp. 188-218.
15. R. Galbraith, M. Boyle, E. Cormany, S. Davison, A. Ginter, J. Ericson, M. Lapitsky, L. Lund, and M. Cooper, Textile Res. J., 39, 329-338 (1969).
16. L. I. Weiner and C. J. Pope, Textile Res. J., 33, 761-764 (1963).
17. E. M. Hicks and A. G. Scroggie, Textile Res. J., 18, 416-423 (1948).
18. H. F. Schiefer and C. W. Werntz, Textile Res. J., 22, 1-12 (1952).
19. J. H. Tait, Rayon Textile Monthly, 26, 171 (1945).
20. S. Backer and S. J. Tanenhaus, Textile Res. J., 21, 635-654 (1951).
21. T. C. Petrie, J. Textile Inst., 28, P206 (1937).
22. G. J. Kyame, J. T. Lofton, and G. F. Ruppenicker, Jr., Textile Bull., 94, 37-39 (1968).
23. G. F. Ruppenicker Jr., G. J. Kyame, and H. W. Little, Proceedings of the Eighth Cotton Utilization Research Conference, New Orleans, La., 1968, pp. 111-122.

24. R. L. Galbraith, paper presented, Division of Cellulose, Wood, and Fiber Chemistry, 153rd Meeting, American Chemical Society, Miami, Fla., April 1967.
25. B. E. Haycook, M.S. Thesis, Univ. of Illinois, Urbana, Ill., 1967.
26. C. L. Warfield and M. J. Fickle, Textile Chemist and Colorist, 5, 181-185 (1973).
27. M. A. Young and M. A. Morris, Textile Res. J., 41, 370-371 (1971).
28. S. U. Douglas, M.S. Thesis, Univ. of Illinois, Urbana, Ill., 1963.
29. C. L. Warfield, M.S. Thesis, Univ. of Illinois, Urbana, Ill., 1967.
30. A. R. Markezich, Am. Dyestuff Reptr., 55, 992-995 (1966).
31. AATCC, Gulf Coast Section, Am. Dyestuff Reptr., 55, 1056-1064 (1966).
32. H. H. Prato and M. A. Morris, Textile Chemist and Colorist, 5, 15-17 (1973).
33. W. R. Goynes and Mary L. Rollins, Textile Res. J., 41, 226-231 (1971).
34. H. J. Ball, Textile Res. J., 8, 134-139 (1938).
35. M. H. Zook, Am. Dyestuff Reptr., 39, 625-630 (1950).
36. C. J. Pope and L. I. Weiner, Textile Chemist and Colorist, 1, 314-322, (1969).
37. American Society for Testing and Materials, Committee D-13, Annual Book of ASTM Standards, Part 24, 1973, pp. 526-530.
38. S. Caldwell, M.S. Thesis, Univ. of Illinois, Urbana, Ill., 1962.
39. E. P. Kaswell, Textile Fibers, Yarns, and Fabrics, Reinhold, New York, 1953, pp. 298-346.
40. W. J. Hamburger, Textile Res. J., 15, 169-177 (1945).
41. R. J. E. Cumberbirch, J. Dlugosz, and J. E. Ford, J. Textile Inst., 52, T513-529 (1961).
42. D. W. Hadley, J. Textile Inst., 54, T409-419 (1963).
43. P. Chippindale, J. Textile Inst., 54, T445-463 (1963).
44. J. D. Reid, R. M. Reinhardt, and A. R. Markezich, J. Home Econ., 58, 189-193 (1966).
45. N. C. Heimbold, Textile World Manmade Fiber Chart 1972, McGraw-Hill, Atlanta, Ga., 1972.
46. H. F. Schiefer, L. Fourt, and R. T. Kropf, Textile Res. J., 18, 18-38 (1948).
47. G. Susich, Textile Res. J., 24, 210-228 (1954).
48. W. E. Morton, J. Textile Inst., 39, P187-192 (1948).
49. M. M. Platt, Textile Res. J., 20, 519-538 (1950).
50. E. Abrams and H. P. Whitten, Textile Res. J., 24, 980-989 (1954).
51. H. F. Schiefer and R. S. Cleveland, Textile Mfr., 60, 182-183 (1934).
52. R. G. Stoll, Melliand Textilber., 22, 125, 235, 308 (1941),
53. W. J. Hamburger, A Technology for the Analysis, Design, and Use of Textile Structures as Engineering Materials, American Society for Testing and Materials, 1955, pp. 15-28.
54. A. C. Walker and P. S. Olmstead, Textile Res. J., 15, 201-222 (1945).

55. L. H. Jameson, B. L. Whittier, and H. F. Schiefer, Textile Res. J., 22, 599-608 (1952).
56. E. R. Kaswell, Textile Res. J., 16, 502-521 (1946).
57. Am. Textile Reptr., 72, No. 20, 19-21 (1958).
58. A. C. Nuessle, Textile World, 104, 92-94 (1954).
59. B. G. Simpson, Am. Dyestuff Reptr., 46, 991-998 (1957).
60. E. A. Murry, Am. Dyestuff Reptr., 44, 141-149 (1955).
61. Midwest Intersectional Technical Paper Committee, AATCC, Textile Chemist and Colorist, 5, 31-37 (1973).
62. E. J. Blanchard, R. J. Harper, G. A. Gautreaux, and J. D. Reid, Textile Ind., 131, No. 1, 116-119, 122, 143 (1967).
63. R. J. Harper, E. J. Blanchard, and J. D. Reid, Textile Ind., 131, No. 5, 172, 174, 177, 181, 184, 186, 233-234 (1967).
64. W. G. Wolfgang, Textile World, 116, No. 1, 104-106, 108, 110 (1966).
65. J. T. Lofton, R. J. Harper, H. W. Little, and E. J. Blanchard, Proceedings of the Eighth Cotton Utilization Research Conference, New Orleans, La., 1968, pp. 123-138.
66. D. D. Gagliardi and F. B. Shippee, Textile Res. J., 31, 316-327 (1961).
67. H. Tovey, Textile Res. J., 31, 185-226 (1961).
68. D. D. Gagliardi and A. C. Nuessle, Am. Dyestuff Reptr., 40, 409-415 (1951).
69. W. A. Reeves, Am. Dyestuff Reptr., 55, 946-950 (1966).
70. J. B. Quig and R. W. Dennison, Textile Res. J., 24, 361-373 (1954).
71. J. B. Goldberg, Textile Res. J., 24, 374-378 (1954).
72. G. G. Whytlaw, J. Textile Inst., 43, P556-583 (1952).
73. W. J. Hamburger, J. Textile Inst., 40, P700-718 (1949).
74. H. Nuding, J. Textile Inst., 43, P352-364 (1952).
75. C. T. Canter, G. D. Jones, and J. W. Weaver, Am. Dyestuff Reptr., 57, 45-49 (1968).
76. A. Kemp and J. D. Owen, J. Textile Inst., 46, T684-698 (1955).
77. R. L. Galbraith, C. N. Sproul, and C. E. Puklin, Textile Chemist and Colorist, 3, 144-148 (1971).
78. D. K. Elias, M.S. Thesis, Univ. of Illinois, Urbana, Ill., 1970.
79. I. V. deGruy, J. H. Carra, V. W. Tripp, and Mary L. Rollins, Textile Res. J., 32, 873-882 (1962).

Chapter 6

RESISTIVITY AND STATIC BEHAVIOR OF TEXTILE SURFACES

S. P. Hersh

School of Textiles
North Carolina State University at Raleigh
Raleigh, North Carolina

I. INTRODUCTION

The generation of static electricity on fibers by rubbing (triboelectrification) leads to many problems in the primary textile and the man-made fiber industries at all stages, from opening and extrusion to fabric finishing. Indeed, trouble can arise even earlier during the ginning of cotton and later in many consumer applications. These problems range from merely mechanical nuisances such as soiling and difficulties in controlling the positioning and lay down of textile materials to extremely dangerous situations such as the failure of parachutes to open and the setting off of explosions of flammable vapors. Areas where soiling causes particular problems include "fog marking" of fabrics during weaving [1]. The pickup of soil, lint, and dirt by garments, especially socks and stockings that are particularly susceptible to charging, is another troublesome problem. One serious annoyance, especially in dyeing and finishing plants where long lengths of dry fabrics must be handled, is the difficulty of folding fabrics into layers so that the material can be stored and transported.

For the consumer, many annoyances in addition to soiling occur frequently. Examples are the clinging of apparel, such as pants to socks, skirts to legs and dresses to the body; the twisting, riding, and bunching of undergarments, particularly slips, under dresses; and the electrical shocks experienced after walking on carpets or sliding over upholstery. Computers and sensitive electronic instruments can also be adversely affected by electrostatic charges on clothing. Of greater significance, however, are the disturbances that frequently result in the loss of life such as the failure of parachutes to open and the setting off of explosions in incendiary vapors and material such as petroleum, food grain, coal dust, and of particular concern, anesthetics in hospitals. In a study of accidents in the chemical industry, Varin has reported that 115 out of 1,600 accidents (7%) were ascribed to static electricity [2].

Not all the consequences of static electrification are bad, of course, and many useful applications can be cited such as purification by electrostatic precipitation; electrostatic separation of minerals and ores; electrostatic painting, printing, and coating; and electrostatic copying [3]. The most successful applications involving fibers have been flocking and spun bonding; much progress has also been made in utilizing electrostatic charges on fibers to aid in fiber cleaning and separation, in controlling fiber transport in open-end spinning, and in singeing and shearing operations [4].

Because of the technological consequences just cited, it becomes important to know and, it is hoped, understand some of the factors influencing the generation and dissipation of static electricity on fibers and how these charges influence the behavior of textile materials.

II. TECHNICAL BACKGROUND

The etymology of the word "electricity" is sufficient in itself to remind us that the static electrification of textile materials has been known since ancient times. The Greek philosopher, Thales of Miletus, around 600 B.C. rubbed amber with cloth and observed that after rubbing the amber attracted light particles [5]. From the Greek word for amber, "elektron," came the word electricity, and from the Greek word "tribein," to rub, comes the term triboelectrification, the generation of electricity by rubbing. It was not until 1733 that Dufay recognized that there were two "kinds" of electricity, which he identified as "vitreous" (glass) and "resinous" (amber). He discovered that objects charged with vitreous electricity repel each other but attract objects charged with resinous electricity. Thirteen years later Benjamin Franklin introduced the terms "positive" and "negative" electricity and identified them with the "vitreous" and "resinous" electricity of Dufay. Franklin also proposed the single-fluid theory which, in effect, infers that negative electricity results from a surplus of negative charge (electrons) and positive electricity results from a deficiency of negative charge. After the existence of positive and negative "electricity" was recognized, the way was open to establishing a "triboelectric" series, that is, a list of materials so ordered that after rubbing any two together, the one listed first would acquire a positive charge and the one lower on the list a negative charge. The first such list was published by Wilche [6] in 1757.

Studies involving the magnitude of charges could not be made, however, until after 1785, when Coulomb established the quantitative relationship between force and charge. Coulomb's law, expressed verbally, states that electrostatic attraction or repulsion between two charged bodies is directly proportional to the product of the charges and inversely proportional to the square of the distance between them. The force is attractive for unlike charges and repulsive for like charges. Since there is a force between charged bodies, energy must be required to establish these charges in proximity to each other in space. As will be described in Section IV-D, this energy is related to the electrostatic "potential" or "voltage" of the system. The forces between charged bodies and the energy associated with them are, of course, the source of the problems arising from the static electrification of materials.

In studying the technological and practical consequences of the static electrification of fibers, the basic question to be answered first is: What is the sign and what is the amount of the charge transferred? In order to understand the phenomenon more completely and thereby have a chance to eliminate it, control it, or put it to constructive uses, many more fundamental questions must be answered such as what molecular factors control the sign of the charge transfer, what particles are transferred, why are they transferred, how many are transferred, how long do transferred particles stay, how many of the particles return as the materials are separated, and many, many others.

Answering these questions, however, is not a simple matter. The study of triboelectrification has long had a reputation for producing discordant, and often nonreproducible and confusing results [7,8]. Some of the reasons why investigating the phenomenon is frought with difficulties are that

1. The observed charge transfer may be limited by the occurrence of electrical breakdown of the atmosphere rather than the electrification process itself (see Section III-B-1).

2. Since classical electrostatic calculations show that only one electron (or charged particle) out of every 100,000 surface atoms must be removed to generate the maximum charge density that can be supported on a flat surface without causing a gaseous discharge in the atmosphere (see Section VII-B-2), minute changes in the chemical composition, the oxidative condition of the surface, and mechanical state of the material can drastically alter the triboelectric characteristics.

3. Since contact and/or rubbing of two objects are required, mechanical disruptions such as elastic deformation, plastic deformation, flow, rupture, creep, and stress relaxation occur, which can readily upset and alter the results. Hard rubbing can produce different results from gentle rubbing since the material surface can be disrupted to different extents and thermal effects can be different in these two situations.

Because of item (1) listed above, the size and shape of the samples also influence the charge transfer. To obtain reproducible results, therefore, very careful control of the chemical and mechanical state of the materials (as well as their geometry), control of ambient conditions including factors affecting the electrical characteristics of the atmosphere, and control of the mechanical contact between the materials are required.

III. THEORIES OF CHARGE GENERATION AND DISSIPATION

A. Generation Processes During Contact

The most common point of departure for explaining triboelectrification is the Volta-Helmholtz hypothesis. Volta in 1779 (according to Perucca [9]) postulated that the source of the frictional electrification of two bodies is in the contact between them and not in the heating, transfer of material, or other effects caused by friction; and that rubbing serves merely to increase the area of contact. Helmholtz [10] realized that the charge Q produced must be present as a double layer when the bodies are still in contact. Upon separation the capacitance C between the materials decreases and the potential V increases correspondingly in accordance with the relation

$$V = Q/C. \tag{1}$$

For contact between metals, it is clear that the contact potential difference will result from a flow of electrons such that their Fermi levels are equalized, i.e., the contact potential will equal the difference in their work functions. In 1917 Frenkel [11] extended this concept to insulators by postulating that insulators possessed an "intrinsic potential" for electrons. Later Frenkel [12] made the first attempt to apply the band theory of solids to contact electrification, followed by Gonsolves [13], Vick [14], Hersh and Montgomery [15], and VanOstenburg and Montgomery [16]. Hersh and Montgomery were able to make certain qualitative predictions of the sign of charge transferred while VanOstenburg and Montgomery extended the treatment to cover the magnitude of charge transferred. Recently Seanor [17,18] has published a list of "work functions" of polymers (attributed to Strella) that correlates with published triboelectric series. A shorter list has been given by Davies [19].

The solid-state treatments enumerated above all depend on the assumption that electrons alone are transferred. Other workers, however, believe that ion transfer is primarily responsible for contact electrification when insulators are involved [20-24].

One logical reason for the discordant and confusing results obtained in triboelectric studies is that perhaps more than one mechanism may be operating. Henry [7,22] has suggested that in addition to charge transfer produced by the Volta-Helmholtz contact potential, additional transfer results from a difference in the abundance of the transferring charge carried on the two surfaces and from a difference in temperature of the two surfaces, effects similar in nature to that of diffusion from a region of high concentration to a region of low concentration or from a higher to lower temperature. This temperature difference can arise from asymmetric rubbing between the two contacting bodies in which a given spot on one material contacts fresh, unrubbed surface on the second (see Section VII-A-3 and 4). The mechanisms just listed are probably the most important ones influencing triboelectrification although others, such as pyro- and piezoelectrification, have been suggested [7,22-25,65].

B. Dissipation Processes Occurring During Separation

1. Mechanisms

Some of the factors influencing the transfer of charge during contact have just been considered. Upon separation, an electric field is produced that exerts a force tending to drive the charged particles back. A portion of these charges therefore will return, so that the charge remaining after the contacting materials are separated is reduced. This leakage through the gap can take place by means of several mechanisms as illustrated schematically in Fig. 1, which shows a sheet of textile material passing through a pair of rollers. These are:

1. quantum-mechanical tunneling across the very small gaps occurring during the initial stages of separation,
2. conduction through the atmosphere,
3. conduction through the contacting bodies, and
4. conduction along the surface of the contacting bodies.

Harper has calculated that for metals electronic tunneling effectively stops after a separation of 25 Å, with about half of the charge flowing back [26]. Although metal-insulator and insulator-insulator contacts have not been studied in detail, it seems likely that the fraction of charge remaining will be reduced from one-half in the metal-metal case to perhaps one-tenth of that initially transferred [15]. For ionic transfer, tunneling would be negligible.

Once the separation distance reaches several times the mean free path of the gas molecules, dissipation through corona discharge can occur if the electric field produced by the charges on the materials is greater than the breakdown strength of the surrounding atmosphere. In general, this dissipation process limits the maximum surface charge density on a planar surface to 30 $\mu C/m^2$, a concentration of charge that produces an electric field of 3,000 kV/m, the generally quoted figure for the dielectric strength of air (760 mm Hg pressure, 20°C, plane parallel electrodes, at separations greater than about 10mm). The

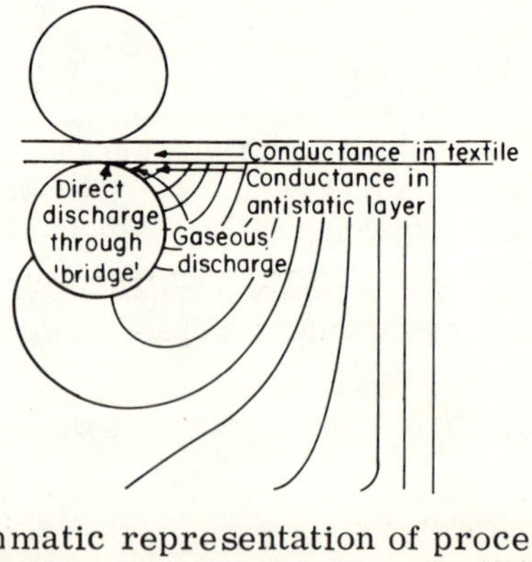

FIG. 1. Diagrammatic representation of processes for dissipating static electrical charges. Reprinted from ref. 171, p. P113, courtesy of the Textile Institute.

breakdown potential for small-diameter cylinders is greater than this, however (see Section VII-B-2).

The final dissipation mechanisms occurring during separation are loss through or along the material itself. Medley [27] has shown that the charge lost through these mechanisms is related to the factor

$$ky/v \tag{2}$$

for dissipation through the bulk of the material, and that the charge decreases exponentially for fiber assemblies coated with a surface finish by the relation

$$q = q_0 \exp(-k\delta/\epsilon\epsilon_0 v). \tag{3}$$

In these equations, k is the conductivity of the bulk fiber in Eq. (2) and of the liquid finishing agent in Eq. (3), y is the thickness of the material, v is the velocity at which the material moves, ϵ is the dielectric constant of the liquid finishing agent, ϵ_0 is the permittivity of free space (see Section IV-A), δ is the length of the wedge of finishing agent formed at the point at which the moving textile assembly departs from the object rubbed (as illustrated in Fig. 15), q is the charge (or charge density) remaining on the fiber after separation to a large distance from the point of contact, and q_0 is the charge (or charge density) transferred upon contact. Cunningham and Montgomery [28] have presented arguments leading to the conclusion that the charge retained on a fiber after rubbing should be functionally related to the term

$$q/q_0 = f(k\ell/\epsilon\epsilon_0 v) \tag{4}$$

where k and v have the same significance as in Eq. (2) and ϵ is the dielectric constant of the bulk material [not the finishing agent as in Eq. (3)], and ℓ is a characteristic length. The dependency indicated in Eq. (4) was derived from the rather simple argument that the fraction of charge remaining on the material should depend on the ratio of a time interval characteristic of the rubbing process t and an average relaxation time for redistribution of charge characteristic of the material τ. The former interval would be given by the time required to traverse a characteristic distance ℓ at a velocity v, i.e., $t = \ell/v$. The time τ is given by electromagnetic theory (see for example ref. 3, p. 26) as $\epsilon\epsilon_0/k$ and the ratio t/τ becomes $k\ell/\epsilon\epsilon_0 v$ as shown in Eq. (4). As pointed out by the authors, relation $k\ell/\epsilon\epsilon_0 v$ can be obtained by a standard dimensional analysis of the selected variables. Although they were unable to develop a theoretical form for the functionality indicated in Eq. (4), Cunningham and Montgomery were able to fit their data, as shown in Fig. 2, by taking $f = [1 - \exp(-\epsilon\epsilon_0 v/k\ell)]$ with some uncertainty in the independent evaluation of ϵ and k.

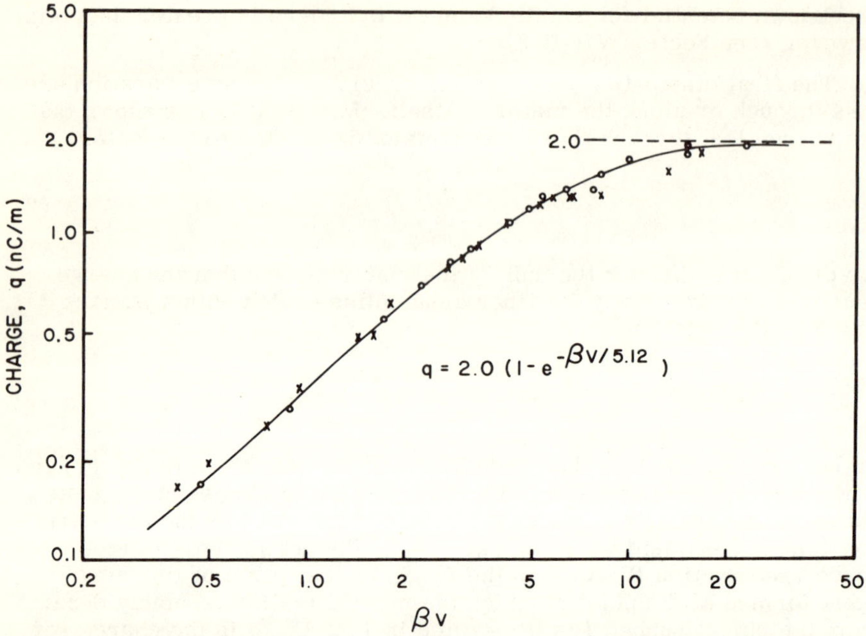

FIG. 2. The fit of data, obtained by rubbing 0.20-mm diameter tantalum wires on 0.20-mm diameter nylon monofilaments at varying temperatures and velocities, to Eq. (4). The value of $\beta/\beta_0(=\epsilon\epsilon_0/k\ell)$ was adjusted for each temperature to give the best fit with $\beta_0 = 5.12$ cm/sec. Reprinted (with change in ordinate units) from ref. 28, p. 976, courtesy of the Textile Research Institute.

Examples of the dependence of the amount of charge remaining on the velocity and conductivity are given in Fig. 3 for single fibers without finishing agents in which the conductivity is varied by altering the relative humidity and in Fig. 4 for cow's tail fiber containing conducting finishes varying in conductivity. The fit of some of the data given in Fig. 4 to Eq. (3) is shown in Fig. 5 by plotting ln q vs. $k\delta/\epsilon v$.

2. Conditions for Complete Dissipation

The technological significance of the last two dissipation mechanisms just described is great, for it permits us to set forth some general rules concerning conditions under which static electrification might be eliminated. Neglecting for the moment the question of just how much charge is transferred by rubbing during contact between two materials, assume that there is charge transfer which increases as the area of contact increases and that part of this charge leaks back

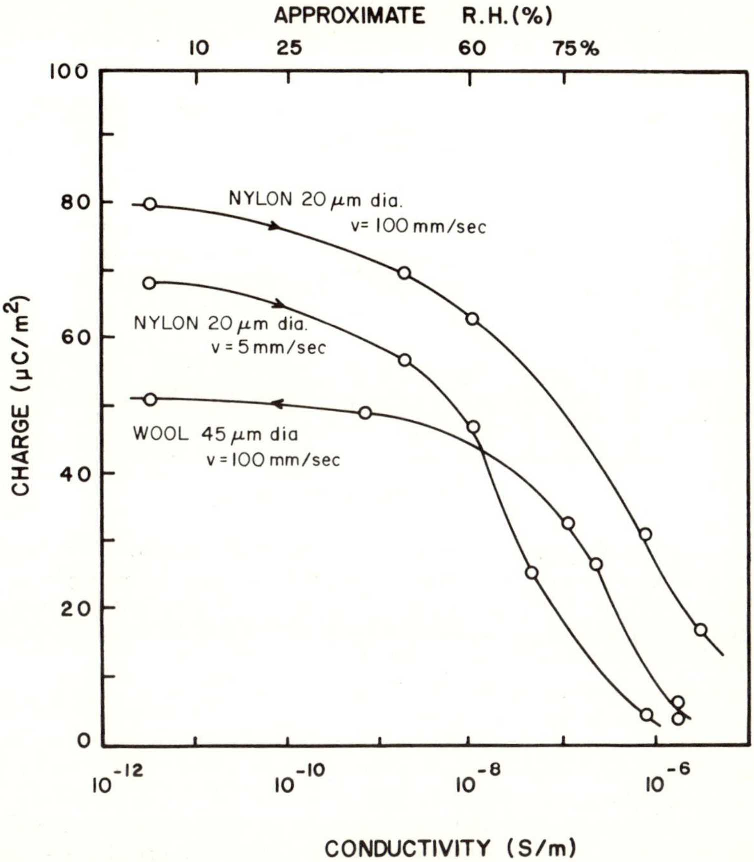

FIG. 3. Net charge transferred to fibers pulled at two velocities v over a platinum wire (1 mm approximate diameter) as a function of fiber conductivity which is varied by changing the relative humidity. Reprinted (with change in units) from ref. 27, p. T129, courtesy of the Textile Institute.

first by tunneling and then is limited by gaseous discharge to a charge density that can be maintained in air. If both contacting materials were conductors, most of the remaining charge would then be dissipated by mechanisms 3 or 4. If one or both of the materials were perfect insulators, no charge could be dissipated by conduction. Somewhere between these extremes, dissipation of charge by these latter mechanisms must become important and the amount of charge dissipated would then depend on two factors: first, how fast the charge flows through the material, and second, the time available for transport of

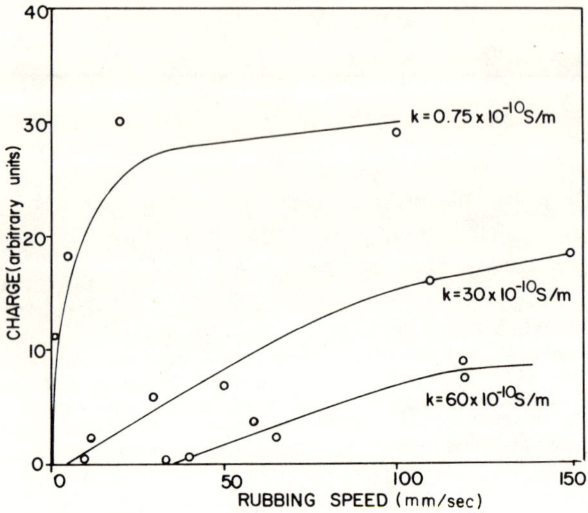

FIG. 4. Net charge transferred to 0.14-mm diameter keratin fiber (cow's tail) coated with solutions of antistatic agent in mineral oil when pulled over a 1.23-mm diameter platinum wire as a function of rub velocity. The volume conductivity k of the solutions is given on the curves. Reprinted (with change in units) from ref. 27, p. T137, courtesy of the Textile Institute.

charge back to the point of contact where it can be neutralized by the charge of opposite sign residing there. Contact time is of course dependent on velocity and charge flow rate is dependent on conductivity. Examples of the tradeoff between velocity and conductivity are indicated in Eqs. (2-4) and in Figs. 2-5. Evaluation of these and other data (see Section VI-D), leads to the conclusion that at speeds encountered in textile processing, say from 1 to 10,000 mm/sec, the charge leaks back too fast to give noticeable effects whenever the resistivity of the more poorly conducting material falls below say 10^7 to 10^8 Ω m^2/m. At higher speeds, static generation is possible even with more conductive materials, and at lower speeds, higher resistivities must occur before static effects can be noticed.

Medley [29] has treated the dissipation problem quantitatively and estimates the ratio of conductivity to velocity required to achieve 50% charge dissipation via mechanism 3 for a sheet of thickness y (after tunneling and gaseous discharge) to be

$$k/v = (2 \times 10^{-11})/y \quad \text{sec}\ \Omega^{-1}\ \text{m}^{-1} \tag{5}$$

or

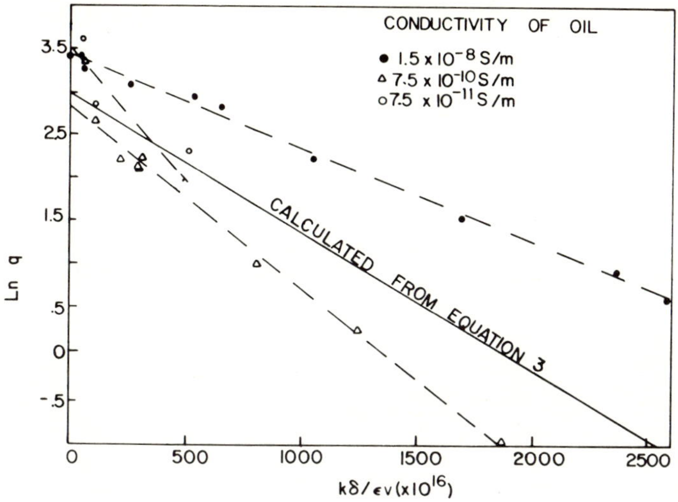

FIG. 5. Test of Eq. (3) for charge transferred to 0.18-mm diameter keratin fiber (cow's tail) coated with three different conducting oils and pulled at varying velocities over a 1.23-mm diameter platinum wire. Reprinted from ref. 27, p. T138, courtesy of the Textile Institute.

$$k_s/v = 2 \times 10^{-11} \ \text{sec}\ \Omega^{-1}\ m^{-1} \tag{6}$$

where $k_s = k\gamma$, the surface conductivity of the material, i.e., the conductance of a sheet of unit length and breadth, as commonly represented in Ω^{-1} (or siemens) per square (see Section VI-A). For dissipation to 10% of the original charge, the conductivity indicated above would have to be doubled. Thus, the greater the relative velocity of the contacting materials, the greater the conductivity (of the material having the lower conductivity) required to eliminate charge formation.

For a cylindrical specimen of diameter D, the critical conductivity/velocity ratio was estimated to be

$$k/v = (8 \times 10^{-11})/D \ \text{sec}\ \Omega^{-1}\ m^{-1}. \tag{7}$$

The relationship between the volume and surface conductivity of a cylinder is

$$k = 4k_s/\pi D. \tag{8}$$

Substituting Eq. (8) into Eq. (7) gives

$$k_s/v = 1.57 \times 10^{-11} \ \text{sec} \ \Omega^{-1} \ m^{-1}. \tag{9}$$

A comparison of Eqs. (9) and (6) shows that the surface conductivity/velocity ratio k_s/v for a cylinder is $\pi/4$ times the value for a sheet. For all practical purposes, they are the same. Manipulation of Eq. (7) or (9) after introducing the appropriate relationships between conductivity, sample geometry, and resistance (see Section VI-A) gives

$$R/L = (1.5 \times 10^{10})/Dv \ \ m \ \Omega \ \text{sec}^{-1} \tag{10}$$

where R is the resistance and L is the length of the cylinder. Equation (10) gives the critical resistance per unit length for a cylinder of radius D moving at velocity v required to reduce the static charge generated to one-half the amount originally transferred at the point of separation. Greater resistance per unit length will generate more static, and lower resistance will generate less. Ward [30], using somewhat different arguments, predicted that the critical resistance for a 200-denier yarn moving at 7.6 m/sec would be about 10^{13} Ω/m, a value essentially in agreement with Eq. (10), which predicts a resistance of 1.5×10^{13} Ω/m.

Calculations of the critical values indicated by Eqs. (6), (9), and (10) required the estimation of one adjustable parameter, the angle of inclination of the electric field generated by the separation of charges to the direction of motion. Although the introduction of a reasonable value for this parameter gave fair agreement with experimental results for materials that did not have a liquid finish applied, predictions of the critical values are quite sensitive to the choice of this parameter. Hence, values given by Eqs. (6), (9), and (10) should be considered only approximations. Validity of these predictions will be considered in Section VI-D.

Three rather important predictions do follow from Medley's treatment, however. First, charge generation decreases with increasing conductivity. Second, for dissipation by mechanism 3, the surface conductivity (or resistivity) is the critical parameter controlling the net charge transferred during triboelectrification even though this leakage mechanism involves conduction through the volume of the material. Third, there is a tradeoff between conductivity and velocity—each order of magnitude of increase in resistivity can be balanced by a reduction of one order of magnitude in the rubbing velocity.

Although as already stated, Eqs. (6), (9), and (10) are in reasonable agreement for materials that do not have a finishing oil added, the conductivity of fibers to which finishing agents have been applied can be one to two orders of magnitude less than that indicated by the above equations to achieve adequate static protection. In this case, charge dissipation should follow from Eq. (3) where k is the conductivity of the surface agent rather than that of the fiber assembly. Medley reported [29] that the application of as little as 3% of a surface agent having a bulk conductivity of between 10^{-7} and 10^{-8} Ω^{-1} m^{-1} is sufficient to eliminate static generation on a wool roving. This result is shown in Fig. 6. In contrast, Fig. 3 shows the conductance required to eliminate charge generation by changing the conductivity of the textile material itself.

Once the textile material is isolated from the rubbing element, any charge remaining can then leak away to ground following the well known principles of R-C circuits (see Section VI-B)

IV. FUNDAMENTALS OF ELECTROSTATICS

Experimentation in static electrification of polymers and textile materials has involved a wide variety of measuring instruments and charging techniques. The results in many instances have not been presented in terms of the quantities that are most desirable to report, that is, the total charge q or charge density, i.e., charge per unit length, area, volume, or mass. Frequently, results are given as electric field strength E or potential V, both of which are proportional to charge. Unless the geometry and the electrical characteristics of the measuring system are known, however, the actual charge cannot be determined from E and V. Another complicating factor is that different systems of units are frequently employed. Because of the great variety of ways that results have been presented in the literature, it would be beneficial to review the fundamentals of electrostatics before describing some of the techniques that have been used to measure static charge (see for example, refs. 3,24,31,32).

A. Coulomb's Law

Coulomb's law states that the force F between two charged bodies carrying charges q_1 and q_2, respectively, is proportional to the product of these charges and is inversely proportional to the square of the distance r between them:*

$$F \propto q_1 q_2 / r^2. \tag{11}$$

*Although mathematical precision would require the use of vector notation here since force and several other quantities to be introduced are vectors, the arguments of this chapter can be readily presented by considering the magnitude of the vectors only. Therefore scalar notation will be used throughout.

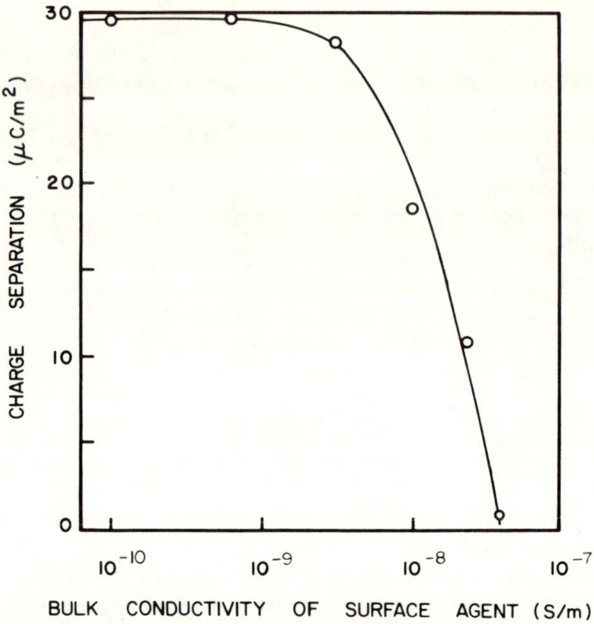

FIG. 6. Charge generated on wool roving in the presence of 3% surface conducting agents varying in volume conductivity. Roving passed between cork rollers at a velocity of 100 mm/sec. Reprinted (with change in units) from ref. 29, p. S26, courtesy of the Institute of Physics.

As shown in Fig. 7a the force acts along a line between the center of the bodies and is repulsive for like charges and attractive for unlike charges.

The proportionality sign may be replaced by an equals sign and a proportionality constant. When this constant is assigned the value unity and the mechanical units involved are in the cgs system, Eq. (11) becomes $F = q_1q_2/r^2$ and serves to define the "electrostatic" system of units (esu). F is in dynes, q is in electrostatic units (called esu or statcoulombs), r is in cm and the unwritten constant has a value of 1 dyne $cm^2/(esu)^2$. For many reasons, however, the electrostatic system of units is being replaced by the rationalized mks system in which Eq. (11) becomes

$$F = \frac{1}{4\pi\epsilon_0} \frac{q_1q_2}{r^2}. \qquad (12)$$

Here F is in newtons (N), q is in coulombs (C), r is in meters (m), and

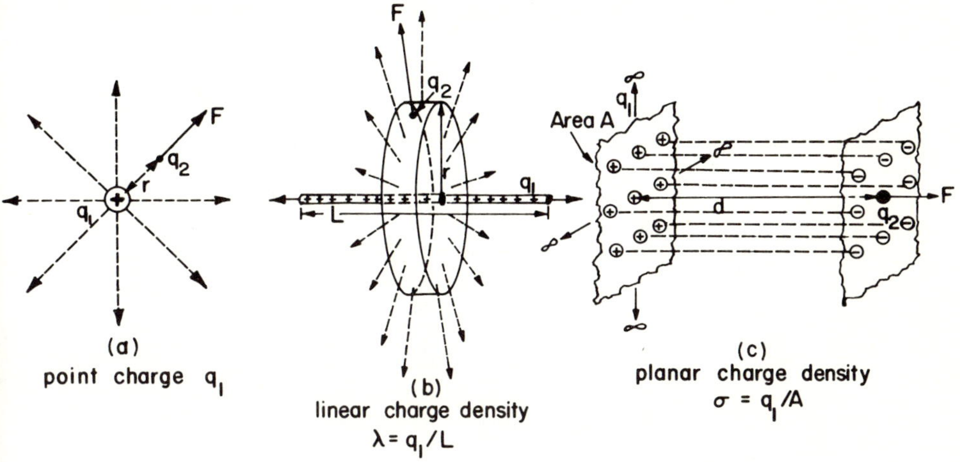

FIG. 7. Force exerted on charge q_2 by (a) a point or spherical charge q_1 at distance r, (b) an infinite line charge of density $\lambda = q_1/L$ at a distance r, and (c) an infinite planar charge of density $\sigma = q_1/A$ at a distance d.

ϵ_0, known as the "permittivity of free space," is equal to 8.85×10^{-12} $C^2/N\text{-}m^2$. If the dielectric constant ϵ of the medium separating the charges is different from 1, the permittivity becomes $\epsilon\epsilon_0$. In the nonrationalized mks system, the constant is $1/\epsilon_0$. By including the quantity 4π in Eq. (12) as is done in the rationalized system, 4π is eliminated from many other, more frequently encountered equations. The mks system has the great advantage of being compatible with the practical system of electrical units, i.e., volt (V), ampere (A), ohm (Ω), and watt (W), which are encompassed by the international system of units (SI). Hence, the rationalized mks and SI systems are used throughout this chapter.

B. Electric Field

A charged body produces an electric field E whose value at any point is defined as the force exerted on a unit charge at that point. If the magnitude of the charged body in the field is q_2, then

$$E = F/q_2. \tag{13}$$

The electric field, like force, is a vector whose direction is the same as that of the force exerted on a positive charge. From Eqs. (12) and (13), it is evident that the electric field produced at a distance r by a point charge q_1 or a sphere having a charge of q_1 is

$$E = q_1/4\pi\epsilon_0 r^2. \tag{14}$$

Point or spherical charges are not often encountered in textile processing, however, and it is necessary to consider the field produced by a line or cylindrical charge (yarn, roving, etc., Fig. 7b), or a planar charge (fabric, film, etc., Fig. 7c). Electrostatic theory shows that for a line charge of $\lambda = q_1/L$ C/m, the electric field produced is

$$E = \frac{1}{2\pi\epsilon_0}\frac{\lambda}{r} = \frac{1}{2\pi\epsilon_0}\frac{q_1/L}{r} \tag{15}$$

and for an infinitely large plane having a charge per unit area $\sigma = q_1/A$ C/m^2, the field is

$$E = \frac{\sigma}{\epsilon_0} = \frac{q_1/A}{\epsilon_0}. \tag{16}$$

Thus, it is evident that the electric field decreases with distance in an inverse square law for a spherical body, decreases inversely for a cylinder, and is independent of distance for an infinitely large sheet. Therefore, for those instruments that measure the electric-field strength, it is necessary to know the geometry of the measuring system in order to determine the electric charge on the test material.

From the definition given by Eq. (13), it is evident that the electric field is expressed in units of force/unit charge which in the mks system is newtons per coulomb, N/C. As will be seen in the following section, the electric field can also be expressed in units of volts per meter, V/m.

C. Electrostatic Potential and Capacitance

An electrically charged body generates an electric field that exerts a force on a second charged body as indicated in the previous section. When a charged particle of charge q_2 is brought from an infinite distance where there is no force on it to a distance r from a charged body, work must be done if the charges are of like sign and energy is released if the charges are of opposite sign. When the force exerted is parallel to the line of movement, the work W required is

$$W = -\int_{\infty}^{r} F(r)\, dr \tag{17}$$

where F(r) is given by Eq. (12). (The negative sign is introduced since the work is done against the electric force for like charges.) This equation may be expressed in terms of the electric field by introducing Eq. (13) to give

$$W = -\int_{\infty}^{r} Eq_2 \, dr. \tag{18}$$

The work W is a potential energy since the only mechanical variable on which the force and the energy depend is the position of the charge q_2 (and of course on the reference position, taken here as $r = \infty$, which merely changes the energy by an additive constant).

The electrostatic potential V at any point may be defined as the ratio of the potential energy of a charged body at that point to its charge, i.e., the energy per unit charge. Thus

$$V = \frac{W}{q_2} = -\int_{\infty}^{r} E \, dr. \tag{19}$$

V represents the difference in potential $(V_r - V_\infty)$ between points r and the reference point. If the reference point is selected at a position where the potential energy is zero, $V_\infty = 0$ and $V = V_r - V_\infty = V_r$. The potential V is obviously a scalar quantity since energy and charge are scalar quantities. From Eq. (19) it is evident that the potential in the mks system is expressed in units of joules/coulomb, J/C, which is equivalent to the volt V in practical electrical units.

Substitution of Eqs. (14), (15), or (16) into Eq. (19) will give the potential produced by a charged body as a function of distance r from the charged body. Consider Eq. (16) for a charged plate and calculate the difference in potential on the plate V_0 to the potential V_d at a distance d from the plate:

$$\Delta V = V_0 - V_d = -\int_{d}^{0} \frac{q_1}{A\epsilon_0} \, dr = \frac{q_1 r}{A\epsilon_0} \Bigg]_0^d = \frac{q_1 d}{A\epsilon_0} = Ed. \tag{20}$$

Two features of Eq. (20) are of great interest. First, the electric field produced between two parallel plates (imagine now the field to be terminated by a second plate at d parallel to the first) is equal to the potential difference between the plates divided by the distance between them, i.e.,

$$E = \Delta V/d. \tag{21}$$

Thus it is evident that the electric field may also be expressed in units of volts/meter. This relationship also follows from the differentiation of Eq. (19) to give $E = -dV/dr$, i.e., the negative of the rate of change of potential with distance equals the electric field at that point. The second important consequence of Eq. (20) is that the charge on the plate q_1 is proportional to the potential difference, i.e.,

$$q_1 = A\epsilon_0 \, \Delta V/d = C \cdot \Delta V. \tag{22}$$

The proportionality constant is called the "capacitance" C and is the ratio of the charge to the potential of any body. In mks units, capacitance is expressed as coulombs/volt (C/V) or in practical units as "farads" F. As shown in Eq. (22), the capacitance between two parallel plates is

$$C = A\epsilon_0/d \tag{23}$$

where A is the area of the plates and d is the distance between them.

The capacitance of any body, i.e., the ratio of its charge to potential, depends on its "surroundings." Only if the surroundings are defined by a relatively simple geometry, such as two parallel plates, two concentric cylinders or spheres, or a cylinder and a plate, can the capacitance be easily calculated. It can be readily measured, however, with an electrometer by using the technique described in Section VI-B for measuring resistance by means of an RC circuit. In this case, it is the capacitance C which is determined rather than the resistance R.

From the relation $q_1 = CV$ (Eq. 22), it is evident that experimental results of triboelectric studies reported in terms of "voltage" cannot be translated into charge measurements unless the capacitance or the geometry of the measuring system is specified.

D. Energy of a Charged System

Equation (19) indicates that the work required to transfer charge from one body at zero potential to another at potential V is $W = Vq$. The work required to transfer an incremental charge would be $dW = V\,dq$. From Eq. (22) $V = q/C$. Hence the energy of a capacitor charged to q_1 will be

$$W = \int V \, dq = \int_0^{q_1} \frac{q \, dq}{C} = \frac{q_1^2}{2C}. \tag{24}$$

Substitution of Eq. (22) into Eq. (24) gives

$$W = CV^2/2. \tag{25}$$

These last two equations indicate the amount of electrical energy possessed by a charged body. Discharge of this energy can cause sparks, explosions, discomfort, and/or death depending on the quantity of energy present and the rate at which it is released. For example, vapors of diethyl ether and cyclopropane, two common anesthetics, can be exploded by the discharge of 0.2 millijoules of electrical energy [33]. This energy is equal to that stored by a man (200 pF capacitance [33,34]) carrying a net charge of 0.28 μC, a charge which can be readily generated on an area of 0.01 m^2 (100 cm^2) (see Section VII-B-2). The man would be at a potential of 1,400 V.

To summarize, relationships have been established in this section for calculating the amount of electrical energy stored by and the force of attraction or repulsion between charged bodies, factors leading to many of the problems associated with static electrification.

V. MEASUREMENT OF CHARGE

This section will describe some of the instruments available for measuring electric charge, some of the difficulties encountered, and some of the experimental arrangements that have been used. A discussion of the problems will be delayed until the electroscope is described.

A. Simple Electroscope

One of the oldest devices for detecting electrostatic charge is the electroscope illustrated in Fig. 8. Two thin metal foils, generally gold, hang side by side in a grounded metal housing. The foils are connected to a metal probe that projects through the housing and is well insulated from it. When the probe is connected to a voltage source, both foils are charged with the same sign and repel each other. The angle of deflection is a direct measure of the charge, and it is possible to calibrate the electroscope in terms of volts.

B. Problems Associated with Charge Measurement

1. Loading or Capacitance of Instrument

When a conductor with charge Q and capacitance C_1 contacts an electroscope of capacitance C_2, the charge is immediately distributed over the connected metal parts. The charge absorbed by the electro-

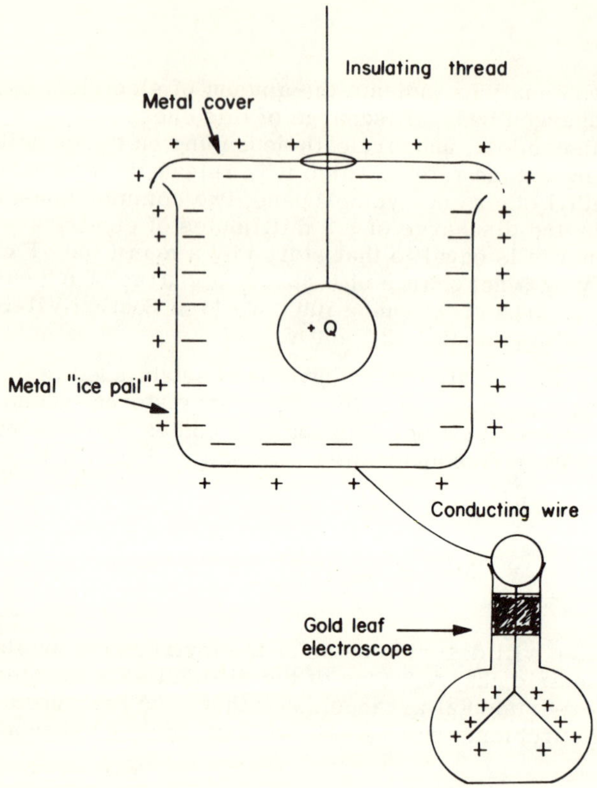

FIG. 8. The principle of inductive charge measurement using a Faraday "ice pail" and a gold leaf electroscope. Reprinted from ref. 3, p. 37, courtesy of John Wiley and Sons.

scope decreases the charge initially on the conductor and the voltage indicated by the electroscope is then less than the voltage V_0 originally on the conductor. According to Eq. (22), $V_0 = Q/C_1$. After contact with the electroscope, the potential falls to $V_f = Q/(C_1 + C_2)$. The greater the capacitance of the electroscope compared with the capacitance of the object, the greater the drop in potential. This phenomenon is commonly referred to as "loading," and in general, the less the capacitance of the instrument, or loading, the greater the sensitivity of the instrument. If both capacitances are known, the total charge on the system can be calculated. The measuring range of electrometers can be extended by connecting a "loading" capacitor across the input to the instrument.

2. Measurement of Charge on Insulators

When the charged object is an insulator, the charge generally cannot be measured simply by touching the probe of the measuring instrument with the insulator. Charges are immobile on an insulator and hence cannot be readily transferred to the probe. Only charges close to the contact area can be transferred, and if the insulator is moved along the probe in an effort to transfer more charge, new charges will be generated, thus further confusing the measurement. For this reason, charges on insulators must be measured by induction, a phenomenon first described by Faraday. When a charged material is enclosed by, but insulated from, a metal container, as illustrated in Fig. 8, the charge on the material attracts a balancing charge of the same magnitude but of opposite sign on the inner surface of the container. If the container is insulated from ground, another balancing charge must appear on the outer surface of the container. For example, if the charge on the insulator is positive, balancing electrons are attracted to the inner surface of the container leaving a positive charge on the outer surface. For negatively charged insulators, electrons will be repelled from the inner surface to the outer surface of the metal. The net result in each case is to induce a charge equal to that on the insulator on the outer surface of the container. If the container is connected to an electrometer, the charge on the insulator can be measured without contact between the container and the insulator. This technique is still referred to as the "ice pail method" or the Faraday "pail" method.

One useful concept of an electric field to consider is that each elementary charged particle generates a line of force which must terminate on a charge of opposite sign. The density of the lines of force (i.e., lines per unit area of "flux") is a measure of the strength of the electric field. In a Faraday pail, each line of force emanating from each elementary charge on the insulator terminates on an opposite charge induced on the container. If the insulator is not completely surrounded by the pail, the lines of force from a portion of the charges might terminate elsewhere. The charge can still be measured, however, if the fraction of the lines of force eventually reaching the electroscope probe is known. This fraction might be estimated theoretically or measured experimentally. It is apparent that the most desirable procedure would be to completely surround the insulator so that all the lines of force terminate on the pail. If it is not possible to determine the fraction of charge reaching the probe, the geometrical relationship of the charged object and its surroundings to the instrument probe must be kept constant in order to maintain quantitative proportionality between measurements.

3. Changes in Capacity

It must be remembered that any movement of the charged object, the probe, the leads, or the surroundings can change not only the fraction of lines of electric flux terminating on the probe, but also the capacitance of the elements and therefore the voltage of the system as indicated by Eq. (22). Therefore when an electrometer is operated at the extreme of its sensitivity, all connections should be rigid and as short as possible to assure accurate and reproducible measurements.

4. Leakage

Another complication results from the fact that the probe cannot be insulated perfectly. Any loss of charge (leakage) by electrical conduction to ground will, of course, change the charge to be measured. Thus the greater the leakage resistance (or input resistance) of the probe, the greater the sensitivity and stability of the instrument. Care must also be taken to insure that the leakage resistance of the connecting cables and components are not reduced by moisture, oils, or other contaminants.

C. Instrumentation for Quantitative Measurements

1. Electrometers

One way of increasing the sensitivity of electroscopes, which is rather low, is illustrated in Fig. 9. A fine wire or "string" fixed at both ends is mounted between two plates maintained at opposite

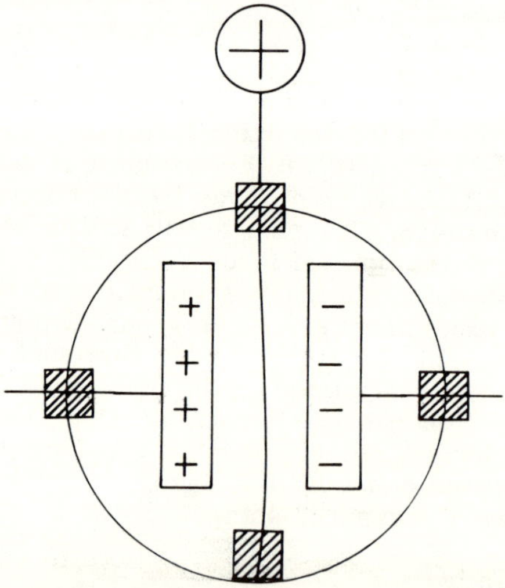

FIG. 9. The operating principle of a string electrometer.

potentials with batteries. After charging, the wire is attracted toward the positively or negatively charged plate, depending on the polarity of the charge. The amount of charge can be determined by measuring the deflection of the wire with a microscope. Instruments of this type are known as string electrometers and are quite delicate. Another example is the quadrant electrometer.

2. "Electronic Electrometers"

Later instruments made use of high-impedance dc amplifiers to detect charge. Instruments of this type are commonly referred to as "electronic electrometers" or more simply (and incorrectly) as electrometers. To achieve the high impedance required, specially designed "electrometer" vacuum tubes, field effect transistors (FET), or vibrating capacitors must be used in the input stage. Electrometer tubes have a "floating" grid, which means that there is no conducting path from the grid to ground. These tubes are also designed to have a small grid or offset current. As indicated in Fig. 10, there is no electrical connection of the input grid to ground via a grid resistor as used in normal dc amplifiers. A high-input resistance is essential, of course, or else any static charge placed on the probe would quickly leak to ground and be lost. High-impedance amplifiers are inherently

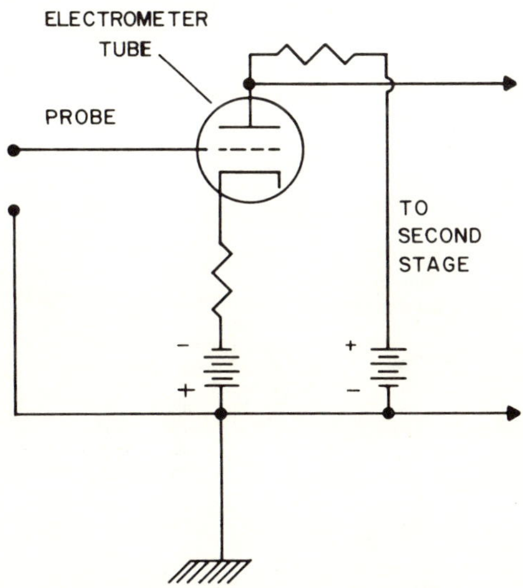

FIG. 10. Input stage of electronic electrometer using an electrometer vacuum tube. Reprinted from ref. 40, p. 19, courtesy of North Carolina State University.

unstable, however, because they do not provide a leakage path to ground for the grid current. Good quality, stable, electrometer amplifiers therefore are relatively expensive. Standard transistors cannot be used in the input stage of an electrometer because their impedance is too low. With the development of the field effect transistors (FET), however, solid-state devices became available with adequate impedance for use in electrometers.

3. Chopper-Type ac Voltmeters

In contrast to dc amplifiers, ac amplifiers have good stability, high gain, and are inherently simpler and less expensive to construct. Electrostatic measurements, however, are essentially dc measurements and hence cannot be measured by direct input to an ac amplifier. By periodically altering the number of flux lines terminating on the electrometer probe, an alternating electric field is created which can then be easily amplified. The alternation can be created by moving the charged region or probe, or by placing a chopper either before or after the probe.

An example of an apparatus in which the charged region is moved is illustrated in Fig. 11 [35]. In this device, the test sample is mounted on a revolving, grounded aluminum wheel and is charged by contact with the rubbing board. The sample then passes in front of the test probe (the grid terminal of an electrometer tube) to generate an ac signal that can be readily amplified and measured. Another example has been described by Sprokel [36] in which the variation in charge along a yarn is measured rather than the average charge by passing the yarn in front of the detector. An instrument of the second type (moving probe) commonly known as a "vibrating reed electrometer" makes use of a metal reed kept in vibration by an electromagnet.

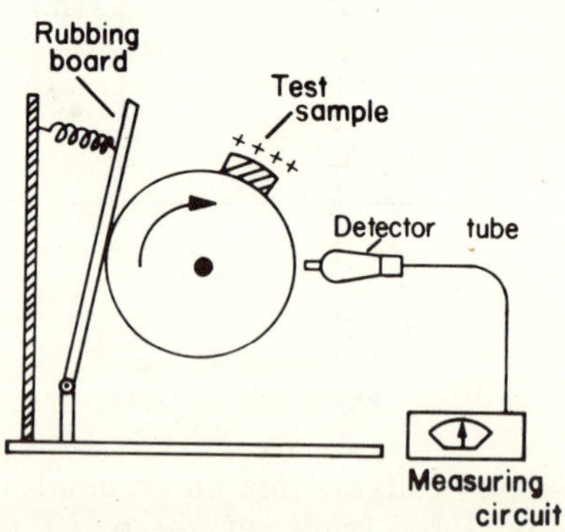

FIG. 11. Operating principle of instrument in which charged material is moved to generate alternating potential (after ref. 35).

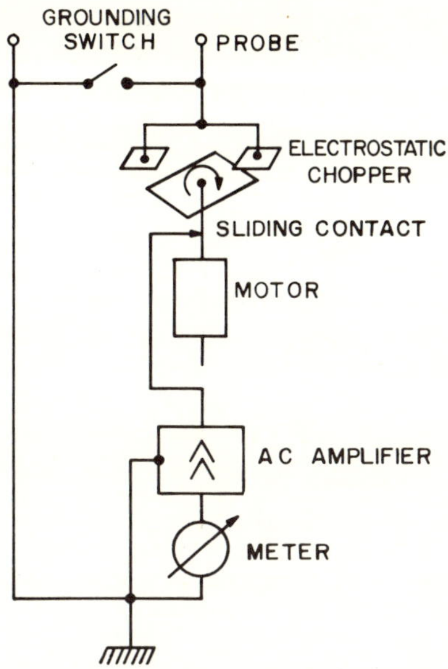

FIG. 12. Operating principle of rotating chopper (dynamic capacitor) type of electrostatic voltmeter in which ac signal is generated by periodically varying the capacitance between probe and amplifier. Reprinted from ref. 40, p. 20, courtesy of North Carolina State University.

The operating principles of instruments using a chopper after the probe and before the probe are illustrated in Figs. 12 and 13, respectively. In the former instrument, sometimes called a "dynamic capacitor" static voltmeter, the probe is connected to two condenser plates. A third plate is mounted on a motor shaft and rotated in a plane parallel to the first two plates to form a varying capacitor. The amplitude of the alternating voltage induced on the rotating plate is proportional to the voltage on the fixed plates and probe. The operating principle of the instrument shown in Fig. 13 is somewhat different. In this device, the electrostatic field between the charged material and the sensing electrode (probe) is chopped by a grounded rotating shield. Since the action involved here is that of chopping or "milling" the electrostatic field, instruments of this type have been given the name "field mill." In this instrument, an ac signal whose amplitude is proportional to the field strength is induced across the resistor R.

Although the operation of the field mill and the dynamic-capacitor instrument shown in Fig. 12 may appear to be similar, there is one very important difference. The field mill is stable over much longer periods of time because leakage is much less of a serious problem. In the dynamic-capacitor type instruments in which an external probe

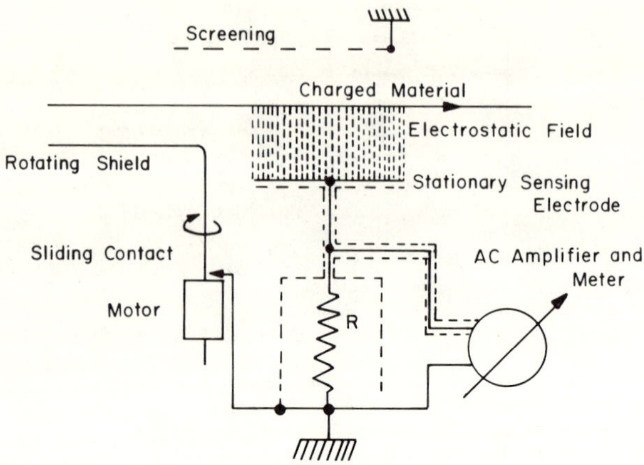

FIG. 13. Operating principle of field mill in which electric field is periodically interrupted to generate ac signal on probe of electrometer. Reprinted from ref. 40, p. 27, courtesy of North Carolina State University.

is employed, care must be taken to minimize leakage of charge from the probe, as discussed in Section V-B-4. For example, consider an instrument of the type shown in Fig. 12, operating for a length of time in which a positive charge on the material remains essentially unchanged. The reading on the meter slowly decreases to zero due to leakage to ground, even though the charge on the material remains. If the charge on the material is then altered, the reading on the meter indicates the change in voltage (positive if the charge increases and negative if the charge decreases) even if the net charge remains positive. As leakage occurs, this type of instrument begins to measure differences rather than the absolute value of charge. To properly re-zero the meter, the probe must be grounded in the absence of an electric field. Either the charged body must be removed or the probe must be shielded from it.

In contrast, the only leakage in the "field mill" is from the sensing electrode through the resistor (labeled "R" in Fig. 13) which only reduces the meter reading—it does not drift to zero. The essential difference in this case is that the sensing electrode is periodically screened from the electric field by the rotating grounded shield and there is no resistance path from charged material to ground through the detection instrument.

4. Portable Instruments

The instruments described above are rather sensitive, delicate, and expensive. For detection of static charges in plant locations, low cost, trouble free, portable instruments are required. Such simple hand-held, pistol-like devices that can be aimed at the suspected charge source for measurement are fairly common. Two instruments operating on different principles are employed. One has a single-stage dc amplifier that uses an electrometer tube and a dry-cell power supply to detect and measure the electric field. The grid of the electrometer tube is shielded by an enclosing barrel, or tube, which can be pointed at the suspected charged body. The sensitivity of the instrument can be changed by altering the size of the tube opening. The other instrument operates on an entirely different principle. A radiation source ionizes the air in the barrel of the instrument. The electrostatic field to be measured penetrates through the front opening and accelerates the ions in the chamber, producing a current that is proportional to the strength of the field. Instrument sensitivity can be adjusted by means of an iris diaphram in the front opening. Care must be taken when making long-term measurements of stationary targets to prevent their gradual discharge by the ions present.

D. Qualitative and Semiquantitative Methods for Detecting Charge

Electric charge may be detected qualitatively by sprinkling or blowing indicating powders, [37-39,172] or even such simple materials as dust, cigarette ashes, or instant coffee [40,41], over an electrified surface. A charged material held over a dish containing these powders also attracts the powders through induction. If certain mixtures differing in color and electrifiability are used, it is even possible to distinguish between positively and negatively charged areas on a fabric or film [38,172].

Another technique is to rub skeins of fibers or yarns with a glass, rubber, or metal rod. If sufficient charge is generated, the skein will balloon out [41,42]. Under carefully controlled conditions, the charges can be estimated by measuring the diameter of the balloon [43,43a].

E. Methods for Generating Charge

1. Laboratory Methods

An almost endless number of ways have been employed for generating charge through contact. These can be separated into those involving rubbing and those that give contact without rubbing. Many of the contact-only studies have been concerned with verifying the Volta-Helmholtz hypothesis. Notable among these are the experiments of Harper [26,44, 45] in which the touching of spheres is carefully controlled through flotation on mercury or by the motion of a lever. Rose and Ward [46] have described an apparatus that provides contact between a spherical insulator and a metal disk at varying pressures by advancing the sphere with a spring-loaded screw. Contact between mercury and solids has also been employed to eliminate rubbing [47]. Rolling contact has been employed by Arridge [48] to study the rub-free contact potential between metals and nylon fiber by contacting the moving fiber with a metal attached to a rotating wheel moving at the same velocity. Davies [19] used a similar arrangement to measure contact potentials of polymer films. Henry [49] and Webers [39] minimized rubbing by sandwiching one sheet of material between two sheets of a different material. Pressure was then exerted, in one case by passing between steel rolls, and in the other with spring clips. After the outer layers were carefully peeled off, the charge on the center layer was measured and was claimed to represent the results of essentially rub-free contact.

The variety of ways for generating charge with rubbing is much larger. These include such simple techniques as squeezing the textile material with a rubber-gloved hand [30,50], or rubbing with a hand-held cloth or bar [51-53]. More complex contacts have been employed such as sliding a powder [54,55] or a fabric wrapped around a box [56] down an inclined plane, rolling a sphere down the sample [57] or in a rotating cylinder [58,59]. The sample might be touched by a spring-loaded plunger [60,61] or rubbed by plates, rollers, or wheels [35, 61-63a]. One common rubbing method for yarns involves passing the sample between rollers into or through a Faraday cage as illustrated in Fig. 14 [29,33,36,64-68]. Another technique frequently employed for measuring charge generated on yarns and rovings is to pass the material over stationary rods or guides connected to an electrometer probe as illustrated in Fig. 15 [27,69-72]. In this arrangement, the charge accumulated on the conductor is assumed to be a direct measure of the charge separated.

In order to eliminate problems associated with contamination of the rubbing surfaces when charging textile materials, especially when using materials containing finishing agents, Shashoua [73] applied an electrical potential directly to the surface of the material in order to charge it. He then studied the rate of charge buildup and decay. A similar technique has been employed by Turer and Smith [170] and in a Federal Test Method Standard for evaluating the electrostatic properties of materials [74].

For more precise control and adjustment of the many rubbing parameters involved, rather elaborate devices have been built. For example, in the device described by Hersh and Montgomery [75], two wires, yarns, or fibers are held in jigs supported in adjustable mountings which permit the two specimens to be rubbed in any geometry from asymmetric to symmetric. The extremes are illustrated in Fig. 16 for two cylinders. Two types of asymmetric rubbing are illustrated in which the same region of one body is exposed to different regions of the other. These are identified as 90°-0° and 0°-90° where the first coordinate refers to the orientation of the fixed fiber and the second to the orientation of the moving fiber. Also shown is a symmetric rub, 45°-45° in which new regions on each body are continually presented

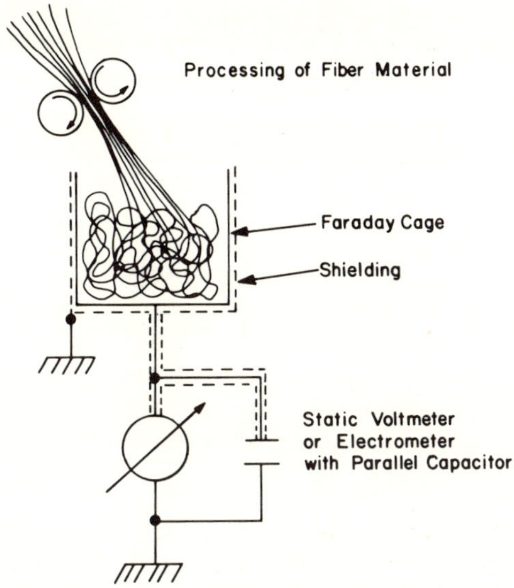

FIG. 14. Batch method for measuring charge generated on textile materials during processing. Reprinted from ref. 40, p. 21, courtesy of North Carolina State University.

to new regions on the other. The normal force between the bodies, length of rub, and velocity of rubbing can be varied. Modification of similar types of apparatus have been made which permit measuring of the frictional characteristics during rubbing [76] and detecting sparks resulting from atmospheric breakdown [28].

2. Tests Simulating Use Conditions

Many anomalies have been observed in which the static properties of textile materials evaluated using either rubbing techniques or resistivity measurements do not correlate well with end-use applications. Therefore many in vivo simulation tests involving wear and use testing have been developed. The first and as yet only one of these to be adopted as a standard test method in the United States is the American Association of Textile Chemists and Colorists (AATCC) test method 134-1969 for measuring the "Electrostatic Propensity of Carpets" [77]. In this test a man carrying a probe connected to an electrometer walks on a carpet for a minimum of 30 sec to a maximum of 60 sec until the measured body voltage builds up to a maximum.

To evaluate the cling rating of various dress-slip combinations, a test has been developed in which women wearing the test garments in a room under controlled environmental conditions go through a prescribed procedure of movements including sitting, walking, putting on and taking off a coat, and sliding on a car seat. The severity of the

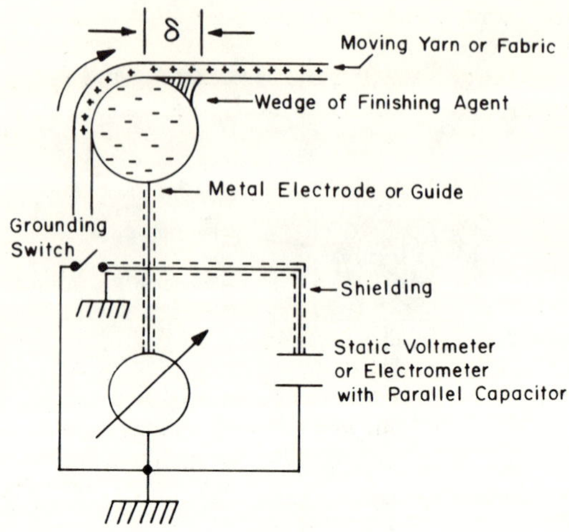

FIG. 15. Method frequently employed for measuring charge generated on yarns passing over metal electrodes. The wedge of finishing agent formed at the point where the yarn leaves the rod has been observed in ref. 27 and used to develop Eq. (3). (After refs. 27 and 40).

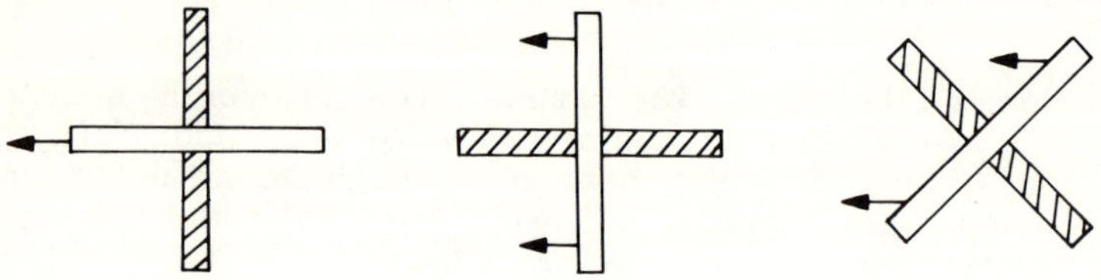

FIG. 16. Sketch of two types of asymmetric rubbing (I and II) and symmetric rubbing (III). Reprinted from ref. 75, p. 284, courtesy of the Textile Research Institute.

resulting clinging, riding, and bunching is then rated on a subjective scale. A somewhat less subjective test used to evaluate clinging of slips and other lingerie is the so-called "sail test" [78], in which the test subject, wearing the garment to be evaluated, charges it by rubbing against the vertically mounted sail in a prescribed manner. After charging, the time required for the test garment to decling from the body is measured.

Other tests concerned with the detonation of explosive mixtures as a result of electrostatic charges generated on clothing have been described [33,34,79]. In these tests, the subject wears the test garment and then generates a charge either by removing the garment or by sitting on a chair momentarily and then touching an electrometer probe or an electrode connected to an explosive mixture. Rizzo [173] has pointed out that in order to understand the static electrical phenomena associated with clothed human subjects, the entire interaction within the materials-clothing-man-environment system must be considered, including the conducting and insulating nature of the platform on which the individual is standing. Although the problem is indeed a complicated one, some efforts have been made to treat the clothing problem quantitatively [79a].

A test method not involving a human test subject called "Electrostatic Clinging of Fabrics: Fabric-to-Metal Test" has been adopted by the AATCC [62]. In this test a fabric clamped at both ends against a stainless steel plate is charged by rubbing in a prescribed manner with a second fabric. The metal plate is then quickly positioned at an angle of 70° from the horizontal with the fabric facing downward. The lower clamp is removed and the lower end of the fabric is pulled away from the metal plate with insulated tweezers for one second and released. The opposite charge induced on the plate by the charged fabric causes it to move back and cling to the plate. After every 30 sec this declinging procedure is repeated until the fabric hangs free of its own accord, and the time required to decling is reported.

VI. ELECTRICAL RESISTANCE OF FIBERS

A. Conductors and Insulators

Electrically, materials have been classified into three basic categories—conductors, semiconductors, and insulators. Conductors are substances, such as metals, containing large numbers of essentially free charge carriers which can move through the material under the influence of an electric field. Insulators are materials in which the charged particles are bound rather strongly and do not move under normal potential gradients. In extremely high fields, however, electrons or ions can be pulled free, and insulators can then start to conduct electricity. When this happens, the field is said to exceed the dielectric strength of the material.

Semiconductors have electrical properties intermediate between conductors and insulators. Actually even the best insulators conduct some current, but the ratio of the conductance in a metal to that in an insulator is so great (10^{20} or so) that it is sufficient to say that insulators are nonconductors. Quantitatively, the ability to conduct an electric current, or rather the lack of this ability, is expressed as the resistance R which is given by Ohm's law as the ratio of the potential applied across the material V to the resulting current flow I,

$$R = V/I. \tag{26}$$

In the rationalized mks and practical systems, V is expressed in volts, I is in amperes, and the unit of resistance is the "ohm" symbolized by Ω.

The resistance of a material depends on its geometry; it increases with length L between the electrodes and decreases with cross-sectional area A. Therefore, in order to express the resistance as an intrinsic property of the material, the resistivity ρ is defined as

$$\rho = \frac{RA}{L} \tag{27}$$

which frequently is expressed explicitly in units of $\Omega\ m^2/m$. It can of course be written more simply as ohm m. For textile materials, Hearle [80] has suggested that it is more convenient to express the results in terms of the mass specific resistance ρ_m, i.e., the resistance in ohms between the ends of a specimen of unit length and mass. The relationship between ρ_m and ρ is

$$\rho_m = \rho d \tag{28}$$

where d is the density of the material expressed in kg/m^3. For most textile materials ρ_m and ρ will differ by less than 50% when d is expressed in g/cm^3.

Under certain circumstances conduction may take place along the surface of a material rather than through its volume. In this case, the appropriate intrinsic property would be the surface resistivity, ρ_s expressed as

$$\rho_s = \frac{Rw}{L} \tag{29}$$

where R is the resistance between electrodes placed on the surface of the material, w is the width of the sample (or perimeter for a cylinder), and L is the distance between the electrodes. Dimensionally ρ_s is Ω m/m, or simply ohm, and equals the resistance of a unit length of material of unit width. To avoid confusion between resistance and surface resistivity, ρ_s is frequently expressed as ohms per square since it represents the resistance measured between the opposite edges of a square piece of material.

There has been much concern about whether conduction in insulators takes place through the interior of the material or along its surface and hence whether Eq. (27) or (29) better expresses the intrinsic resistivity of an insulator. For large specimens such as blocks, films, or fabrics, it is possible to arrange electrode systems in standard ways to separately measure the volume and surface resistance [81,82]. In principle the conductance of a fiber can be separated into surface and volume components by measuring the effect of the diameter on the conductance. Conflicting results have been reported, however, with some results tending to support volume [83,84] and others surface [85,86] conductance. The discrepancies may perhaps result from difficulties in assuring that the composition, orientation, and morphological structures of the fibers do not differ from sample to sample. In general, it appears that in hygroscopic fibers volume conduction is the dominant mechanism, while surface conduction is likely to be more important in hydrophobic fibers having negligible moisture absorption [84]. When conducting surface finishes are applied, surface conductivity undoubtedly predominates.

The conductivity k is the reciprocal of resistivity and can be expressed in units of $ohm^{-1}\ m^{-1}$. The reciprocal of resistance is conductance K which is measured in units of reciprocal ohms, "siemans" in the SI system, and frequently called "mhos." To conform to the SI system, conductivity will be expressed in siemens/meter, S/m. Although there may be some disagreement about the exact values, the resistivity of conductors is generally placed in the range of from 10^{-8} to $10^{-5}\ \Omega$ m, and that of insulators from 10^{8} to $10^{18}\ \Omega$ m. The intermediate range of resistivity between 10^{-5} and $10^{8}\ \Omega$ m is considered to be characteristic of semiconductors.

The theory of electrical conductivity in organic and polymeric materials concerns the number of and nature of the charge carriers. Specifically, one seeks the answers to such questions as: Are the carriers electrons or ions? How many charge carriers are generated? How are they generated? What are their mobilities? What is the mechanism of charge transport? Answers to these questions are beyond the scope of this chapter and therefore will not be considered further here. An introduction to the theory of electrical conduction in organic materials, however, may be obtained from a number of review articles and books (for example, refs. 18,87-92).

B. Measurement of Resistance

The basic method for determining resistance is to place a potential across a specimen and measure the resulting current passing through it. The resistance would then be given by Ohm's law, Eq. (26). While the application and measurement of the applied voltage involves no difficulty, the currents encountered in textile materials and other insulators are usually so small, as low as 10^{-17} A, that only highly sensitive ammeters can detect them. It has been necessary, therefore, to employ certain voltage divider type techniques such as the circuit shown in Fig. 17 used by many workers (i.e., in refs. 93-96). With the advent of stable and accurate high-value standard resistors (up to $10^{13}\ \Omega$), it became possible to simply measure the voltage drop V_s across the standard resistor R_s using a high-impedance voltmeter, i.e., an electrometer. The input impedance of the electrometer should be at least one and preferable several orders of magnitude greater

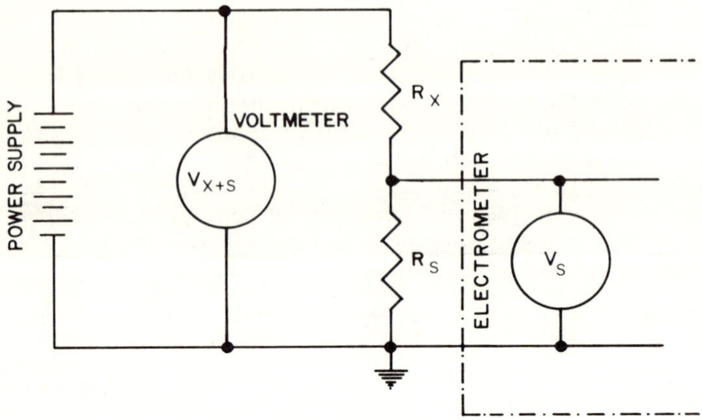

FIG. 17. Voltage divider circuit for measuring resistance of insulating materials. See Eq. (30).

than that of the standard resistor to insure that the current flowing through the two resistances is essentially equal, i.e., little current flows through the voltmeter measuring V_s. If this condition holds, the ratio of the voltages across each resistor equals the ratio of their resistances:

$$\frac{R_x}{R_s} = \frac{V_{x+s} - V_s}{V_s}. \tag{30}$$

The voltage V_{x+s} across both resistors can be measured by a conventional low-resistance voltmeter. The circuitry illustrated in Fig. 17 has been adopted by The American Society for Testing and Materials (ASTM) [82] and AATCC [97] for measuring the resistivity of insulating and textile materials. For routine measurements, several teraohmeters are available commercially. A partial listing is given in the AATCC Standards.

A very sensitive bridge-type circuit employing an electrometer as a null indicator and in which the leakage resistance is not a critical factor has been described by Hersh and Montgomery [83].

A second method commonly used to measure the resistivity of insulating materials is based on the rate of discharge (or sometimes charging) of an RC circuit. Consider the circuit illustrated in Fig. 18. It is well known from electrical circuit theory that the rate of discharge of the condenser of capacitance C through the resistor of resistance R is

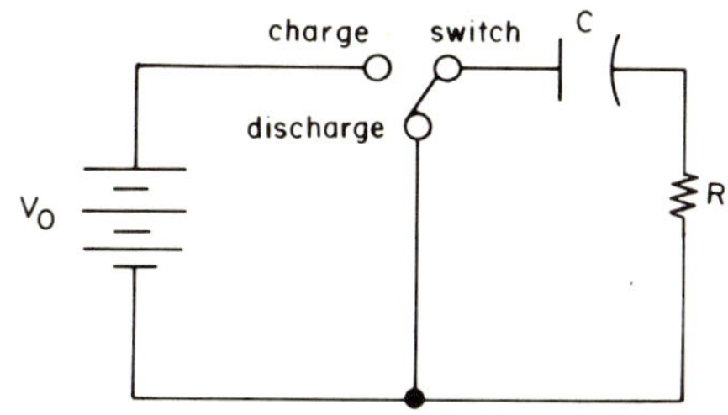

FIG. 18. Simple RC circuit for measuring resistance or capacitance by rate of charging or discharging. See Eqs. (31) and (32).

$$Q(t) = Q_0 e^{-t/RC} \tag{31}$$

or

$$V(t) = V_0 e^{-t/RC} \tag{32}$$

where Q(t) is the charge on the capacitor at time t, V(t) is the potential difference across R or C at time t, Q_0 and V_0 are the initial charge and voltage, respectively, and $Q_0 = V_0 C$, Eq. (22). Thus there is an exponential decay of charge and voltage with a time constant $\tau = RC$, i.e., the charge and voltage will decay to 1/e of their original values after a time interval equal to RC. The rate of charging or discharging of a high-resistance material should therefore depend on its capacitance and leakage resistance to ground, the valves of which will generally be that of the high-resistance material itself. Hence, if the capacitance of the material is known, the resistance can be calculated and vice versa. If similar geometries are used in a series of tests, it can be assumed that C remains constant. Therefore in many cases the time to decay to a certain value, which is proportional to R, is reported rather than the value of R itself [74,98].

The correlation between the time constant τ and R has been studied by several workers [56,63,99], who have found that Eqs. (31) and (32) hold reasonably well. The deviations that have been observed are relatively small and are generally ascribed to the occurrence of additional leakage of charge to the atmosphere through corona discharge or neutralization by ions present in the atmosphere [63,73,100]. Arridge [101] reported that the charge decay on nylon at relative humidities below 35% is exponential as predicted by Eq. (31). Above 35% RH, however, he found that the charge decay was proportional to the square root of time, that is, the charge spread along the yarn surface following a Fickian diffusion process.

C. Typical Values

The resistance of a material was given quite simply and definitively by Ohm's law as the voltage/current ratio, Eq. (26). The value of this ratio for any particular sample, however, may be dependent on a great number of variables including such factors as the potential difference (voltage applied), potential gradient (voltage/electrode spacing), time of application of potential (this dependence results from a wide variety of polarization phenomena), relative humidity, temperature, internal impurities, external impurities and additives, molecular orientation, morphological structure, voltage polarity, the direction of measurement with respect to molecular or fiber axis, geometry of sample, type of electrodes, and pressure of electrodes. It would be far beyond the scope of this chapter, and indeed would require at least a complete volume, just to cover these many facets of charge conduction in fibers, fiber assemblies, and polymeric materials and their theoretical implications. For this reason, only those aspects of resistance having the greatest impact on triboelectrification will be considered.

Of the variables just listed, RH has by far the greatest effect on fiber resistivity, and it is not at all uncommon for the resistance of fibers to vary by 10 orders of magnitude as the RH ranges from 10 to 90%. Fiber resistivity is actually dependent on fiber moisture content and generally follows a power law over a significant portion of the moisture range of the form

$$\rho = BM^{-n} \tag{33}$$

or

$$\log \rho = \log B - n \log M \tag{34}$$

where B and n are constants and M is the moisture content in percent, calculated as 100 times the weight of moisture divided by the wet weight of the fiber [80,83,96,102]. The value of the exponent n has been found to range from a low of 2.3 to a high of 18.8. However, values below 8 are rare. Because of the wide range in resistivity, only the value of the exponent is of great concern when relating resistivity with static behavior. Hence resistance and resistivities are frequently reported in powers of ten or as logarithms. To provide a frame of reference, Figs. 19 and 20 are presented to show some examples of the dependence of fiber resistivity on relative humidity and moisture content. These values should not be taken as definitive for a particular fiber type since the exact values are highly dependent on specific samples and test conditions as pointed out quite strikingly in a 1952 review by Hearle [80]. Other resistivity benchmarks are

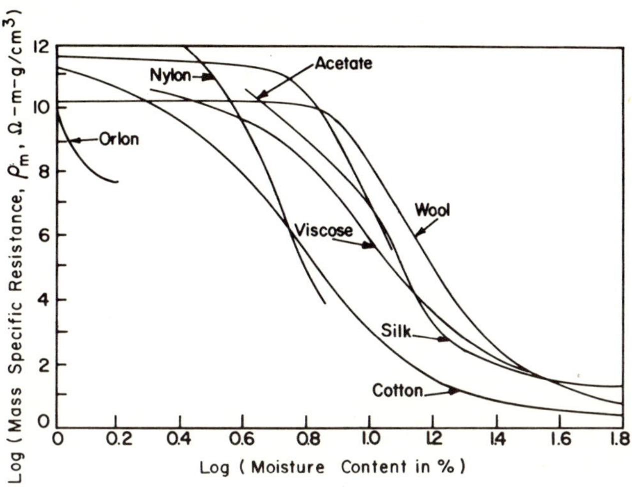

FIG. 19. Variation of resistance of fibers with moisture content. Reprinted (with change in ordinate units) from ref. 84, p. 462, courtesy of Butterworth and Co., Ltd.

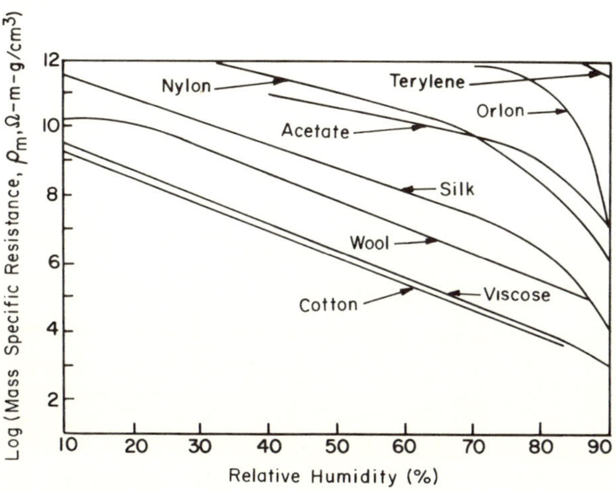

FIG. 20. Variation of resistance of fibers with relative humidity. Reprinted (with change in ordinate units) from ref. 84, p. 463, courtesy of Butterworth and Co., Ltd.

given in Table 1, which lists the values of the constants given by Eqs. (33) and (34) and the resistivity at 65% RH for a variety of fibers and fiber assemblies [83,84].

TABLE 1

Electrical Resistance of Fibers and Fiber Assemblies [83,84]

Material	Constants of Eq. (33) n	Constants of Eq. (33) (Log B - n) = log ρ (in Ω m) at M = 10%	Log ρ (in Ω m) at 65% RH
Cotton yarn	11.4	4.1	5.6
Washed cotton yarn	10.7	4.8	6.0
Mercerized cotton yarn	10.5	5.6	6.0
Washed cotton fiber[a]	7.5	5.8	3.6
Washed cotton yarn[a]	7.9	5.2	5.0
Flax yarn	10.6	4.6	5.7
Hemp yarn	10.8	5.8	5.9
Ramie yarn	12.3	5.1	6.3
Viscose rayon yarn	11.6	6.8	5.8
Washed viscose rayon yarn	12.0	7.8	6.3
Washed viscose rayon fiber[a]	5.3	6.6	4.9
Washed viscose rayon yarn[a]	10.5	7.4	5.1
Washed cuprammonium rayon[a] yarn	11.8	7.6	5.3

Fortisan yarn	11.0	6.4	6.2
Acetate rayon yarn	-	7.9	10.6
Washed acetate rayon yarn[a]	5.8	8.9	9.1
Silk yarn	17.6	7.9	8.7
Washed silk yarn[a]	16.8	8.5	7.7
Wool yarn	15.8	9.3	7.3
Washed wool yarn	14.7	10.8	8.8
Washed wool fiber[a]	12.6	9.6	6.7
Washed wool yarn[a]	14.5	10.1	7.1
Washed human hair[a]	16.8	8.5	7.0
Fibrolane yarn	16.0	9.4	8.1
Nylon yarn	-	-	8 to 11
Washed nylon fiber (30 denier)[a]	8.7	6.2	8.5
Washed nylon tire cord[a]	9.6	7.0	9.8
Orlon yarn (as received)	-	-	7.6
Purified orlon yarn	-	-	13
Terylene yarn (as received)	-	-	6.9
Purified terylene yarn	-	-	13

[a]Data from ref. 83. Other data from ref. 84, p. 464.

Like most semiconducting materials, the conductivity of fibers increases with temperature as the increasing thermal energy leads to an increase in the number of available charge carriers [84,86-88]. The conductivity of nylon yarns follows an Arrhenius relationship (linear plot of log k vs. 1/T) with an activation energy of approximately 40 kcal/mole around room temperature [86,102]. This dependence translates to about a 10-fold change in conductivity for a 10°C temperature change, i.e., 100% per degree Celsius. Other temperature coefficients reported for fibrous materials range from 1.2 to 19.9% [84,96]. Since static electrical problems of greatest importance occur near room temperature, the effect of temperature on resistivity is of much less concern than the influence of humidity. The other parameters listed above do not exert a critical influence on static properties, and they therefore will not be considered here. These factors are discussed in several of the general review articles already mentioned [80,88,90,92].

D. Resistivity Criteria for Static-Free Materials

In Section III-B-2, theoretical arguments were presented indicating that static generation could be eliminated if the surface conductivity of the rubbing elements were great enough. According to Eq. (6), the conductivity required to reduce the charge generated to one half its original value for a rubbing velocity of 100 mm/sec would be about 2×10^{-12} S/sq. That value would have to be doubled to about 4×10^{-12} S/sq to reduce the charge to 1/10, which corresponds to a resistivity of about 2×10^{11} Ω/sq. Equations (6) and (7) were derived by Medley and data obtained by him using wool roving and fabric were consistent with the values predicted by these equations.

Much experimental evidence has been accumulated since then which confirms this conclusion. Sprokel [36], working with finishing oils on cellulose acetate yarns, reached the conclusion that no charge would be detected if the surface resistivity were less than 10^{10} or 10^{11} Ω/sq. For fabrics treated with antistatic agents, Steiger [103] placed the limit at 10^{11} Ω/sq. Henry et al. [63] have measured the time for the charge generated on a number of fabrics to decay to one-half of the original value. Their results, reproduced in Fig. 21, show a linear relationship between decay time and log surface resistivity, with a charge half life of 0.01 sec for fabrics having a surface resistivity of about 2×10^{10} Ω/sq and a decay time of 0.1 sec for fabrics having a resistivity of about 2×10^{11} Ω/sq. Decay times at higher resistivities are 1 sec, 2×10^{12} Ω/sq; 10 sec, 2×10^{13} Ω/sq; 100 sec, 2×10^{14} Ω/sq; 1,000 sec, 2×10^{15} Ω/sq; and 10,000 sec, 2×10^{16} Ω/sq.

More recent studies continue to reaffirm the concept. Henry [34], in 1971, studied the risks of igniting explosive vapors by static generated on clothing. He concluded that in order to have only one chance in twenty of igniting ordinary fuel vapors (i.e., generating 0.2 mJ of

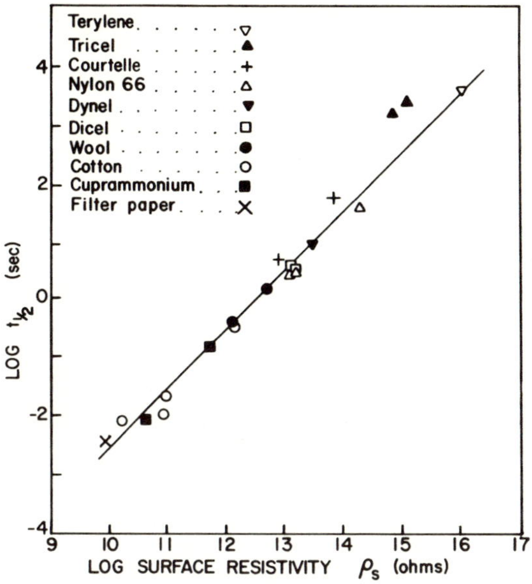

FIG. 21. Time for electric charge to decay to one-half plotted logarithmically against surface resistivity in Ω/sq. Reprinted from ref. 63, p. 76, courtesy of the Textile Institute.

energy), 10^{11} Ω/sq was the maximum surface resistivity that could be tolerated for the most sensitive fabric. With some fabrics, this value could be increased to 10^{12} Ω/sq. Such a resistance standard failed, however, for a worn cotton-polyester blend fabric. In a study of carpets, Wilson [104] concluded that carpets having a surface resistivity of less than 5×10^{11} Ω/sq would have a low static propensity. Some exceptions were noted however. Lobel [79a] concluded that charging is highly improbable if the surface resistivity is below 10^{12} Ω/sq and is practically excluded below 10^{10} Ω/sq.

Ramer and Richards [56] also studied the correlation between fabric resistivity and its ability to develop electrostatic charges. Although it is difficult to interpret their results in a manner that specifies a critical resistivity, they concluded that "there is probably no more meaningful a test than electrical resistivity for measuring the static propensity of fabrics." The correlation between resistivity and charging tendency has also been confirmed by Williams et al. [105].

From a study of charge decay rates, Wilson [99] suggested the antistatic rating scale shown in Table 2. Similar ratings have been suggested by Valko and Tesoro [106] who reviewed efforts to correlate surface resistivity with antistatic protection and concluded that the ratings shown in Table 2 generally agree with experience.

TABLE 2

Static Propensity and Surface Resistivity [99,106]

Log surface resistivity ($\log \rho_s$, Ω/sq)	Antistatic protection	
	Wilson [99]	Valko and Tesoro [106]
>13	Nil	Nil
12-13	Poor	Poor
11-12	Moderate	Moderate
10-11	Fairly good	Good
<10	Good	Excellent

In terms of volume resistivity, Hersh and Montgomery [75] reached the conclusion that a measurable electric charge cannot be generated on a fiber when the volume resistivity is reduced below 10^8 Ω m. Their rubbing speeds ranged from 10 to 120 mm/sec. For insulators in general, Cooper [107] placed the limit at 10^9 Ω m. For contact charging, Rose and Ward [46] observed little charge formation for resistivities less than 10^{10} Ω m. It is interesting to note that Bulgin [108] reported automobile tires running on grounded steel drums at 30 mi/hr (135 m/sec) generated negligible charge when the resistivity of the rubber was reduced to 10^6 Ω m. This value is consistent with the predictions of Eqs. (3-10) that a decade reduction in resistivity is required to offset every decade increase in rubbing velocity.

Thus it can be concluded that most static electrical problems arising from the use of textile materials are reduced to manageable levels if the volume resistivity of the material is lowered to 10^7-10^8 Ω m or the surface resistivity is reduced to 10^{11}-10^{12} Ω/sq. The resistivity values shown in Table 1 indicate that the hygroscopic fibers, notably cotton, viscose, and wool should have little tendency to generate static charge at relative humidities above 60-65%.

Since it is so much easier to determine the resistivity of a textile material and the measurements are much more reproducible than those of charge generation, the earliest standard tests established for characterizing the static properties of textile materials were based on resistivity values [97]. In fact, the National Fire Prevention Association standards for sheeting, film, and textiles intended for use in hospitals are based on resistivity measurements [98]. In areas where anesthetics are present, such materials must have a surface resistivity of less than 1×10^{11} Ω/sq, a value apparently arrived at as a result of the study by Henry cited above [34]. The resistivity can also be specified by the rate of decay of a dc potential imposed on the material. Another reason for the use of the resistance measurements is that they have the added advantage (or perhaps more correctly, the convenience) of being specific to the sample itself and not dependent on the choice of a second rubbing surface, as would be required for a test based on charge generation.

The reasons just cited for assessing static protection on resistance measurements are admittedly based more on expediency than on what might be a more realistic but less reproducible generation test. A number of workers [33,51,81,109] have pointed out some of the shortcomings of resistivity measurement used without additional supporting tests. Indeed, over the past several years many anomalous results have arisen, and it is becoming evident that specifications based only on resistivity measurements will not always be adequate. These anomalies usually involve composite materials such as coated fabrics [34] or fiber blends such as cotton-acrylic [33], cotton-polyester [34], or organic-metallic [104].

VII. CHARGE GENERATION

A. Qualitative Studies

1. Symmetric Rubbing of Different Materials

a. The Triboelectric Series. Much of the early work in the field of triboelectrification was concerned with establishing the so-called triboelectric series, a list of materials arranged so that a given substance will become charged positively when it is rubbed with another substance listed below it. The first such list was published by Wilcke in 1757 [6] and contained only one textile material, "wool flock," which occupied a position near the positive end of the series, just below "polished glass" and two positions above "paper" and three above "sealing wax." As will be evident shortly these items follow the same order given in many recent lists of wool, cotton, and polyethylene.

Shaw [110-113] and his co-workers studied the problem of generating a triboelectric series quite thoroughly in the 1920's and found that the relative position of materials in the series could be changed by rubbing pressures, roughness of surface, previous heat treatments, surface strain, and the presence of organic films. They also showed that it was possible to arrange some materials in a circle, that is, if material A is positive to B and B is positive to C, then A might be negative to C instead of positive as would be expected in a linear series [111]. This example is exceptional, however, and remains unexplained.

In spite of these complications Shaw concluded that a triboelectric series could be formed if sufficient care is taken in controlling the condition of the samples and the mechanics of rubbing. Some triboelectric series that have been published since the introduction of man-made fibers are shown in Table 3. Except for Grüner's series which was obtained under conditions thought to give contact without rubbing, the others all involve rubbing of some type. As discussed in Section III-A, several mechanisms may play a role in static electrification, and thus it would not be too surprising if there were some differences between triboelectric series generated by contact and those generated by rubbing. In spite of this possibility, there are few inversions between the Grüner series and the others. At least five other electrostatic series containing some polymeric materials based on contact alone have been published [18,19,39,46,49], but these include only a few textile materials. An extensive tabulation of contact electrification studies have been given by Inculet in ref. 3, pp. 102-105.

Although the various series listed in Table 3 are not completely consistent, they are more so than might be expected considering the bad reputation acquired by static electrical studies over the years. One noteworthy feature of the series listed is the consistent arrangement by chemical types. Generally the polyamides head the list at the positive end (wool, silk, nylon), the cellulosics are in the middle, and the halogenated hydrocarbons are found at the negative end. There are some inversions within these general groups from series to series, but few changes between the groups. It should be noted that the first published series of Wilcke in 1757 was consistent with this arrangement in that it had wool, paper (cellulose), and wax (polyethylene) arranged in the list from positive to negative.

b. Theories Proposed. Even though it will not be possible to treat the theories of charge transfer in detail, it would be valuable to briefly note some that have been suggested for explaining the position of materials in the triboelectric series. Coehn [116] in 1898 formulated an empirical rule stating that when two insulators are rubbed together, the one with the higher dielectric constant will become positively charged. Coehn's rule can be somewhat naively rationalized by arguing that the material having the higher dielectric constant is more easily polarized making it easier for an electron to leave, thus rendering the material positively charged. Many efforts have been made to verify Coehn's law, but suffice it to say that although some correlations have been observed, they are thought to be fortuitous [23] and hence Coehn's law has almost been abandoned [117].

Sippel [118] has established a correlation between the lightfastness and position of materials in a triboelectric series. He related static susceptibility to the tendency of the materials to lose electrons upon absorbing light quanta. Thus, materials with the poorest lightfastness are found at the positive end of the series and those with the best are at the negative end. Undoubtedly poor lightfastness is associated with the tendency of materials to polarize, which is in turn related to the dielectric constant. Still another example of the possible influence of the dielectric constant of a material on its position in the triboelectric series is given by the effect of RH on the series of Frotscher listed

TABLE 3

Some Triboelectric Series Containing Textile Materials (Positive at Top)

Lehmicke [50]	Grüner [65] (58% RH)	Ballou [114] (15% RH)	Hersh and Montgomery [75] (33% RH, 30°C)	Frotscher [115]	
				(35% RH)	(65% RH)
Glass	Wool	Wool	Wool	Glass	Glass
Human hair	Nylon 6[a]	Nylon	Nylon	Wool	Wool
Nylon yarn	Cuprammonium and viscose rayon	Silk	Viscose rayon	Polyamide	Polyamide
Nylon polymer	Nylon 6[b]	Viscose rayon	Cotton	Steel	Viscose rayon
Wool	Silk	Cordura rayon	Silk	Viscose rayon	Cotton
Silk	Nylon 6[c]	Cotton	Acetate rayon	Cotton	Acetate rayon
Viscose rayon	Terylene	Fiberglas	Lucite[e]	Acetate rayon	Poly(acrylonitrile)
Cotton	Paper	Spun ramie	Poly(vinyl alcohol)	Polyester	Steel
Paper	Steel	Acetate rayon	Dacron	Poly(acrylonitrile)	Polyester
Ramie	Glass	Dacron	Orlon	Poly(ethylene)	Poly(ethylene)
Steel	Nylon 66	Orlon	Poly(vinyl chloride)		
Hard rubber	Aluminum	Poly(ethylene)	Dynel		
Acetate rayon	Cotton	Saran	Velon[f]		
Synthetic rubber	Leather		Poly(ethylene)		
Orlon	Brass		Teflon		
Saran	Orlon				
Poly(ethylene)	Hard rubber				
	PeCe[d]				
	Synthetic rubber				

[a]Undrawn as received.
[b]Drawn as received.
[c]Drawn washed in CCl_4 3-4 hr.
[d]Chlorinated poly(vinyl chloride).
[e]Poly(methyl methacrylate).
[f]Copolymer of vinylidine chloride and vinyl chloride.

in Table 3. The presence of moisture greatly increases the dielectric constant of a material because of the high dielectric constant of water. In terms of Coehn's rule, increasing moisture content would mean a shift of a material toward the positive end of a triboelectric series, the effect actually observed for the more hygroscopic viscose and polyacrylonitrile, which move upward past cotton and polyester, respectively. In contrast, Ballou reported that little change occured in his series given in Table 3 when the RH was increased from 15 to 60%. The only major change was a shift in Fiberglas to a position between wool and nylon, a move rather difficult to explain on the basis of increasing moisture content and dielectric constant since glass has such a small moisture regain.

The third approach taken to explain the position of materials in the triboelectric series is based on the band theory of solids. In essence these theories accept the basic premise of the Volta-Helmholtz hypothesis, outlined in Section III-A, that an equilibrium transfer of electrons takes place upon contact such that the Fermi levels of the contacting bodies are equalized. Thus in order to predict the polarity of charge transfer resulting from contact, it is necessary to know the value of the Fermi levels in the materials. The first attempt to assign such values to insulators in order to explain their position in the triboelectric series was made by Hersh and Montgomery [15]. Some quantitative predictions were made by VanOstenburg and Montgomery [16]. Harper [24], however, has disputed this approach since by using a somewhat different assumption he calculates that only negligible electron transfer can take place. Consequently Harper suggests that triboelectrification must result from the transfer of ions. The different approaches taken in these last two papers has been analyzed by Seanor [17].

Although the exact transfer mechanism is in dispute, contact potentials between metals and insulators do exist and can be measured. These measurements are usually made by plotting the charge transferred in insulator-metal contacts against the work function of the metal (which corresponds to its Fermi energy). An example is shown in Fig. 22. The value of the metal work function which results in no charge transfer is taken as the contact potential of the insulator. Results have been reported for glass [119,120], polyethylene [119], nylon 66 [19,48], PET [19], and a number of other polymers [17-19]. A triboelectric series can be formed from such a list.

2. Symmetric Rubbing of Similar Materials

When similar materials are stroked under symmetrical conditions such as those illustrated by rub-type III in Fig. 16, the charge transferred is small and of random sign [46,117]. In some instances molecular orientation has been reported to influence the position of chemically identical materials in the triboelectric series. For

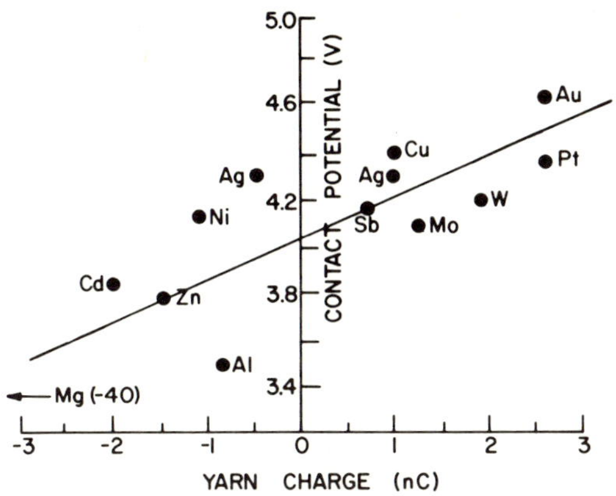

FIG. 22. Charge generated on undrawn nylon 66 20 denier monofilament by contact with various metals as a function of contact potential (contact potential of gold taken to be 4.70 V). Reprinted (with change in abscissa units) from ref. 48, p. 1314, courtesy of the Institute of Physics.

example, Grüner [65] has reported that an undrawn nylon 6 is positive to a drawn sample. In contrast, Hersh and Montgomery [75] examined a series of nylon 66 filaments drawn from the same melt to ratios of 1 through 6 in unit increments and found that in symmetric rubbing, the undrawn nylon was always negative to a drawn one. The drawn nylons of draw ratio 2 through 6, however, were triboelectrically indistinguishable from each other. Similar measurements made on Dynel and Dacron polyester yarns showed no differences between drawn and undrawn samples.

3. Asymmetric Rubbing of Similar Materials

Electric charges can be generated by contacting similar or even chemically identical materials if the nature of their surfaces is different or they are rubbed asymmetrically. For example, it has long been known that polishing or etching can render bodies of the same material sufficiently different so that a charge can be generated between them [110,121]. Examples of asymmetric rubbing are illustrated in Fig. 16 for two cylinders in which first the entire length of the top cylinder is rubbed over one spot on the bottom (90-0) and the converse in which a single spot on the top cylinder is stroked over the entire length of the bottom (0-90). Asymmetric rubbing of this type on like materials has been studied by Shaw [110], Shaw and Hanstock [112], Grüner [65], Hersh and Montgomery [75], Ballou [114], and Henry [22,49].

Shaw and Grüner both found that the specimen rubbed on a single spot becomes negative. Henry noted that the sample rubbed on one spot usually becomes negative, while Hersh and Montgomery found that it usually becomes positive. Henry also observed the occurrence of both polarities and noted that reversals sometimes take place with changing relative humidity. Shaw explained his results on the basis of differential surface strain, while Grüner suggested a differential heating effect. He proposed that the surface ions on the rubbed materials diffused or migrated in the thermal gradient produced by the rubbing from the hot surface to the colder surface. The sample that was rubbed in one spot will, of course, be heated to the higher temperature. Grüner assumed that most of the ions present were H^+ and OH^- derived from the water on the fibers, and since the mobility of H^+ is greater than the mobility of OH^-, there would be a net migration of positive charge to the colder surface rubbed along its entire length.

Henry used a similar argument but in a more general way by reasoning that the polarity would depend on the most mobile ion present, which might be positive or negative. The presence of a thermal gradient was also employed by Hersh and Montgomery to explain their results, but in this case, the migrating charged particles were assumed to be electrons. The only two instances observed by the latter authors that could not be explained by the transfer of electrons involved the two most hygroscopic materials used, cellulose acetate on cellulose acetate and viscose rayon on viscose rayon. The hydrogen ion migration mechanism hypothesized by Grüner might have predominated in these two cases. Ballou ascribed charge formation in asymmetric rubbing to a difference in dielectric constant in the axial and transverse directions of the fiber.

Another example of charging by asymmetric rubbing of chemically identical materials is given by wool and human hair. Martin [25] and Hersh and Montgomery [122] have shown that when the root ends of these animal fibers are lined up in opposite directions and rubbed, essentially no charge is transferred. If the tips are aligned in the same direction, however, the polarity of the charge generated will be positive when a fiber is pulled in one direction and negative when pulled in the opposite direction.

4. Asymmetric Rubbing of Different Materials

Chemical differences normally overwhelm the thermal effects of asymmetric rubbing. For certain combinations of materials close together in the triboelectric series, however, the direction of charge transfer may be reversed by the thermal gradient produced by asymmetric rubbing. Examples have been reported for polyethylene and velon, polyethylene and Dynel [75], and cellulose acetate and glass [22].

B. Quantitative Studies

1. Historical Review

It would be quite a difficult task to describe concisely and without distortion the many quantitative studies of charge generation on fiber surfaces that have been reported. Since a wide variety of materials and test procedures have been employed, many of the studies are not compatible and/or comparable with each other. Likewise it would not be feasible to present all the quantitative material available because there is simply too much of it. Rather only a representative picture of the variables involved will be offered along with a guide to some of the relevant literature where the experimental details may be consulted in order to eliminate the distortions that inevitably arise when summarizing the work of others. In general, the material presented will be essentially limited to the triboelectrification of fibers and fiber assemblies and will concentrate on work reported after 1945 when data covering the man-made fibers first became available in significant quantities.

First, however, to provide a frame of reference for the work of the past half century, consider briefly some of the studies made prior to 1930. The earliest systematic quantitative investigation of frictional electrification was made by Peclet [123] who studied the influence of velocity, pressure, extent of surface of contact, state of surface, thickness of bodies, nature of the materials, and in a later work [124], the effects of carbon dioxide instead of air for the surrounding atmosphere. Later studies of a similar nature were made by Owen [125], Jones [126], McClelland and Power [121], French [127], Macky [128], Kluge [129], and Wolf [130]. The apparatus used by all these investigators was basically the same. An arm covered with one of the materials to be rubbed was held at fixed pressure against a rotating wheel covered with the other material. The wheel was turned for a fixed length of time or given a certain amount of kinetic energy to be expended during the rubbing, and the charges generated were then measured. In all these studies the reproducibility was good.

It would be impractical to bring out all the facts and speculations presented in these papers, but the following general results will be mentioned. Peclet found that charge was independent of velocity, pressure, and thickness of the substances rubbed. The charge was found to depend on the nature and curvature of the materials. Changing the atmosphere from air to carbon dioxide did not change the amount of charge generated. Owen, Jones, and French found that charge increased with energy expended in rubbing, but that a maximum limiting charge was reached, which was independent of pressure. French reported that charge decreased with increasing relative humidity. McClelland and Power found that pressure, velocity, temperature, relative humidity, and surrounding gas pressure affected the charges produced in a manner that depended on the polarity of the charge transfer. Macky confirmed the earlier work of Peclet and found also that the charge generated decreased with decreasing atmospheric pressure. Kluge found that in a vacuum of the order of 10^{-5} mm of mercury, the charge was greater than at atmospheric pressure, and upon heating, the charge increased further. In general, the authors agreed that no charge transfer takes place at zero rubbing velocity (that is upon mere contact with no rubbing). The fact that rubbing was necessary to produce a measurable charge transfer indicates that the sensitivity of their measurements was not too high.

Though these studies were all quite carefully made, the significance of the results that charge is independent of pressure, velocity, and energy expended in rubbing is uncertain since it is highly probable that in most cases the materials rubbed were charged to a maximum potential limited by air breakdown. The truth of this supposition was indicated by Macky who found that the maximum charge developed on rubbing was proportional to the breakdown strength of the surrounding atmosphere. Medley [131], studying the electrification of wool rubbed on nylon and polyethylene, was perhaps the first to point out that some earlier quantitative triboelectric studies may have been of doubtful value because atmospheric gaseous discharge may have limited the charge buildup in many instances. The surprisingly small amount of charge transfer required to cause atmospheric breakdown was mentioned in Section II and will be discussed further in the next section.

2. Magnitude of Charge Transferred

In Section III-B-1, it was noted that the dielectric strength of air limits the maximum surface charge density σ that can be maintained on a planar surface to about 30 $\mu C/m^2$. This value of charge density is obtained from Eq. (16) by setting $E = 3{,}000$ kV/m with $\epsilon_0 = 8.85 \times 10^{-12}$ $C^2/N\text{-}m^2$. Since the electronic charge q_e is 1.6×10^{-19} C, the maximum charge density is created by a deficiency of $\sigma/q_e = 3 \times 10^{-5}$ $C/m^2 \div 1.6 \times 10^{-19}$ C/electron $= \sim 2 \times 10^{14}$ electronic charges per m^2 assuming the atoms are singly charged. Taking the surface area occupied by an atom to be approximately 5 $Å^2$, it is evident that there are about 2×10^{19} surface atoms per m^2. Hence, only $(2 \times 10^{14})/(2 \times 10^{19})$ surface atoms, i.e., one out of every 10^5, need be charged to produce the maximum electrification that can be sustained on a planar surface in air.

If a crystal of sodium chloride could be fractured such that chloride ions remained on one face and sodium ions remained on the other, the charge formed would be 10^5 times that which could be sustained in air since every surface atom would then carry a charge. When two materials having a contact potential of 1 V are separated by 1 Å, the charge transferred according to Eq. (16) would be $\sigma = \epsilon_0 E = \epsilon_0 V/d$ (8.85×10^{-12})

$C^2/N\text{-}m^2 \times (1\ V/10^{-10}\ m) = 0.1\ C/m^2$, about 3,000 times greater than that which can be finally sustained in air. The excess is quickly dissipated by mechanisms (1) and (2) listed in Section III-B-1. The highest charge density that has ever been measured was about 1,000 times the maximum value for a flat surface [45]. This value was obtained by contacting two different crystal faces of quartz with each other. By using a technique designed to reduce leakage, Medley [47] was able to form a charge about 50 times greater than the maximum for a flat sheet by contacting mercury with nylon, mica, and alkathene films.

The charge density which can be maintained on single fibers and roving of small radius is greater than that which can be formed on flat sheets since the electric field surrounding a cylinder decreases with distance [Eq. (15)] [132]. For example, Medley [27] reports that a maximum charge density of 150 $\mu C/m^2$ can be formed on single fibers such as 20 μm diameter nylon (3 denier). This finding is essentially confirmed by Liddy and Downes [132a] who observed maximum charge densities of about 200 $\mu C/m^2$ on 20-μm diameter wool fibers. Although the surface charge density may be high, the total charge is of course small because of the small radius. The theoretical maximum charge density on a planar surface of 30 $\mu C/m^2$ has frequently been observed on relatively small areas; however, the maximum charge density that can be produced on areas greater than 1 or 2 cm^2 is about half this [133]. Charge densities of 1/300 of the maximum (0.1 $\mu C/m^2$) can cause problems in textile processing [134].

A summary of the charge densities corresponding to the situations just described is given in Table 4.

One consequence of the fact that the net charge transferred is generally limited by atmospheric breakdown is that the amount of charge transferred will not depend on the distance between the materials in a triboelectric series [49,75]. As long as there is any difference in positions, the charges transferred in all but the lightest contacts are of the same order of magnitude limited by dielectric breakdown. The additivity of contact potentials, however, has been verified for metals by Harper [26].

3. Influence of Experimental Variables

a. Reproducibility. Several studies have considered the reproducibility of the charge generated in triboelectric experiments [26,28, 44,45,49,69,75,123,135-137]. In general, the reproducibility for repeat runs on the same two materials is about ±5% when good control of the mechanical and ambient conditions is maintained. For different specimens of the same material, the variation between samples generally runs about ±20%.

TABLE 4

Magnitude of Charge Transferred Under Certain Limiting Conditions

Limiting conditions	Fraction of surface atoms charged	Elementary charges per m^2	Charge density ($\mu C/m^2$)
Separation of double layer	1	2×10^{19}	3,000,000
Highest ever observed (quartz/quartz) [45]	1/100	2×10^{17}	30,000
Highest ever observed on polymer under special conditions [47]	1/2,000	1×10^{16}	1,500
Highest on fiber in air (20 μm diameter, ~3 denier) [27]	1/20,000	1×10^{15}	150
Limiting charge on flat sheet in air	1/100,000	2×10^{14}	30
Charges frequently observed on textiles [133]	1/200,000	1×10^{14}	15
Charges found to be critical in processing [134]	1/30,000,000	7×10^{11}	0.1

The variability just cited refers to repeat runs on samples after they have reached a "steady state" which might take anywhere from none to sometimes as many as 60 preliminary rubs [75]. If rubbing then continues for thousands of rubs, Levy et al. [76] have shown that the charge generated slowly drops for 5,000 rubs or so until a level is reached which then remains constant for several thousand more rubs. An example is shown in Fig. 23.

b. Effect of Atmosphere. The most common atmospheric variable that has been studied is relative humidity [27,29,36,52,64,70,73,75,114, 134]. Some typical results are given in Fig. 24 for charge generated on card slivers leaving the calender rolls as a function of moisture regain. The charge drops rather dramatically as the moisture content increases and the resistivity decreases. Although these authors noted that the same maximum charge was generated on all these fibers, they did not recognize that the limit reached might be controlled by

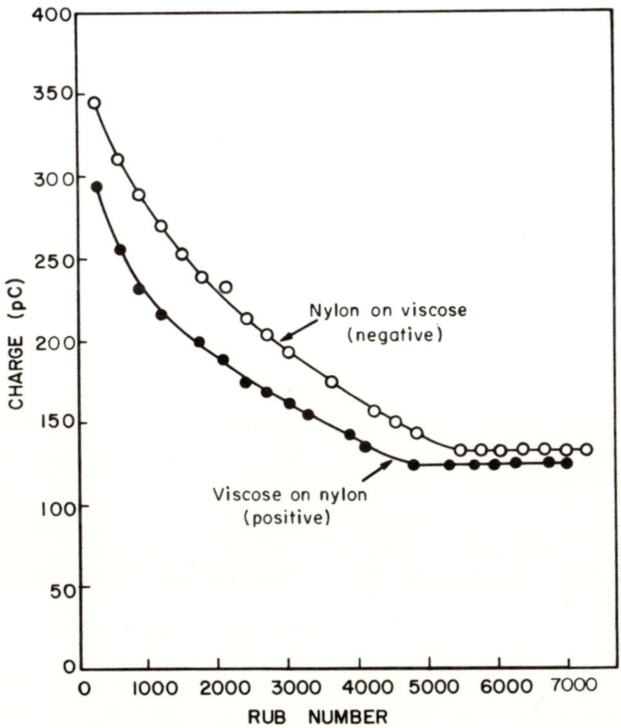

FIG. 23. Charge generated on rubbed fiber as a function of rub number. Reprinted from ref. 76, p. 902, courtesy of the Textile Research Institute.

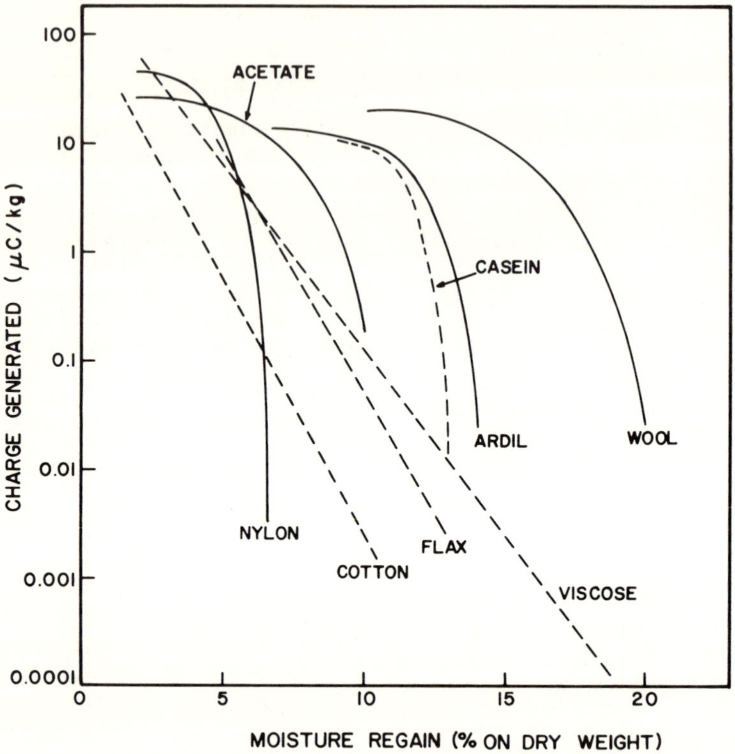

FIG. 24. Electric charge on fibers after carding at various moisture regains. Reprinted (with change in ordinate units) from ref. 64, p. T706, courtesy of the Textile Institute.

the onset of atmospheric breakdown. Rather they suggested that the maximum might be characteristic of the carding process or the carding engine itself.

In all except one of the reports cited above, charge generation increased with decreasing relative humidity and moisture content of the fiber as would be expected from the associated decrease in conductivity. Sereda and Feldman [52], however, observed that the charge generated on four fabrics (cotton, wool, nylon, and polypropylene) rubbed against a nickel-plated brass surface went through a maximum at relative humidities corresponding to the moisture content at which a monomolecular layer of water is formed on the fiber surface. Their results for wool, shown in Fig. 25, can be contrasted with the results reported by Medley in Fig. 3, which shows a maximum charge at the lowest humidity examined (unspecified, but less than 10%). None of the other authors reported results at low enough RH to show the

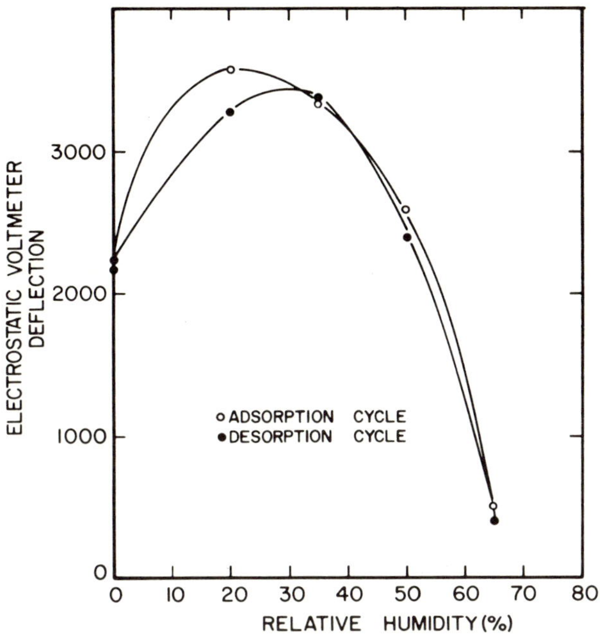

FIG. 25. Maximum electric charge generated on wool fabric after rolling 10 times with a nickel-plated brass roller at different relative humidities. Reprinted from ref. 52, p. T294, courtesy of the Textile Institute.

possible existence of a charge maximum at the RH reported by Sereda and Feldman. These authors felt that the initial buildup to a monomolecular layer of water aids in the transfer of charged particles from one surface to the other. After the monolayer of water is formed, they hypothesized that additional moisture serves to decrease the charge transferred as a consequence of the increased conductivity of the fiber.

Other atmospheric variables that have been examined are composition [27,124,128,131] and pressure [27,121,137]. The usual conclusion reached in these studies is that all observations could be explained by the effect of the change on the breakdown strength of the atmosphere.

c. Effect of Temperature. The influence of ambient temperature on charge generation has been examined only in a limited range close to room temperature. A typical result is shown in Fig. 26 in which the charge decreases with increasing temperature. The effects produced by small ambient temperature changes are generally interpreted as the result of changes in conductivity [117]. Undoubtedly, local heating at the contact region also influences both the character of the

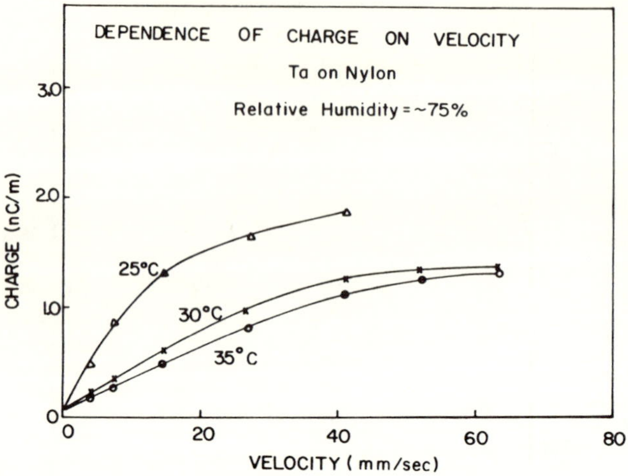

FIG. 26. Net charge transferred to 0.20-mm diameter nylon 66 monofilament rubbed symmetrically with 0.20-mm diameter tantalum wire as a function of velocity at various temperatures. Reprinted (with change in ordinate units) from ref. 28, p. 976, courtesy of the Textile Research Institute.

materials at that point and the thermal diffusion of charges from the hotter to the cooler surface by the mechanism described in Sections III-A and VII-A-3.

An example of the effect produced when the temperature of one of the contacting materials is varied has been given by London [138] and is illustrated in Fig. 27 for polyester fiber passing over a ceramic cylinder. The charge generated on the fiber (measured as electric field strength) ranges in sigmoidal fashion from -0.1 kV/m to -6.5 kV/m as the temperature increases from 27 to 93°C. London also reported that the charge generated on a polyester yarn rubbing on a copper tube at 24°C was -1.3 kV/m. When the tube was cooled to 0°C, the polarity of the charge on the fiber reversed and produced a field of +1.5 kV/m. These changes are consistent with those expected if the thermal diffusion of electrons is controlling, i.e., as the temperature of the ceramic or metal increases, the tendency to drive off electrons increases, thus changing the polarity of the polyester fiber from positive to negative.

McClelland and Power [121] reported the opposite effect, however, when rubbing metals on silk. Heating the metal to 100°C led to a reduction in the algebraic value of charge formed on the metal, i.e., a reduction in charge when the metal became charged positively and an increase in charge when it was charged negatively.

d. Effect of Velocity. A theoretical treatment of the influence of rubbing velocity on charge generation has been given in Section III-B-1,

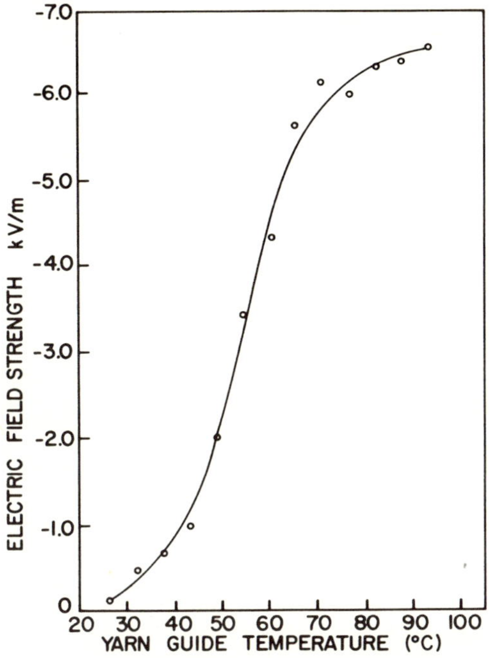

FIG. 27. Static charge generated on 250-denier 50-filament Dacron polyester yarn passing over 12.7-mm diameter Heanium rod as a function of rod temperature [138].

and some selected results are shown in Figs. 2 through 5 and Fig. 26. These and other studies in which the effect of velocity was measured [68,70,75,103,139] at moderate speeds (of the order of 100 mm/sec) indicate that charge tends to increase with velocity until a limiting value is reached. Although in some cases saturation may occur as a consequence of dielectric breakdown of the atmosphere (ref. 24, pp. 334-336; ref. 75), it is likely that the plateau results most often from the mechanism described in Section III-B-1, i.e., a decrease in the amount of charge lost by leakage back to the contact point because of the shorter time available. In contrast, Levy et al. [76] have reported one situation in which charge generation is decreased as the rubbing velocity increases. After a pair of fibers are rubbed for several thousand times at a fixed velocity and then discharged, the charge measured in the next single rub (made at the same test velocity) is found to decrease as the testing velocity is increased. Their explanation is that because of the higher temperatures reached at the higher rubbing velocities, more material is transferred from one fiber to the other, which results in a reduction in the charge transferred after extensive rubbing. At high speeds ranging from about 0.1 to 10 m/sec, however, the dependence of charge on velocity is very erratic, sometimes going through maxima and minima and even changing polarity [71,138].

e. Effect of Force Between Bodies, Energy Expended in Rubbing, and Area Rubbed. Since 1945 there have been only a few studies of charge generation as a function of the normal force between the contacting bodies [28,75,136]. Much more data has been accumulated in which the tension of a yarn wrapped around a friction post is varied [36,70,71,138,140]. In such an experimental arrangement, the yarn tension is not constant but increases as the yarn passes around the post, according to the well-known capstan formula. The resulting normal force does, of course, increase as the tension in the yarn increases, but it is not constant along the yarn path. Another approach taken is to correlate charge generation with the energy expended during rubbing [76,141-143].

It has generally been observed that the charge generated q increases with increasing normal force F between the contacting bodies. In some cases, a maximum value is reached that might be associated with the onset of atmospheric breakdown. Rose and Ward [46] measured the charge transferred when spheres of insulators were pressed against metal disks without rubbing and concluded that the charge was directly proportional to the area of contact between the sphere and disk. The area, of course, increased with increasing pressure. The most careful study involving the rubbing of filaments was made by Montgomery et al. [136] who measured the charge generated on undrawn nylon filaments varying in diameter D made from the same polymer and run at a velocity high enough to minimize charge dissipation by conduction. Their results for tantalum wire rubbing on nylon are shown in Fig. 28. From a log-log plot of these data for q vs. (FD), it becomes evident that the charge transferred is proportional to the square root of the product of the normal force between the filaments and the diameter of the nylon. For poly(ethylene) rubbed on these same nylon samples, the charge transferred was proportional to the square root of the normal force, but independent of the diameter. Because of the nature of the yarn geometry employed, these findings were consistent with the hypothesis that the charge transferred is proportional to the true area of contact made with the object having the greatest resistivity and tends to confirm that part of the Volta-Helmholtz hypothesis which states that rubbing serves merely to increase the area of contact between the bodies. As shown by many studies of friction and elasticity [144,145], pressure will, of course, increase the true area of contact (i.e., the area over which interpenetration of molecular fields occurs) between the bodies even though the apparent area of contact remains constant. When filaments are rubbed at constant pressure, it has been shown that the charge increases linearly with the length of yarn rubbed, again indicating that the charge is proportional to contact area.

Charge measurements made by pulling a yarn around a pin also show that the charge increases with the frictional force [36,70,71]. In

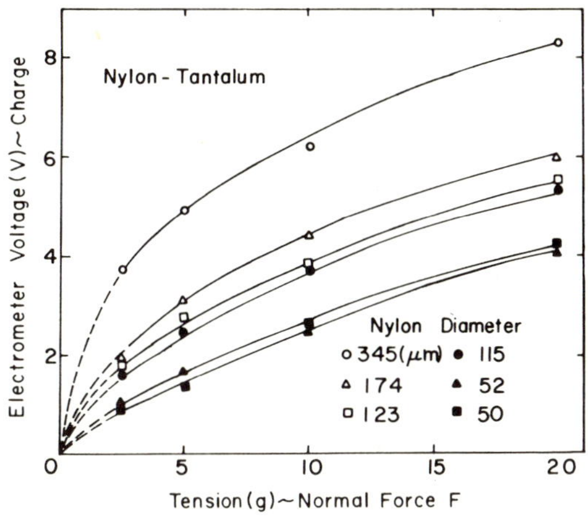

FIG. 28. Electric charge transferred (in terms of electrometer voltage) as a function of normal force (proportional to filament tension) to nylon 66 filaments of various diameters rubbed against a 315-μm diameter tantalum wire. Reprinted from ref. 136, p. 27, courtesy of the Textile Research Institute.

some instances, it has been observed that the positive charge on a fiber passing over a metallic friction pin can first increase as the output tension is increased by increasing the angle of wrap around the pin, pass through a maximum and then become negative [70]. The opposite case in which a fiber, first charged negatively, reverses polarity and becomes positive with increasing tension was not observed. These results were explained on the basis of solid state band theory [13].

Several workers have tried to relate the charge generated with energy W expended in rubbing. In 1915 Jones [126] reported that the relation was exponential:

$$q = q_m(1 - e^{-bW}) \tag{35}$$

where q_m is the maximum charge obtained at high values of W, and b is a constant. Levy et al. [76] in 1958 found the charge to increase with the square root of the energy expended, and in 1964 Cunningham [141] reported the relationship to be linear. Finally in 1966 Cunningham [142] studied the problem over a much greater range of energy levels and concluded that the q-W relationship was indeed exponential as expressed by Eq. (35). A typical result is shown in Fig. 29 for a polyester film rubbing on hardened, stainless steel. An exponential curve can, of course, be fit over a limited range at the low end by a linear relation and at intermittent levels by a square root relation, so

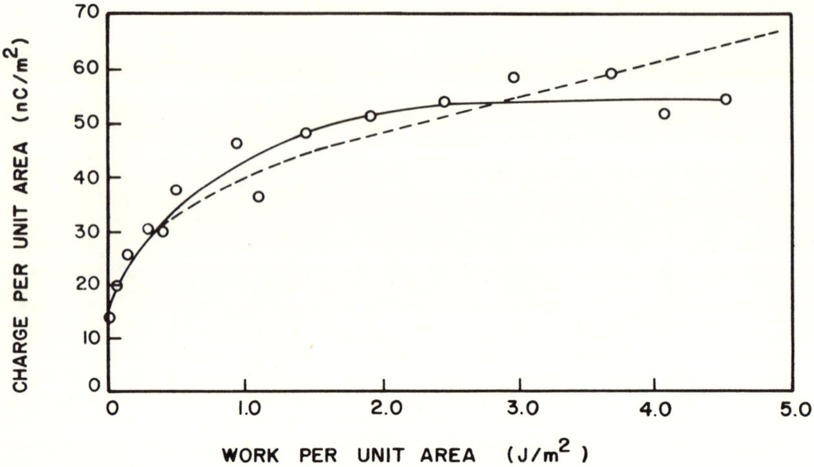

FIG. 29. Electric charge generated on a belt of polyester film slipping on a hardened stainless steel roller as a function of work expended. The solid line is the best exponential fit and the dashed line is the best square root fit. Reprinted from ref. 142, p. 1735, courtesy of the American Institute of Physics.

in effect the earlier results reported by Cunningham and Levy were actually different parts of the same phenomenon. In Fig. 29 the best exponential fit is given by the solid line and the best square root fit is given by the dashed line.

f. Effects of Electric Field.

1. Influence on Charge Transfer. The charge generated can be altered by the application of an external electric field. Depending on the direction of the field, the charge transfer can be increased or decreased, and if high enough, the polarity of the charge transfer might even be reversed. In fact, Murasaki [146] has suggested the use of such a technique as a method for eliminating static problems. An example showing how the charge generated during rubbing has been reduced or eliminated by adjusting the potential of one of the materials has been reported by Hersh [4] for yarns passed over metal and ceramic pins at varying velocities. Figure 30 shows that the charge varies with the dc potential applied to the pin and that conditions can be selected that prevent charge from being generated on the textile material. Other examples have been given by Murasaki [146] for textile fabrics passing between two rollers, one grounded and the other at a varying dc potential, as well as for other configurations.

It has also been demonstrated that the presence of an electric field can increase, neutralize, or reverse the charge picked up by particles of ebonite, sand, or glass rolling down a metal slide [55,147]. Similar

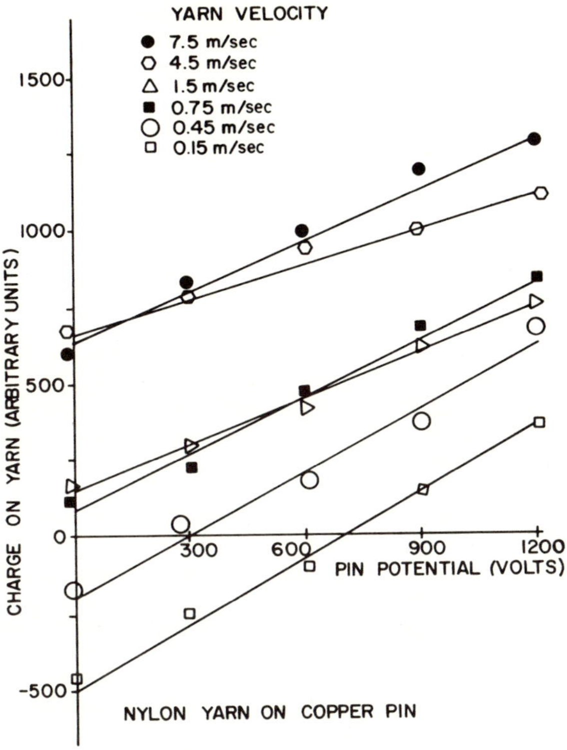

FIG. 30. Electric charge generated on 200-denier 34-filament nylon 66 yarn after passing over a copper pin maintained at various dc potentials. Reprinted (with change of units) from ref. 4, p. 32, courtesy of Marcel Dekker, Inc.

effects have been shown for contacts that do not involve rubbing such as those between nylon film and mercury [47], gold and nickel contacting glass and quartz [148], and for measuring the contact potential of gold and chromium with brass [149]. The influence of an applied potential on the electric field within a tank filled with water vapor has also been reported [150].

2. Behavior of Fibers in Electric Fields. Many applications of electrostatics in textiles are based on controlling the orientation or motion of fibers in an electric field. As mentioned in Section I, the most successful applications have involved flocking and spun bonding. The influence of the electric field on the orientation, velocity, and momentum of fibers during flocking has been discussed by Lion [151], Kassack et al. [152], and Bershev [152a]. In the spun-bonding process, a potential is applied to the fibers as they leave an aspirator jet after cold drawing in order to keep them separated as they move to the

collector belt to form the web [153]. An electric field is also maintained between the jet and the belt to attract the fibers to the belt. The separation of fibers by electrostatic charging has been studied quantitatively by Uchiyama et al. [43a].

Other potential applications that have not yet achieved any significant commercial use include cleaning and separation of fibers [61,154] and control of fibers during open-end spinning. In the latter application, electrostatic forces can be used to transport, orient, and condense fibers [155], and perhaps to twist them [156]. A comprehensive survey and study of the behavior of fibers in an electric field has been given by Gregoric [157].

g. Fiber Blends. By blending fibers from opposite ends of the triboelectric series, it is possible to neutralize the net electric field measured at some distance from the surface of the textile material [114,158,159]. The most extensive work in this area has been carried out by Ballou [114]. He was able to demonstrate an essentially linear relationship between the magnitude of charge generated and the composition of binary blends of nylon 66 with either poly(ethylene terephthalate) or an acrylic fiber rubbed against cotton, wool, and a chrome-plated surface. An example is given in Fig. 31. This technique has been suggested many times as a method for eliminating the detrimental effects of static electrification [40]. As Ballou pointed out, however, this method offers only a very limited solution to the static problem because as illustrated in Fig. 31, the composition of the optimum blend can vary with the specific rubbing surface and, no doubt, with other rubbing conditions.

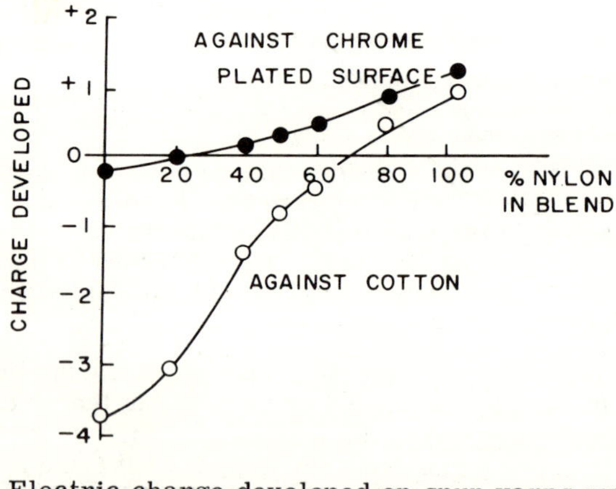

FIG. 31. Electric charge developed on spun yarns made from nylon 66/acrylic fiber blends after contacting chrome-plated and cotton surfaces. Reprinted from ref. 114, p. 153, courtesy of the Textile Research Institute.

One apparent exception to this generalization has recently been claimed by Crawley [160] who reported that static charging in carpets can be reduced to an acceptable level by blending yarns of nylon (from the positive end of the triboelectric series) and polypropylene (from the negative end). Results seem satisfactory for contact with both leather and synthetic shoe soles. If so, this development represents the first successful application of the blending technique as a practical remedy for a static problem in the textile area.

The blending of fibers having different tendencies to generate static electricity usually produces a material having intermediate charging characteristics. Such a result is expected because of the correlation between conductivity and charging rate discussed in Section VI-D. The conductance of a fiber blend is simply the sum of the conductance of each of the fibers in the cross section, and the volume conductivity would thus be the sum of the conductivities of the components weighted by their volume fractions (assuming the composition is uniform along the cross section). Hence, the charging characteristics of the blend should also show the same linear averaging of its components. When one of the components is a metal, the conductance goes up dramatically even if only a small amount of metal fiber is included. Thus fabrics and carpets containing from 0.2 to 2% of metal fibers have been reported to be sufficiently static resistant for many end uses [2,79,161, 162].

As a result of the increasing use of cotton-polyester blends, several studies of the electrical characteristics of these blends have appeared. They generally report, as expected, that the addition of the higher conducting cotton to polyester moderates the severe electrification characteristics of the polyester. Care must be exercised, however, since the cotton can be lost with wear or washing and a material originally safe for use in certain critical areas such as hospital operating rooms and refueling depots may become unsatisfactory with use [34]. Other problems associated with the electrical properties of textile materials made from fiber blends were mentioned in Section VI-D.

h. Effect of Antistatic Agents. In order to eliminate the many static problems that arise in processing and use, much effort has been made to increase the conductivity of textile materials. These efforts have concentrated on developing suitable internal (e.g., in refs. 163-167) and external (e.g., in refs. 106,168) antistatic agents, coating the fiber with conductors such as silver or carbon [169,169a], and incorporating fine metal fibers in the textile material [2,79,161,162]. Until about five years ago, most successful antistatic treatments were based on the application of conducting surface finishes to the fibers in order to increase their surface conductivity. These finishes were generally of two types: hygroscopic materials such as polyethylene oxide adducts of glycols and amines or ionized materials such as quaternary ammonium compounds or surface-active sulfonates and phosphate esters. An example of the effects of finishes of this type was given in Figs. 4 and 5.

Since many comprehensive studies and excellent reviews have appeared over the past 20 years describing the mechanism and use of surface antistatic agents (e.g., see refs. 106,168), this aspect of the static problem will not be considered further here. The so-called permanent antistatic fibers, containing internal antistatic agents, also depend on increasing conductivity for their static protection. Recent reviews have been given by Hersh [4] and Grady [86].

REFERENCES

1. D. Finlayson, J. Textile Inst., 48, P55 (1957).
2. R. R. Varin, Mod. Textile Mag., 52(2), 28 (1971).
3. A. D. Moore, ed., Electrostatics and Its Applications, Wiley-Interscience, New York, 1973.
4. S. P. Hersh, Polymer-Plastics Technol. Eng., 3(1), 29 (1974).
5. S. M. Edelstein, Am. Dyest. Rep., 42, 70 (1953).
6. J. C. Wilcke, Disputatio Physica Experimentalis de Electricitatibus Contrariis, Academy of Rostock, 1757.
7. P. S. H. Henry, Brit. J. Appl. Phys. Suppl. No. 2, 4, S6 (1953).
8. L. B. Loeb, Science, 102, 573 (1945).
9. E. Perucca, Z. Physik, 51, 268 (1928).
10. H. Helmholtz, Ann. Phys. Chem., 7, 337 (1879).
11. J. Frenkel, Phil. Mag., 33, 297 (1917).
12. J. Frenkel, J. Phys. (USSR), 5, 25 (1941).
13. V. E. Gonsolves, Textile Res. J., 23, 711 (1953).
14. F. A. Vick, Brit. J. Appl. Phys. Suppl. No. 2, 4, S1 (1953).
15. S. P. Hersh and D. J. Montgomery, Textile Res. J., 26, 903 (1956).
16. D. O. VanOstenburg and D. J. Montgomery, Textile Res. J., 28, 22 (1958).
17. D. A. Seanor, in Electrical Properties of Polymers (K. C. Frisch and A. Patsis, eds.), Technomic, Westport, Conn., 1972, pp. 37-58.
18. D. A. Seanor, in Polymer Science, A Materials Handbook (A. D. Jenkins, ed.), Vol. 2, North-Holland, Amsterdam, 1972, pp. 1187-1280.
19. D. K. Davies, Brit. J. Appl. Phys. (J. Phys. D), 2, 1533 (1969).
20. J. A. Medley, Nature, 171, 1077 (1953).
21. P. Boning, Z. Angew. Phys., 8, 516 (1956).
22. P. S. H. Henry, J. Textile Inst., 48, P5 (1957).
23. L. B. Loeb, Static Electrification, Springer-Verlag, Berlin, 1958.
24. W. R. Harper, Contact and Frictional Electrification, Oxford University Press, London, 1967.
25. A. J. P. Martin, Proc. Phys. Soc., 53, 186 (1941).
26. W. R. Harper, Proc. Royal Soc., A205, 83 (1951).
27. J. A. Medley, J. Textile Inst., 45, T123 (1954).
28. R. G. Cunningham and D. J. Montgomery, Textile Res. J., 28, 971 (1958).

29. J. A. Medley, Brit. J. Appl. Phys. Suppl. No. 2, 4, S23 (1953).
30. G. R. Ward, Am. Dyest. Rep., 44, 220 (1955).
31. W. Lobel, W. Schumann, O. Lochmuller, H. Schiedewitz, H. Friedel, and F. Kux, Static Electricity in The Processing of Man-Made Fibers, Fachbuchverlag, Leipzig, 1963.
32. J. R. Reitz and F. J. Milford, Foundations of Electromagnetic Theory, Addison-Wesley, Reading, Mass., 1960.
33. A. M. Crugnola and H. M. Robinson, "Measuring and Predicting The Generation of Static Electricity in Military Clothing," U.S. Army Quartermaster Res. and Eng. Command, Textile, Clothing, and Footwear Division, Textile Series Report No. 110, Natick, Mass., 1959.
34. P. S. H. Henry, Proc. 3rd Conf. on Static Electrification: Inst. Phys. Conf. Ser. No. 11, 212 (1971).
35. M. Hayek and F. C. Chromey, Am. Dyest. Rep., 40, 164 (1951).
36. G. J. Sprokel, Textile Res. J., 27, 501 (1957).
37. K. Burker, Ann. Phys. Leipzig, 1, 474 (1900).
38. H. H. Hull, J. Appl. Phys., 20, 1157 (1949).
39. V. J. Webers, J. Appl. Polymer Sci., 7, 1317 (1963).
40. J. Gaylor, R. E. Wiggins, and J. B. Arthur, Static Electricity-Generation, Measurement, and Its Effects on Textiles, School of Textiles, North Carolina State University, Raleigh, N. C., 1965.
41. R. D. Fine, Am. Dyest. Rep., 43, 405 (1954).
42. R. G. Cunningham and D. J. Montgomery, Am. J. Phys., 24, 54 (1956).
43. W. Sprenkmann, Melliand Textilber., 34, 971, 1076 (1953).
43a. S. Uchiyama, Y. Miyagawa, and M. Kobayashi, J. Textile Mach. Soc. Jap., 19(3), 57 (1973).
44. H. R. Harper, Proc. Roy. Soc., A218, 111 (1953).
45. H. R. Harper, Proc. Roy. Soc., A231, 388 (1955).
46. G. S. Rose and S. G. Ward, Brit. J. Appl. Phys., 8, 121 (1957).
47. J. A. Medley, Brit. J. Appl. Phys. Suppl. No. 2, 4, S28 (1953).
48. R. G. C. Arridge, Brit. J. Appl. Phys., 18, 1311 (1967).
49. P. S. H. Henry, Brit. J. Appl. Phys. Suppl. No. 2, 4, S31 (1953).
50. D. J. Lehmicke, Am. Dyest. Rep., 38, 853 (1949).
51. H. T. McLean, Am. Dyest. Rep., 44, 485 (1955).
52. P. J. Sereda and R. F. Feldman, J. Textile Inst., 55, T288 (1964).
53. H. F. Richards, Phys. Rev., 22, 122 (1923).
54. D. E. Debeau, Phys. Rev., 66, 9 (1944).
55. E. W. B. Gill and G. F. Alfrey, Nature, 163, 172 (1949).
56. E. M. Ramer and H. R. Richards, Textile Res. J., 38, 28 (1968).
57. R. S. Havenhill, H. C. O'Brien, and J. J. Rankin, J. Appl. Phys., 17, 338 (1946).
58. J. W. Peterson, J. Appl. Phys., 25, 907 (1954).
59. P. E. Wagner, J. Appl. Phys., 27, 1300 (1956).
60. R. S. Havenhill, H. C. O'Brien, and J. J. Rankin, J. Appl. Phys., 15, 731 (1944).
61. C. E. Kirkwood, Jr., O. H. Bellamy, N. S. Kenrick, and H. M. Brown, Textile Res. J., 24, 853 (1954).

62. American Association of Textile Chemists and Colorists, Technical Manual, 48, Research Triangle Park, N. C., 1972, p. 204, Test Method 115-1969, "Electrostatic Clinging of Fabrics: Fabric-to-Metal Test."
63. P. S. H. Henry, G. R. Livesey, and A. M. Wood, J. Textile Inst., 58, 55 (1967).
63a. E. R. Frederick, Am. Dyest. Rep., 57, 550 (1968).
64. F. J. Kleggin, G. Morris, and A. M. Yuill, J. Textile Inst., 40, T702 (1949).
65. H. Grüner, Faserforch. Textiltech., 4, 249, 275 (1953).
66. A. G. Beaumont, J. Textile Inst., 48, P100 (1957).
67. J. Lunenschloss and K. Schnaidt, Melliand Textilber., 43, 674, 807, 1029 (1962).
68. W. Lobel, Faserforch. Textiltech., 19, 110 (1968).
69. G. W. Graham, Nature, 168, 871 (1951); Can. Textile Semin. Book Pap. 3rd, 32 (1952); Textile Mercury Argus, 128 (3336), 16 (1953).
70. V. E. Gonsalves and B. J. van Dongeren, Textile Res. J., 24, 1 (1954).
71. D. A. Zaukelies, Textile Res. J., 29, 794 (1959).
72. A. Jeziorny and G. W. Urbanczyk, Faserforsch. Textiltech., 24, 151 (1973).
73. V. E. Shashoua, J. Polymer Sci., 33, 65 (1958).
74. U.S. General Services Administration, Federal Supply Bureau, Federal Test Method Std. No. 101B, U. S. Government Printing Office, Washington, 1969, Method 4046, "Electrostatic Properties of Materials."
75. S. P. Hersh and D. J. Montgomery, Textile Res. J., 25, 279 (1955).
76. J. B. Levy, J. H. Wakelin, W. J. Kauzmann, and J. H. Dillon, Textile Res. J., 28, 897 (1958).
77. American Association of Textile Chemists and Colorists, Technical Manual, 48, Research Triangle Park, N. C., 1972, p. 199, Test Method 134-1969, "Electrostatic Propensity of Carpets."
78. P. R. Wilkinson, Mod. Textile Mag., 51(12), 35 (1970).
79. R. E. Thompson, Safe Eng., 3 (Oct./Nov.), 17 (1969).
79a. W. Löbel, Faserforsch. Textiletech., 24, 195 (1973).
80. J. W. S. Hearle, J. Textile Inst., 43, P194 (1952).
81. N. A. Teixera and S. M. Edelstein, Am. Dyest. Rep., 43, 195 (1954).
82. American Society for Testing and Materials, 1972 Annual Book of ASTM Standards Part 29, ASTM, Philadelphia, 1972, p. 176, ASTM Designation D 257-66, "D.C. Resistance or Conductance of Insulating Materials."
83. S. P. Hersh and D. J. Montgomery, Textile Res. J., 22, 805 (1952).
84. W. E. Morton and J. W. S. Hearle, Physical Properties of Textile Fibres, Butterworths, London, 1962.
85. R. V. C. Roche, M.S. Thesis, Textile Research Institute and Princeton University, Princeton, N. J., 1972.
86. P. L. Grady, Ph.D. Thesis, North Carolina State University, Raleigh, N. C., 1973.

87. H. A. Pohl, in Modern Aspects of the Vitreous State (J. D. Mackenzie, ed.), Vol. 2, Butterworths, 1962, pp. 114-151.
88. D. A. Seanor, Adv. Polymer Sci., 4, 317 (1965).
89. J. E. Katon, ed., Organic Semiconducting Polymers, Dekker, New York, 1968.
90. R. H. Norman, Conductive Rubbers and Plastics, Elsevier, Amsterdam, 1970.
91. K. C. Frisch and A. Patsis, eds., Electrical Properties of Polymers, Technomic, Westport, Conn., 1972.
92. M. E. Baird, Electrical Properties of Polymeric Materials, Plastics Inst., London, 1973.
93. J. B. O'Sullivan, J. Textile Inst., 39, T368 (1948).
94. J. W. S. Hearle and E. H. Jones, J. Textile Inst., 40, T311 (1949).
95. M. Hayek and F. C. Chromey, Am. Dyest. Rep., 40, 225 (1951).
96. J. W. S. Hearle, J. Textile Inst., 44, T117, T144, T155, T177 (1953).
97. American Association of Textile Chemists and Colorists, Technical Manual, 48, Research Triangle Park, N. C., 1972, p. 193, Test Method 76-1972, "Electrical Resistivity of Fabrics," p. 195, Test Method 84-1969, "Electrical Resistivity of Yarns."
98. National Fire Protection Association International, Standard for The Use of Inhalation Anesthetics 1972, NFPA No. 56A, Boston, 1972, pp. 43-45.
99. D. Wilson, J. Textile Inst., 54, T97 (1963).
100. W. T. Morris, Plastics Polymers, 38, 41 (1970).
101. R. G. C. Arridge, Brit. J. Appl. Phys., 11, 202 (1960).
102. E. P. Sharman, S. P. Hersh, and D. J. Montgomery, Textile Res. J., 23, 793 (1953).
103. F. H. Steiger, Textile Res. J., 28, 721 (1958).
104. N. Wilson, Textile Inst. Ind., 10, 235 (1972).
105. M. J. Williams, D. R. Lawson, and M. L. Staples, Can. Textile J., 88(1), 63 (1971).
106. E. I. Valko and G. C. Tesoro, in Encyclopedia of Polymer Science and Technology, Vol. 2, Wiley-Interscience, New York, 1965, pp. 204-229.
107. W. F. Cooper, Brit. J. Appl. Phys. Suppl. No. 2, 4, S71 (1953).
108. D. Bulgin, Brit. J. Appl. Phys. Suppl. No. 2, 4, S83 (1953).
109. P. S. H. Henry, Proc. 3rd Conf. on Static Electrification: Inst. Phys., Conf. Ser. No. 11, 195 (1971).
110. P. E. Shaw, Proc. Roy. Soc. (London), A94, 16 (1917); 39, 449 (1927).
111. P. E. Shaw and C. S. Jex, Proc. Roy. Soc. (London), A111, 339 (1926); A118, 97, 108 (1928).
112. P. E. Shaw and R. F. Hanstock, Proc. Roy. Soc. (London), A128, 474, 480 (1930).
113. P. E. Shaw and E. W. L. Leavey, Proc. Roy. Soc. (London), 138, 502 (1932).
114. J. W. Ballou, Textile Res. J., 24, 146 (1954).
115. H. Frotscher, Z. Gesamte Textilind., 62, 746 (1960).

116. A. Coehn, Ann. Phys. Chem., 64, 217 (1898); 66, 1191 (1898).
117. D. J. Montgomery, Adv. Solid State Phys., 9, 139 (1959).
118. A. Sippel, Textile Prax., 8, 1030 (1953).
119. D. K. Davies, Proc. 2nd Conf. on Static Electrification: Inst. Phys. Conf. Ser. No. 4, 29 (1967).
120. I. I. Inculet and E. P. Wituschek, Proc. 2nd Conf. on Static Electrification: Inst. Phys. Conf. Ser. No. 4, 37 (1967).
121. J. A. McClelland and C. J. Power, Proc. Roy. Irish Acad., 34, 40 (1917).
122. S. P. Hersh and D. J. Montgomery, Textile Res. J., 24, 755 (1954).
123. E. Peclet, Ann. Chim. Phys. (2), 57, 337 (1834).
124. E. Peclet, Ann. Chim. Phys. (2), 71, 83 (1839).
125. M. Owen, Phil. Mag., 17, 457 (1909).
126. M. W. Jones, Phil. Mag., 21, 261 (1915).
127. N. R. French, Phys. Rev., 9, 151 (1917).
128. W. A. Macky, Proc. Roy. Soc. (London), A119, 107 (1928).
129. W. Kluge, Ann. Phys. Leipzig, 1, 1 (1929).
130. L. Wolf, Ann. Phys. Leipzig, 1, 260 (1929).
131. J. A. Medley, Nature, 166, 524 (1950).
132. S. Whitehead, Dielectric Phenomena, Vol. 1, Electrical Discharges in Gases, Ernest Benn, London, 1927.
132a. D. T. Liddy and J. G. Downes, J. Textile Inst., 49, T467 (1958).
133. N. Gibson and F. C. Lloyd, Brit. J. Appl. Phys., 16, 1619 (1965).
134. M. Kato and K. Yoshida, J. Textile Mach. Soc. Jap., 4(2), 27 (1958).
135. M. M. Bredov and I. Z. Kshemianskaia, Zh. Tekhn. Fiz., 27, 921 (1957) (in Russian); Soviet Phys. Tech. Phys. (English Transl.), 2, 844 (1957).
136. D. J. Montgomery, A. E. Smith, and E. H. Wintermute, Textile Res. J., 31, 25 (1961).
137. R. G. Cunningham, Ph.D. Thesis, Michigan State University, East Lansing, Michigan, 1957.
138. J. F. London Jr., M.S. Thesis, North Carolina State University, Raleigh, North Carolina, 1966.
139. K. Gotze, W. Brasseler, and F. Hilgers, Melliand Textilber., 34, 548, 658 (1953).
140. M. H. Ghanam, M.S. Thesis, North Carolina State University, Raleigh, North Carolina, 1972.
141. R. G. Cunningham, J. Appl. Phys., 35, 2332 (1964).
142. R. G. Cunningham, J. Appl. Phys., 37, 1734 (1966).
143. G. Quintelier, M. Warzee, and R. Sioncke, J. Textile Inst., 48, P26 (1957).
144. F. B. Bowden and D. Tabor, The Friction and Lubrication of Solids, Clarendon, Oxford, 1950.
145. S. P. Timoshenko, Strength of Materials, 2nd ed., Van Nostrand, New York, 1941.
146. N. Murasaki, J. Textile Mach. Soc. Jap., 7(1), 37 (1961).
147. J. W. Peterson, J. Appl. Phys., 25, 501 (1954).

148. I. I. Inculet and W. D. Greason, Proc. 3rd Conf. on Static Electrification: Inst. Phys. Conf. Ser. No. 11, 23 (1971).
149. W. R. Harper, Brit. J. Appl. Phys., 4, 111 (1953).
150. J. M. Van De Weerd, Proc. 3rd Conf. on Static Electrification: Inst. Phys. Conf. Ser. No. 11, 158 (1971).
151. P. Lion, Rayonne Fibres Synth., 13, 445 (1957).
152. F. Kassack, F. Reich, A. Schmitz, G. Heyl, and G. Luttgens, Flocking Bayer Farben Revue, Special Edition No. 2, Dyestuffs Sales Division of Farbenfabriken Bayer, Leverkusen, W. Germany.
152a. E. N. Bershev, Tekstil. Prom., 25(3), 75 (1965); 26(2), 64 (1966).
153. E. Parrish (nee Grandique) and M. Katz (to E. I. Du Pont de Nemours and Co.), U.S. Pat. 3, 117, 055 (1964).
154. J. J. LaFranka Jr., M. Mayer Jr., and H. W. Weller Jr., Textile World, 116(6), 58 (1966).
155. J. Shaw, in New Ways to Produce Textiles, The Textile Institute, Manchester, England, 1962
156. P. R. Lord, Textile Mfr., 88(1055), 490 (1962).
157. A. Gregoric, Das Verhalten von Textilfasern im elektrischen Feld, Mitteilungen aus dem Institut Für Textilmaschinenbau v. Textilindustrie an der ETH, Nr. 14, Juris Verlag, Zurich, 1973.
158. F. Ecochard, Rayonne Fibres Synthetiques, 10, 85 (1954).
159. A. Sippel, Melliand Textilber., 35, 831 (1954).
160. W. P. Crawley, quoted in Floor Covering Weekly, 21(32), pp. 1, 12 (Aug. 7, 1972).
161. G. F. Barry, Mod. Textile Mag., 49(5), 64 (1968).
162. H. F. Lewis, Textile Inst. Ind., 10, 199 (1972).
163. J. Boyd and D. Bulgin, J. Textile Inst., 48, P66 (1957).
164. E. E. Magat and D. Tanner (to E. I. Du Pont de Nemours and Co.), U.S. Pat. 3,329,557 (1967).
165. W. K. Wilkenson (to E. I. Du Pont de Nemours & Co.), U.S. Pat. 3,341,627 (1967).
166. G. T. Barnes, Mod. Textile Mag., 49(5), 64 (1968).
167. E. E. Magat and W. H. Sharkey (to E. I. Du Pont de Nemours & Co.), U.S. Pat. 3,475,898 (1969).
168. L. R. Kumar, Silk Rayon Ind. India, 12, 315 (1969).
169. T. Okuhashi (to Teijin, Ltd.), U.S. Pat. 3,582,445 (1971).
169a. V. S. Ellis, Textile Manufact., 101(1193), July 19 (1974).
170. J. Turer and J. K. Smith, Am. Dyest. Rep., 51, 41 (1962).
171. J. A. Medley, J. Textile Inst., 48, P112 (1957).
172. H. Bertein, Proc. 2nd Conf. on Static Electrification: Inst. Phys. Conf. Ser. No. 4, 11 (1967).
173. F. J. Rizzo. "Electrostatic Phenomena in Textiles and Clothing Systems," Technical Report 74-2-CE, Series C&PLSEL-111, U.S. Army Natick Laboratories, Natick, Mass., 1973.

Chapter 7

SCANNING ELECTRON MICROSCOPY OF SYNTHETIC FIBERS

H. R. Billica and R. D. Van Veld

Textile Fibers Department
Fiber Surface Research Section
E. I. du Pont de Nemours & Co., Inc.
Kinston, North Carolina

I. INTRODUCTION

The scanning electron microscope (SEM) and fiber-textile science . . .one of those rare weddings where bride and groom seem truly destined for one another! Although the instrument was developed at the University of Cambridge for use with microcircuitry, its great potential for use in the fiber field was recognized in the early 1960's by J. Sikorski at Leeds University and fiber scientists at the Du Pont Company. This recognition led to development of the first commercial instrument, the Stereoscan,* the original prototype being delivered to the Du Pont Company. Now there are numerous scanning electron microscopes and it is difficult to imagine conducting adequate fiber-fabric studies without the use of such an instrument.

The imaging system of the scanning electron microscope, with its wide range of magnification and great depth of focus, is remarkably well suited to the needs of textile technology for the following reasons:

1. Fibers are small but are not microscopic and hence may be imaged easily and quickly.
2. Fiber features that determine performance in manufacturing and conversion to end-use are of the same size range and are easily imaged.
3. Garment appearance is vital to consumer acceptance; fiber features that determine appearance are easily imaged.

There are now a number of good reference works (e.g., see refs. 1,2,3) that detail the principle and general techniques of scanning electron microscopy. There are also numerous publications that adequately demonstrate the range of "pretty" pictures that can be made of fiber systems. Accordingly, it does not seem appropriate to repeat such material in this chapter. Rather, the authors will point out special techniques, interpretation methods, and possible problem areas where one can go astray. Such items, amply illustrated, should be of value in enhancing the use of the instrument and obtaining maximum correct information.

II. SEM-SPECIMEN INTERACTION

Fibers share with other specimens—ceramics, electrically conductive metals, semiconducting metals, and biological specimens—the same general considerations for sample preparation and SEM operation. However, the extremely low electrical conductivity of fibers makes metal coating a requirement for good imaging and this coating, because of the labile materials often of interest on filament and yarn surfaces,

---

*Trademark owned by Cambridge Instrument Company, Ltd.

can cause image artifacts. Also, even when they are carefully prepared in a metal evaporator, textile objects of complex geometry, such as carpet tufts or double knit garments, can display artifacts caused by electrical charge buildup in elements of the object not grounded electrically. A discussion of the factors contributing to "charging" artifacts is presented in the following section.

A. Charge in Beam

In the SEM, an electron beam of current I is scanned over a specimen's surface. The response of the specimen to this beam is such that the negative charge input from the beam is balanced by the combined effect of charge conduction to ground from the specimen and secondary emission at the specimen's surface. The area density of the charge approaching the specimen at every point during the time of one frame, F (in seconds), with the instrument set at magnification M, and with the area of the image viewing screen as A (cm^2) is [4]

$$\sigma_{beam} = \frac{M^2 IF}{A}.$$

Typically, $\sigma = 5 \times 10^5$ electrons per square micron per frame.

B. Specimen Charge Loss

Assuming that simple mechanisms dominate the electrical conduction of charged filaments, the decay of charge resulting from conduction to ground may be stated in terms of Ohm's law

$$\sigma = \sigma_0 \exp(-t/\eta\epsilon)$$

where resistivity η and permittivity ϵ are specimen properties and are the only parameters used to describe the conduction to ground. The specimen also loses charge by emission of secondary electrons. The secondary emission coefficient δ, which is the ratio of secondary current to beam current is dependent on beam penetration and electron absorption effects [2] and also, importantly, on the energy of the beam electrons penetrating the surface. Hence, in the SEM-specimen interaction there is the circuitous mechanism in which the dissipation of charge by secondary emission depends on the effective voltage of the beam, which is related to the specimen's charge; the specimen's charge in turn depends on secondary emission.

C. Charging

A quantitative study of this SEM-specimen interaction [5] for a glass filament confirmed an earlier report [6] that synthetic fibers could reach negative potentials in the 1 to 10 kV range. The study also proposed that the concomitant large electrostatic repulsive forces from the bulk charge in the specimen increase secondary electron emission to cause image artifacts such as abnormally bright "charged" areas on the synthetic fiber specimens. The increased emission has been identified as the primary mechanism for charging in poorly conducting specimens other than synthetic fibers [7], and there are other mechanisms proposed as well for nonfiber specimens, such as those involving specimen collection and field emission [8]. However, the mechanistic models involving increased emission and mirroring [4] of beam electrons from highly charged areas have been the most satisfactory in applying the SEM to fibers.

III. SAMPLE AND MICROGRAPH PREPARATION PROCEDURES

The high resistivity of synthetic fibers causes their charge buildup in the SEM, and of course this accumulation, or at least the effects of it, can be largely eliminated by depositing an electrically conductive coating over the specimen. It should be recognized, however, that much useful detail can be imaged without adding this coating. Reducing beam current I, magnification M, or frame time F, will equilibrate the charged condition without eliminating the charge. Reducing the beam accelerating voltage can, in many cases, produce a neutral specimen [5]. Obviously, these adjustments sacrifice resolution because of beam instability and weaker signal current. However, manufacturers of instruments that photograph at low frame time (TV scan rates) and also use low accelerating voltages report high resolution [9].

In selecting a coating material and a coating procedure, it has become apparent to SEM users that there exists no single combination that is good for all SEM specimens, or even for a simple majority of specimens in the much narrower field of synthetic fibers. For a general discussion of important variables, reference may be made to several recent, and excellent, review articles that collectively tend to cover all possible factors [3,10-12]. In addition to these, special note may be taken of techniques used to protect volatile materials, which may comprise part of the filament surface features, from the vacuum of metal evaporative methods. A good example is the application of a technique developed to replicate water drops and ice crystals. The technique was subsequently used to locate volatile liquid-smoke components on filter fibers [13].

All specimens shown in the figures in this chapter (see Appendixes) were attached to SEM stubs with conductive cement such as that used in electronic circuitry. The consistency of the cement was adjusted with solvent so that a light contact could be dabbed at the specimen extremities without causing the cement to become thin enough to wick into the specimen. After 10 min drying time, the stubs were metallized by evaporation at 10^{-4} to 10^{-5} Torr. To provide rotation during metal deposition, the stubs were placed at the edge of a 5-cm disk that rotated about its axis at 10 rpm. Arrays of filaments such as those in fabrics or in loose bundles were metallized first from a source 20 cm overhead at 30° incidence and then again from 15 cm at 70° incidence. 60/40 gold/palladium alloy was used for most specimens. Aluminum has commonly given noticeably better results with specimens like those shown but the aluminum-coated specimens have a relatively short shelf life, which prevents reexamination after one day.

Specimens that remained too nonconductive after the second metallization were dipped in a dilute solution of antistatic agent such as Duron [14] according to the methods prescribed in the references cited on specimen preparation. Alternatively, for low-magnification studies ($<200\times$), a common antistatic spray marketed for phonograph records was used.

The SEM used to produce most of these micrographs was of conventional design. Conventional operational parameters (beam current, accelerating voltage, working distance) were used. The micrograph record time, 20-25 sec, was possibly unconventionally short. Polaroid T-42 film was used exclusively.

Typical image deficiencies and artifacts, which can be caused either by preparation faults or by SEM misoperation, will be discussed in the following section.

IV. COMMON IMAGING PROBLEMS

A. Charging

Figures 1a and b show the familiar effects of high-negative charge accumulation, in this case on polyester staple fibers teased from spun yarn (Fig. 1a) and on a continuous filament polyester taffeta (Fig. 1b). These overly bright spots completely obscure all detail, even the gross feature of fiber shape. A loose array of fibers under no tension like those in Fig. 1a, has been found to be particularly difficult to render conductive through metallization alone. For low-resolution work conductive liquids may be applied. For higher resolution, the array may be remounted on the specimen stub, still in a relaxed state but with the precaution of fixing both ends of all fiber segments.

The charging in Fig. 1b can be prevented by an initial coating of carbon or by post-treatment with antistatic solution. It is interesting to note the images of fibers immediately adjacent to the ultrabright spots in this figure. No distortion in the linearity of their shape is detectable, which indicates that the electric field around the charged bright spots is too small to deflect the incident beam. This is consistent with the "increased emission" mode of image charging, which requires only a few volts to assist the escape of low-energy secondary electrons from the sample surface.

B. Beam Damage

The rectangular pattern on the filament of Fig. 2 is the familiar mark produced by too much exposure to a fairly intense scanning beam. A mark as obvious as this cannot mislead image interpretation and is thus of no concern. What is of concern is the question of beam damage to small protrusions, which may be thinly metallized and consequently have lower heat capacity. In addition, the surface detail may consist of lower melting-point material. This concern points to the need for cross checking, by using variations in the coating preparation and also by studying surface replicas with a transmission microscope. Some quantitative beam damage studies have been reported [15-18]; however, the question of beam damage needs much additional examination because of the large number of fiber-forming synthetic polymers, the many states of fiber crystallinity and molecular orientation, and the large number of materials common to fiber surfaces.

Within the beam damage mark of Fig. 2 appears a second type of fiber-image artifact. Note the series of jagged dark lines traversing the mark. These are indeed cracks in the surface of the object being viewed but are not necessarily cracks in the surface polymer of the fiber. Figure 3 shows that these cracks can occur in larger size and greater frequency. Cases have been observed in which several fibers in the same field contain cracks over all their surface. A useful criterion for distinguishing fiber cracks from metal coating cracks is to determine whether the edges or bottom of the cracks show as much detail as the fiber surface adjacent to the cracks. If they do not, the cracks should be considered to be in the coating, as they are in Fig. 3, and hence they are an artifact in the fiber surface structure image.

In addition to merely cracking, metal coatings can also buckle, giving the appearance of a fiber in the process of shedding a skin, as in Fig. 4. If the coating vs. fiber surface criterion is applied to this micrograph, the skin is shown to be merely the coating and not the fiber.

To confirm the existence of coating cracking, it is obvious that one should not remetallize for subsequent reexamination of the specimen in question. Also, coating cracks that go undetected in an examined specimen, which for some reason is remetallized or sprayed with antistat for reexamination, are a particularly insidious source of artifact.

C. Coating Problems

1. Fast Evaporation of Metal

Although "fast" evaporation is usually recommended to reduce agglomeration accompanying a "slow" buildup of coating on the specimen, Fig. 5 shows that the evaporation of Au/Pd alloy (and perhaps other alloys and metals) can be made too fast. Here, a thick specimen coating was formed in the initial stages of metal evaporation but the metal remaining in the molten bead formed a large agglomerate which plowed into the yarn specimen. For this specimen the action resulted in an obvious artifact, because the projectile buckled the metal coating and also disrupted three filaments. Projectiles considerably smaller than their targets may not leave such positive clues.

2. Antistatic Liquids

Figure 6a shows two sections of a polyester continuous-filament yarn sampled upstream of the spindle in a commercial false-twist texturing process. Both sections were in the field of view for the SEM photograph; the left-hand yarn section contains only a sprayed-on coating of antistat and the right-hand section a conventional metal coating. Both images reveal the highly twisted and contorted nature of the yarn with sufficient clarity. However, the greater brightness in the metallized specimen, owing to the higher emission of its coating is obvious. This directly benefits the resolution of small surface detail, as shown by comparing Fig. 6c with 6b. In Fig. 6b, not only do the small protrusions appear unencumbered by a coating but also there are no strong signal variations over the field of view. This latter feature is not always recognized as an advantage of metal coatings as compared to antistat liquids.

V. APPLICATION OF SEM TO FIBERS TECHNOLOGY

The micrographs shown in Appendixes B-E are intended to illustrate the range of applicability of the SEM to fibers technology and also to acquaint workers new to the field of synthetic fibers with the range of shapes and structures found in fibers. The micrographs should be taken as representative of fiber technology but not all inclusive.

Appendix B. Fiber shape features (Figs. 7-11).
Appendix C. Fiber structure features (Figs. 12-16).
Appendix D. Yarn structure features (Figs. 17-22).
Appendix E. Fabric structure features (Figs. 23-31).

VI. SPECIAL TREATMENT OF SECONDARY EMISSION SIGNALS

To this point, the discussion has dealt exclusively with the use of the SEM to form images of textile objects via secondary electron emission. Mention should be made of methods being used to enhance the topographic information provided by the secondary electron image. Among these are gamma controls, which permit display of signal strengths proportional to some fractional power of the raw signal. The data treatment methods also include Y modulation, which shifts the display raster in proportion to signal strength. Y modulation also provides a more quantitative measure of contrast range. Another procedure that assists topographic studies is the modulation of the display raster intensity with the time derivative of the raw signal fluctuations, the so-called derivative amplifier.

VII. PARTICLE COUNTING AND SHAPING

Apart from operations on the raw signal to extract the most information from secondary electron images, there are also devices directly coupled to the SEM to assist in processing the signal, such as auxiliary equipment to provide fast, quantitative information on particle size and shape [19], as seen in fiber sections or on surfaces.

VIII. DYNAMIC EXPERIMENTS

The great interest in conducting dynamic experiments with fibers within the specimen chamber should be noted. The first mention of this possibility described in-situ viewing of the stretching of wool fibers, crack development in the cuticular cells, and the removal of these cells both by friction and by dissection with a razor blade [14]. Another application described twisting, compression, and loop and tensile fractures of synthetic fibers [20]. Subsequently, most instrument manufacturers began to offer as stock items special stages equipped for stressing the specimen in a wide variety of modes. As an auxiliary item, a particularly elegant micromanipulator consisting of two probes, each independently positionable to 0.1 μm, has been described [21]. This device and the manipulators that are now becoming available as stock items permit microdissection of features at 1000 to 5000× magnification.

To assist dynamic examination, TV scan rates are now common. They allow easier viewing and, more importantly, real-time recording of events to replace the time consuming, pseudodynamic time-lapse photographs [22] used before fast scanning became available.

IX. OTHER SIGNAL SOURCES

The literature has just recently begun to contain reference to the other signal sources (elastically scattered primary electrons, induced voltage, cathodeluminescence) available in the SEM-specimen interaction. Indeed, most SEM-equipped fiber laboratories appear to be making good use of modern X-ray detection and data-handling instrumentation. This capability has wide application in analysis of fiber coatings and discoloration, polymeric inhomogeneities, deposits on textile processing equipment, and textile machinery failures. Development of systems to increase detection sensitivity to elements lighter than fluorine promises to widen this capability.

APPENDIX A: COMMON IMAGING AND COATING PROBLEMS

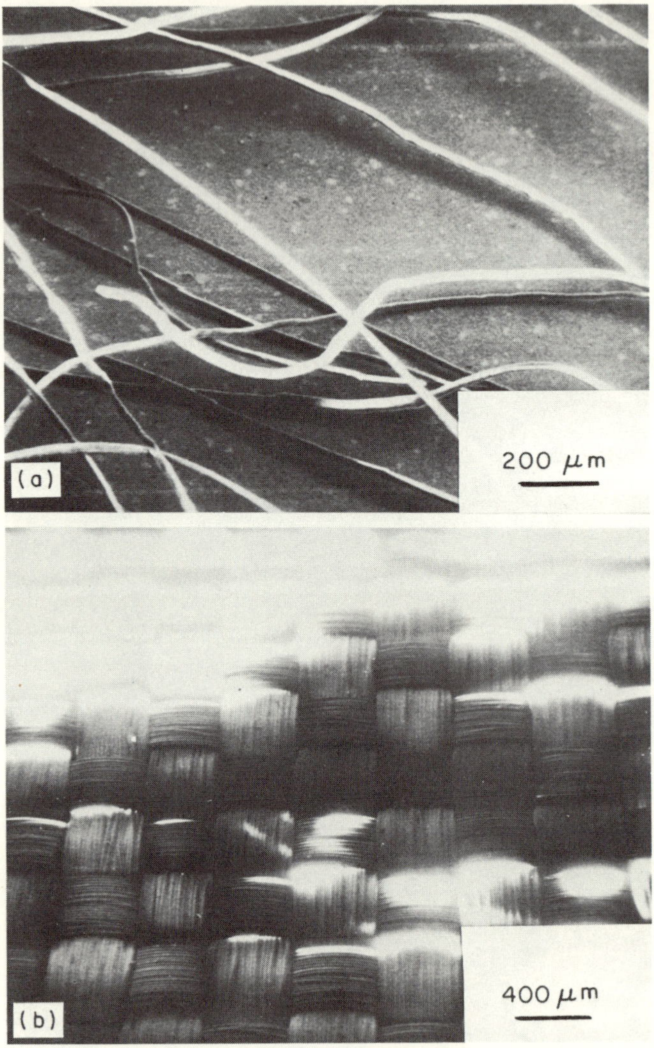

FIG. 1. Typical charging artifacts. (a) Overly bright filaments in a loose cluster, (b) overly bright areas on a woven fabric.

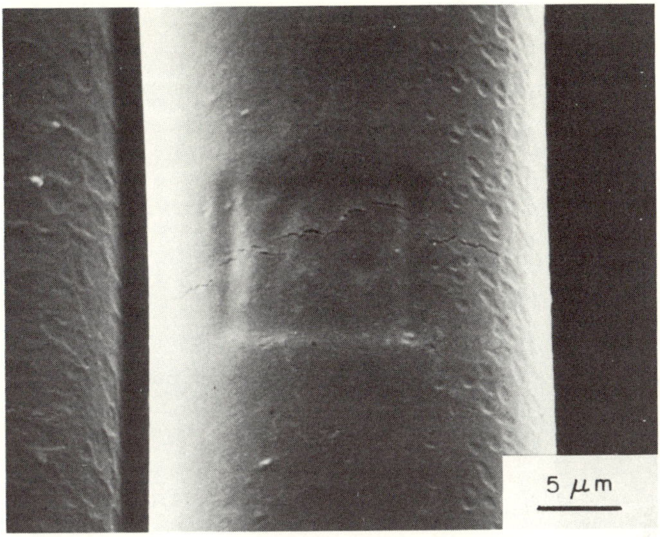

FIG. 2. SEM beam raster damage on synthetic fiber. Fiber melting point is approximately 285°C.

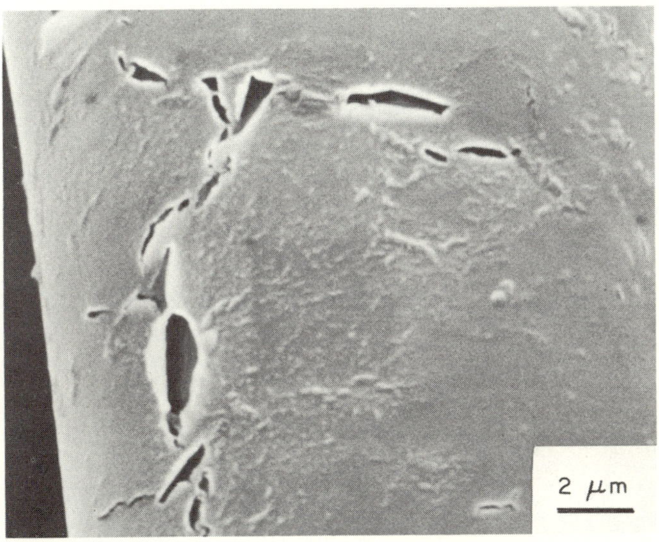

FIG. 3. Beam damage cracks in metal layer on metallized fiber.

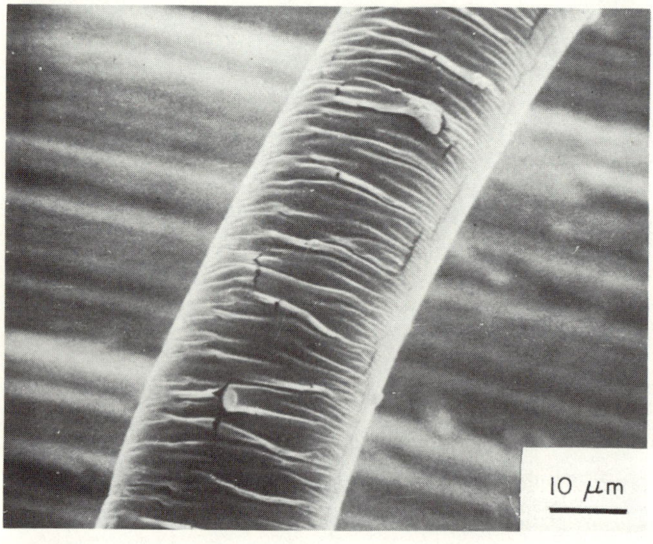

FIG. 4. Buckled metal layer on metallized fiber.

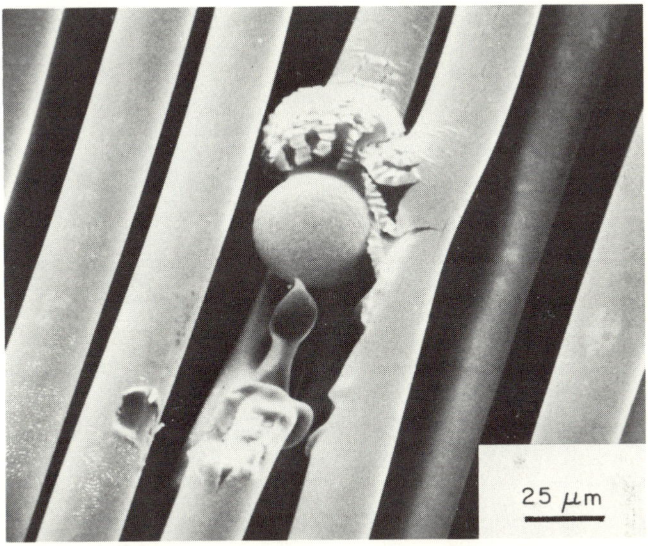

FIG. 5. Au/Pd ball shot into filaments during too fast metallization.

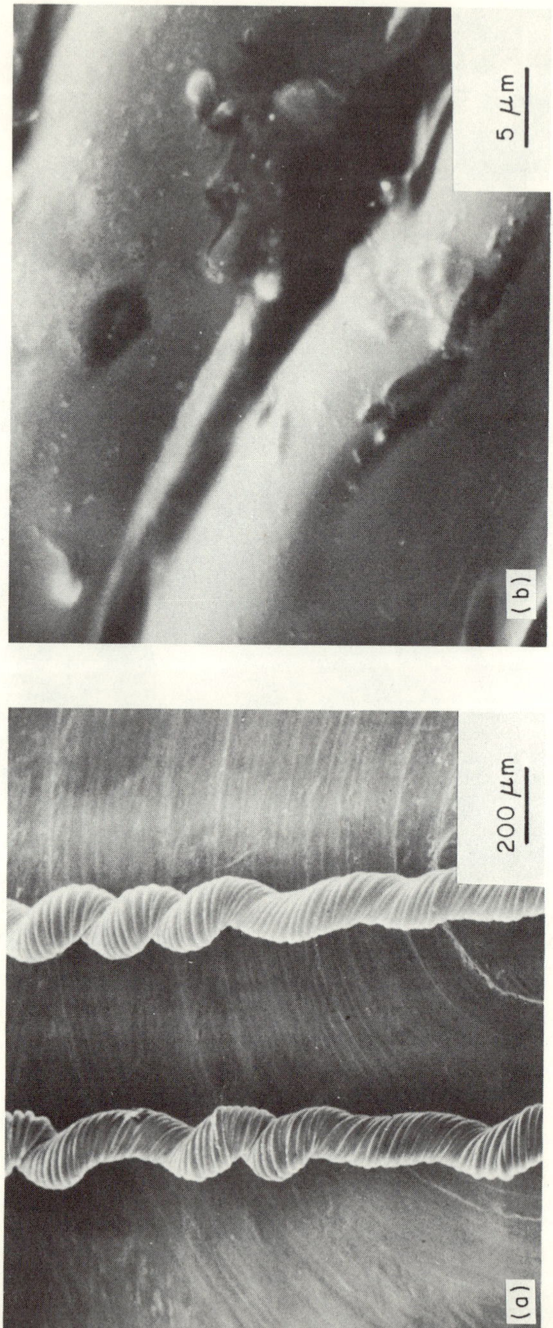
(b)
5 μm
(a)
200 μm

FIG. 6. Yarn sections from false-twist texturing process. (a) Liquid antistat coating on left, metal coating on right gives bright image, (b) image detail obscured by antistat liquid, (c) as in b, but metal coated.

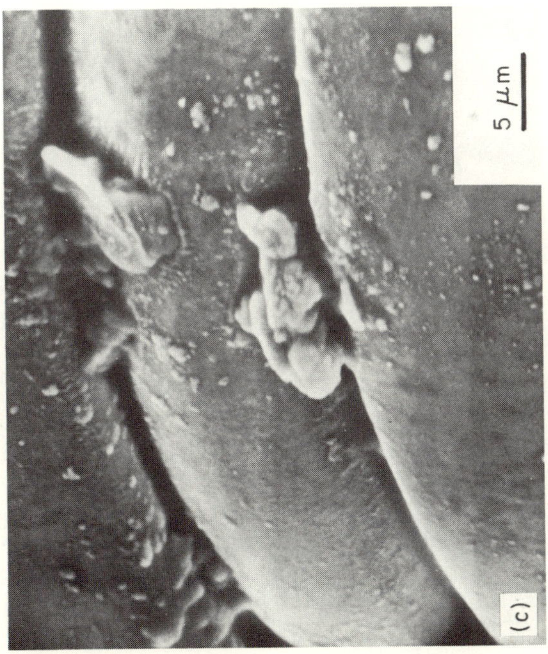

APPENDIX B: FIBER SHAPE FEATURES

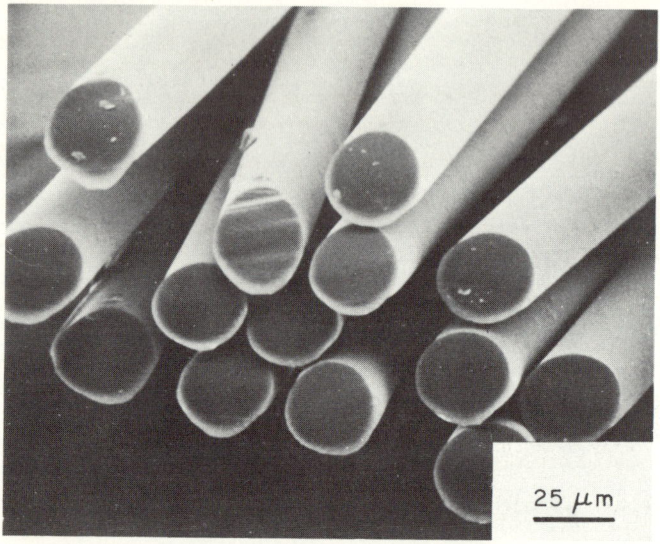

FIG. 7. Round filaments

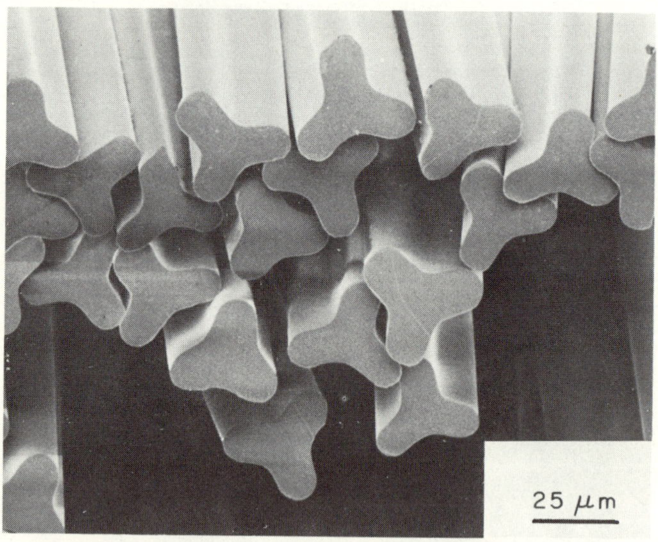

FIG. 8. Trilobal filaments.

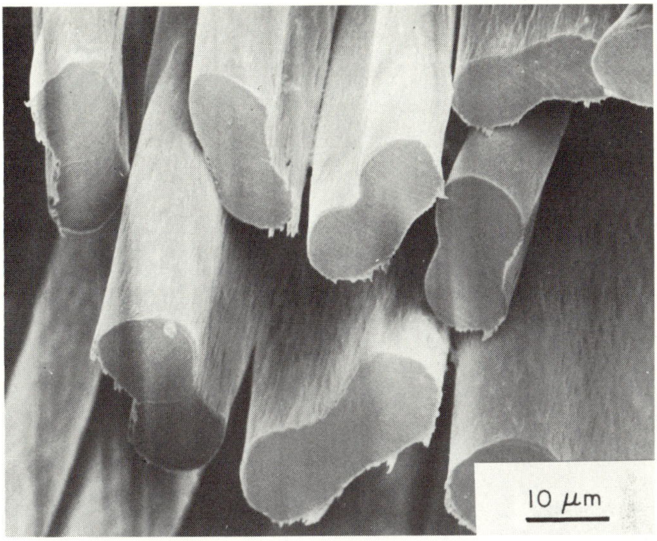

FIG. 9. Acrylic filaments with "dog-bone" cross section.

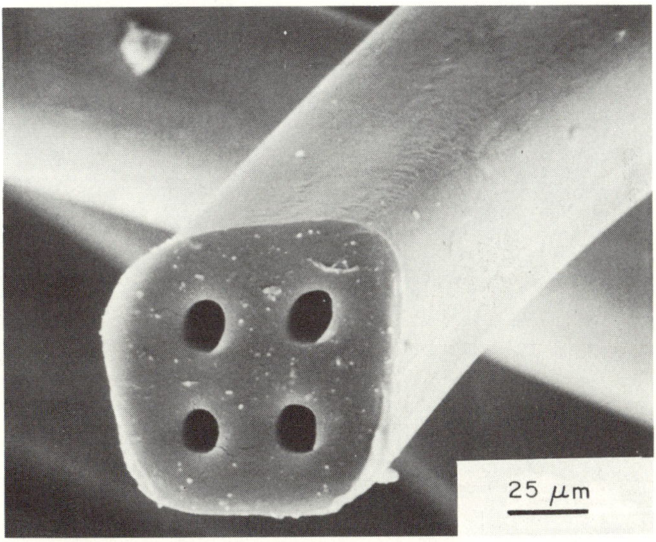

FIG. 10. Nylon carpet yarn filament. Note internal void channels intended to affect light transmission.

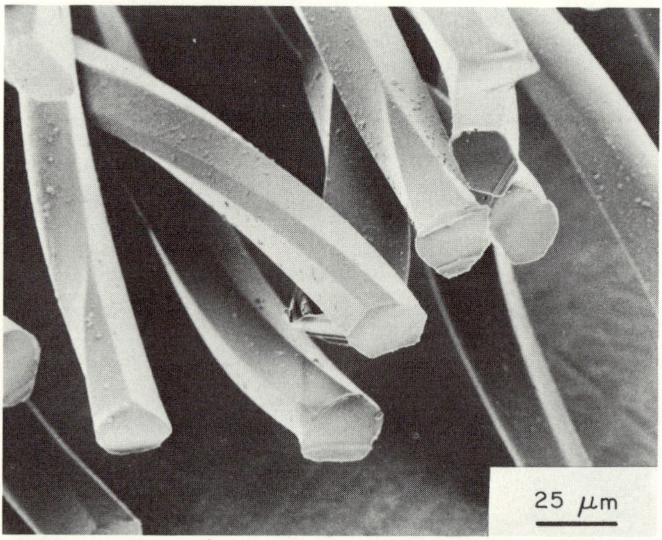

FIG. 11. Round polyester filaments after false-twist texturing.

APPENDIX C: FIBER STRUCTURE FEATURES

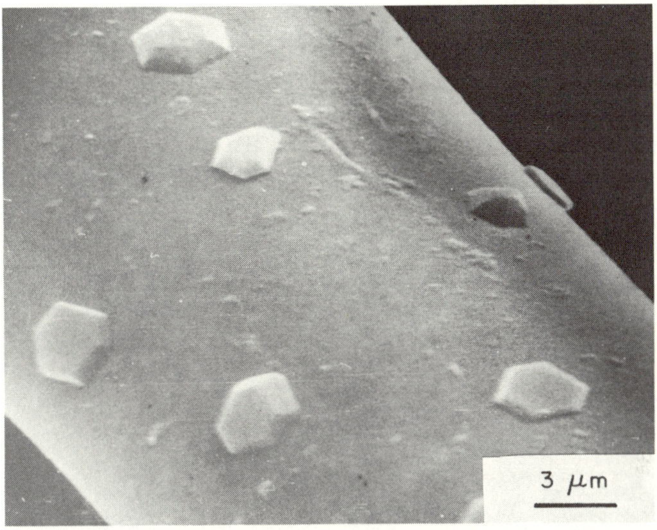

FIG. 12. Cyclic trimer sublimate on surface of a polyester fiber.

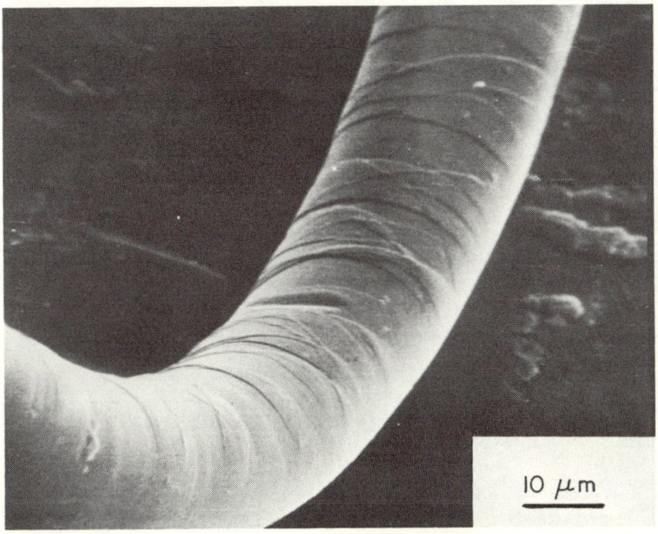

FIG. 13. Bent polyester fiber with structure resembling Lueder's bands.

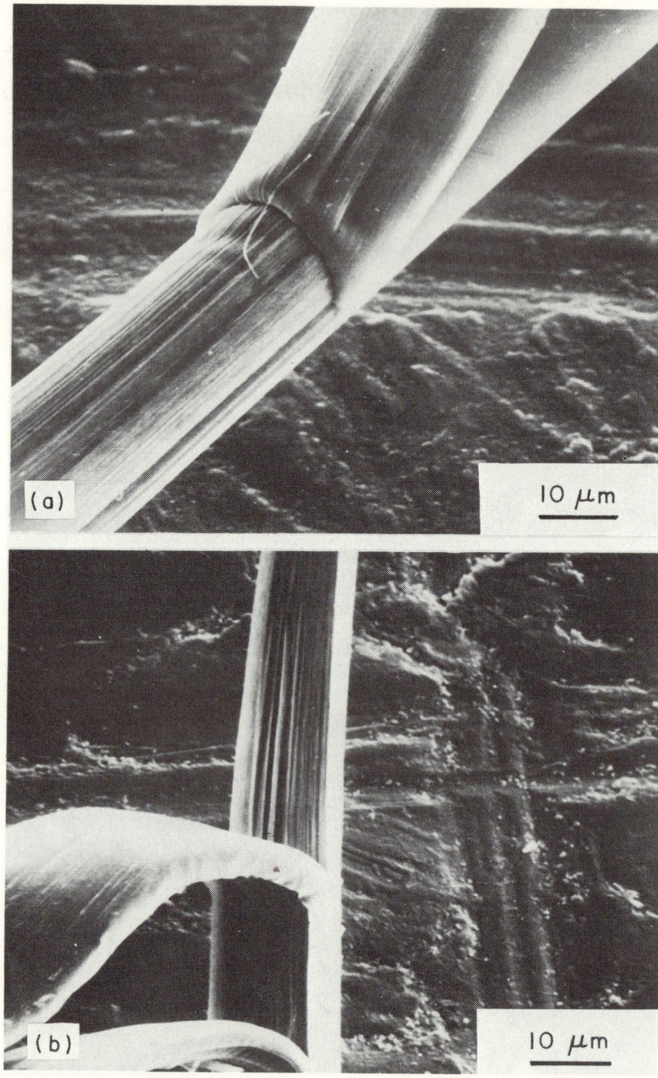

FIG. 14. Peeling characteristics of polyester fibers (a) skin peeling (b) chord peeling.

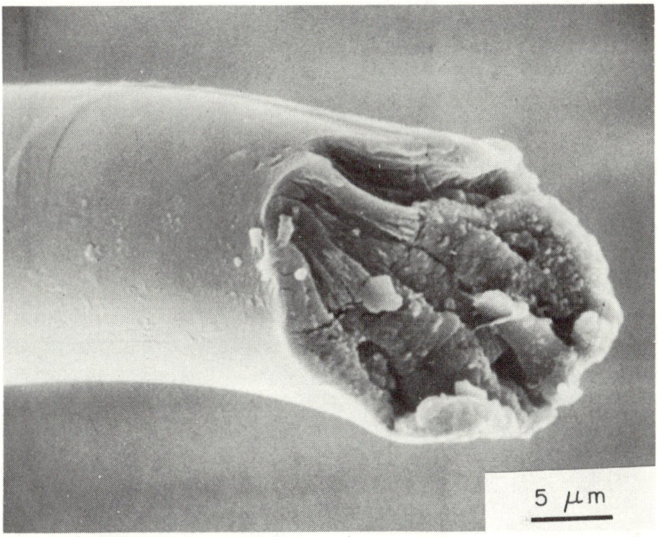

FIG. 15. High-strength polyester fiber broken in testing device.

FIG. 16. Surface of acrylic fibers from two different manufacturing sources.

APPENDIX D: YARN SURFACE FEATURES

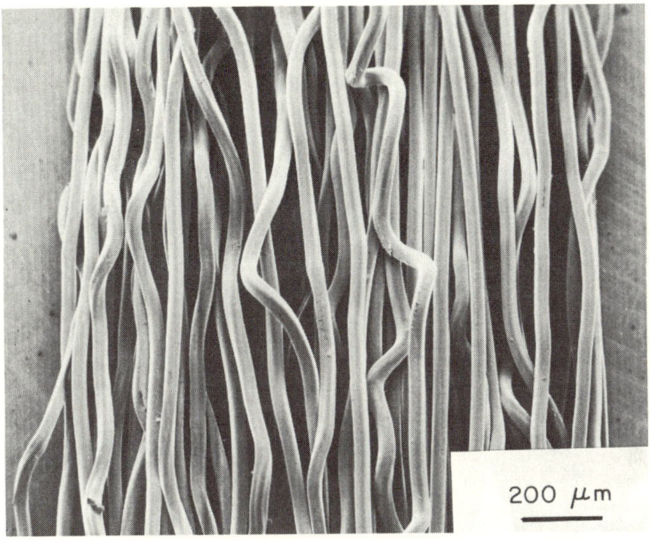

FIG. 17. Bulked continuous-filament nylon carpet yarn.

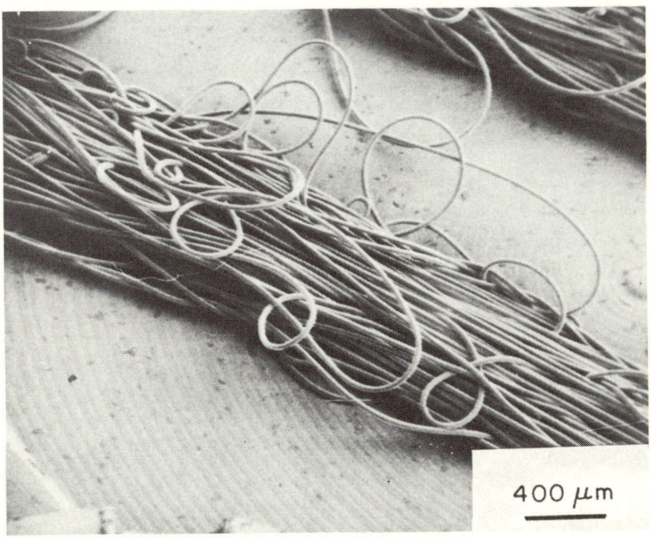

FIG. 18. Taslan* textured nylon yarn.

*Trademark owned by E. I. du Pont de Nemours and Company, Inc.

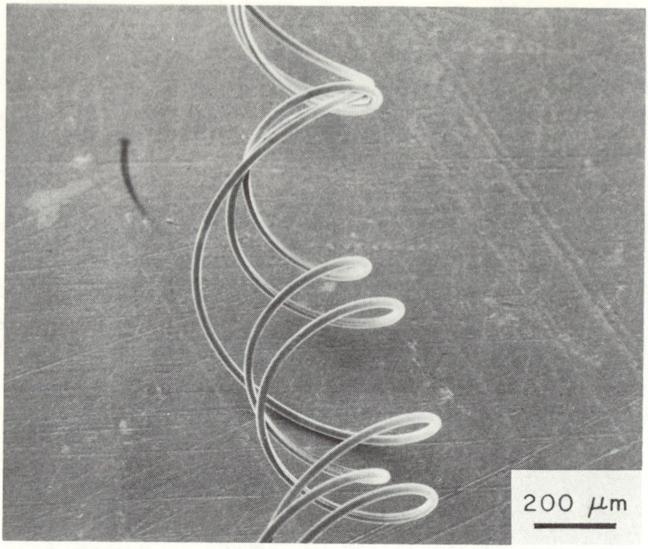

FIG. 19. Agilon* textured nylon yarn.

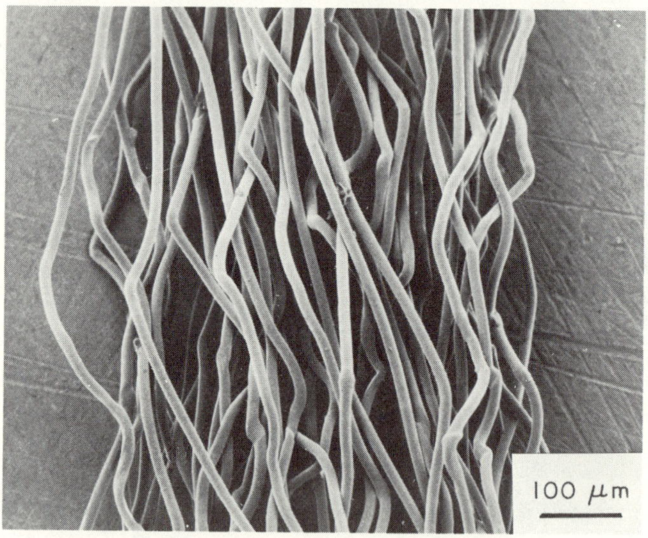

FIG. 20. Ban-Lon† textured nylon yarn.

*Trademark owned by Deering-Milliken Research Corporation.
†Trademark owned by Bancroft Licensing.

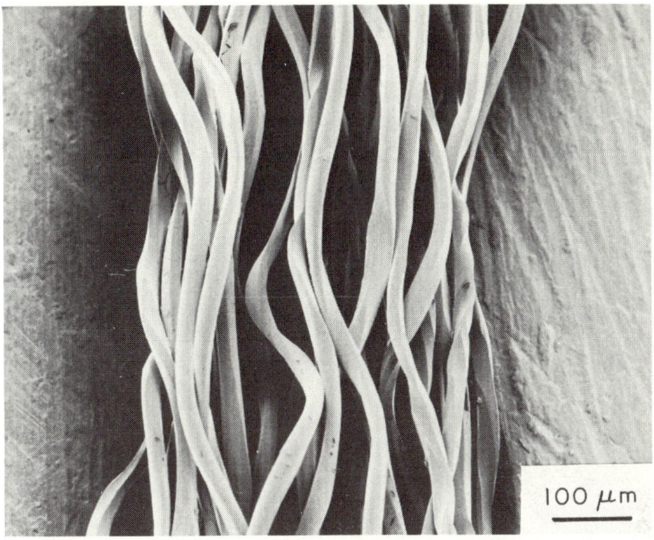

FIG. 21. False-twist textured polyester yarn.

FIG. 22. Nylon tire cord damaged in cord-forming process. Note flattened sides of filaments.

APPENDIX E: FABRIC STRUCTURE FEATURES

FIG. 23. Nylon zipper in spun yarn woven fabric.

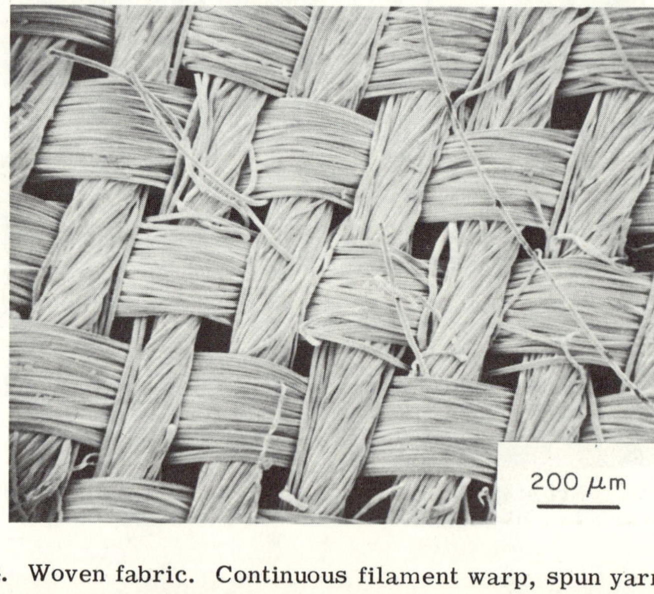

FIG. 24. Woven fabric. Continuous filament warp, spun yarn filling.

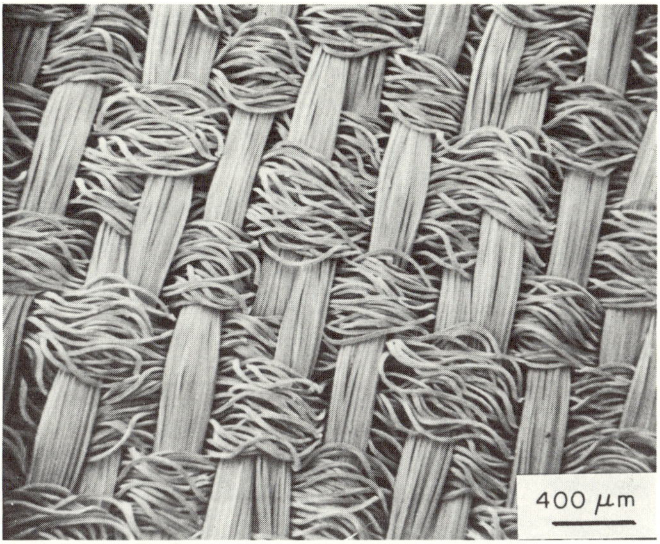

FIG. 25. Woven fabric from continuous filament yarn. Filling yarn is textured.

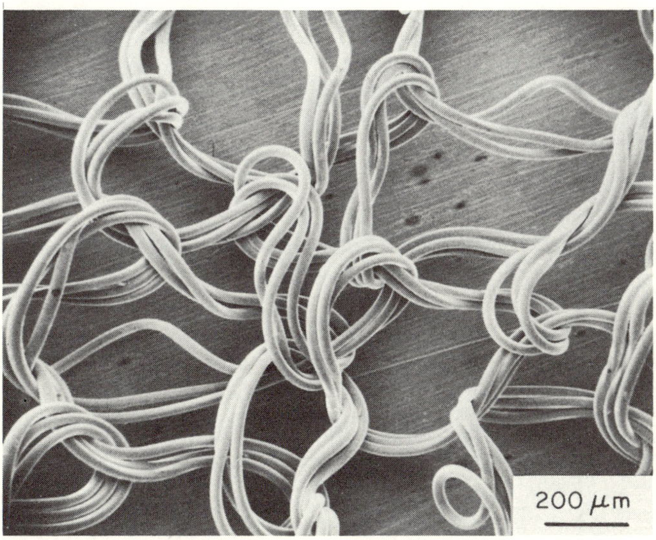

FIG. 26. Nylon hosiery fabric.

FIG. 27. Support garment from nylon-wrapped spandex yarn. (a) Low SEM magnification, (b) higher magnification.

FIG. 28. Jersey knit tricot from continuous-filament nylon yarn.

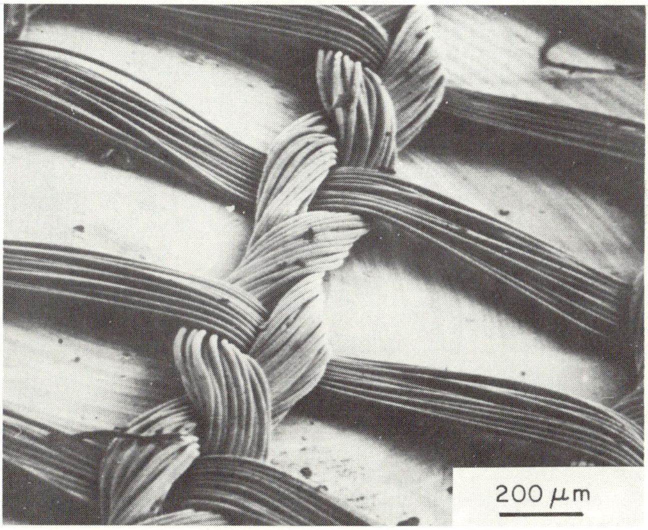

FIG. 29. Plain gauze woven fabric.

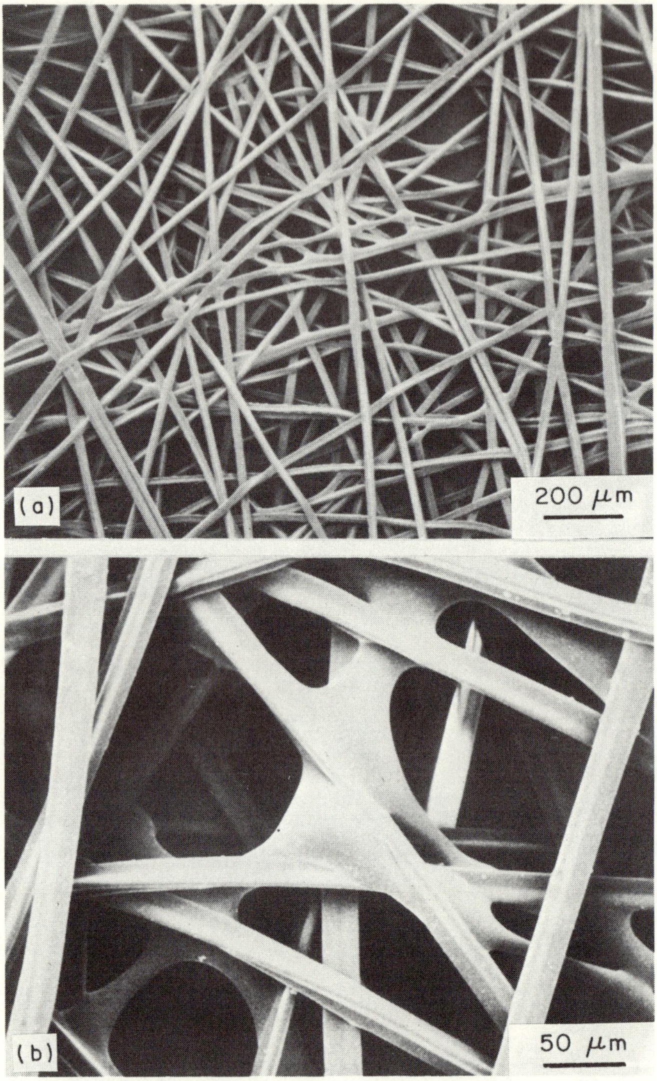

FIG. 30. Nonwoven fabric from polyester continuous-filament yarn. (a) Low SEM magnification, (b) higher magnification showing bonding.

FIG. 31. Example of appearance defect in spun yarn fabric. Melted and flattened fiber end forms an undesired glitter point. (a) Low SEM magnification, (b) glitter point at higher magnification.

REFERENCES

1. C. W. Oatley, The Scanning Electron Microscope, Cambridge University Press, London, 1972.
2. P. R. Thorton, Scanning Electron Microscopy, Chapman and Hall, London, 1968.
3. J. W. S. Hearle, J. T. Sparrow, and P. M. Cross, The Use of the Scanning Electron Microscope, Pergamon Press, Oxford, 1972.
4. R. D. Van Veld and T. J. Shaffner, J. Phys. (E: Scientific Instruments), 4, 633 (1971).
5. T. J. Shaffner, Scanning Electron Microscopy/1973, Proceedings of the Sixth Annual Scanning Electron Microscope Symposium (O. Johari, ed.), Illinois Institute of Technology Research Institute, Chicago, 1973, p. 293.
6. L. A. Weitzenkamp, J. Phys. (E: Scientific Instruments Series), 2, 561 (1969).
7. A. Boyde and C. Wood, J. Microscopy (Pt 3), 90, 245 (1969).
8. J. B. Pawley, Scanning Electron Microscopy/1972, Proceedings of the Fifth Annual Scanning Electron Microscope Symposium (O. Johari, ed.), Illinois Institute of Technology Research Institute, Chicago, 1972, p. 153.
9. L. M. Welter, and A. N. McKee, Scanning Electron Microscopy/1972, Proceedings of the Fifth Annual Scanning Electron Microscope Symposium (O. Johari, ed.), Illinois Institute of Technology Research Institute, Chicago, 1972, p. 161.
10. P. Echlin and J. W. Hyde, Scanning Electron Microscopy/1972, Proceedings of the Fifth Annual Scanning Electron Microscope Symposium (O. Johari, ed.), Illinois Institute of Technology Research Institute, Chicago, 1972, p. 138.
11. G. E. Pfefferkorn, H. Gruter, and M. Pfautsch, Scanning Electron Microscopy/1972, Proceedings of the Fifth Annual Scanning Electron Microscope Symposium (O. Johari, ed.), Illinois Institute of Technology Research Institute, Chicago, 1972, p. 147.
12. G. E. Pfefferkorn, Scanning Electron Microscopy/1973, Proceedings of the Sixth Annual Scanning Electron Microscope Symposium (O. Johari, ed.), Illinois Institute of Technology Research Institute, Chicago, 1973, p. 752.
13. V. Peck, in J. Appl. Polymer Sci., Applied Polymer Symposium No. 16 (L. H. Princen, ed.), Wiley, New York, 1972, p. 19.
14. J. Sikorski, J. S. Moss, A. Hepworth, and T. Buckley, Proceedings of the Second Annual Stereoscan Colloquium, Engis Equipment Co., Morton Grove, Ill., 1969, p. 25.
15. J. W. Heavens, A. Keller, J. M. Pope, and J. Rowell, J. Mat. Sci., 5, 53 (1970).
16. J. Dlugosz and A. Keller, J. Appl. Phys., 39, 5776 (1968).
17. L. F. Thompson, R. L. Venable, and K. G. Mayhan, J. Appl. Polymer Sci., Applied Polymer Symposium No. 16 (L. H. Princen, ed.), Wiley, New York, 1971, p. 51.

18. O. T. Grubb and A. Keller, J. of Mat. Sci., 7, 822 (1972).
19. J. L. Brown, J. L. Hubbard, and L. H. Glassman, Proceedings of the Fifth Annual Stereoscan Colloquium, Kent Cambridge Scientific Co., Morton Grove, Ill., 1972, p. 95.
20. A. N. McKee and C. L. Beattie, Textile Res. J., 40, 1006 (1970).
21. J. B. Pawley and T. L. Hayes, Scanning Electron Microscopy/1971, Proceedings of the Fourth Annual Scanning Electron Microscope Symposium (O. Johari, ed.), Illinois Institute of Technology Research Institute, Chicago, 1971, p. 107.
22. R. D. Van Veld, F. D. Bryant, and H. R. Billica, Textile Res. J., 40, 1138 (1970).

Chapter 8

APPLICATIONS OF THE ATR TECHNIQUE TO THE STUDY OF FIBER AND FABRIC SURFACES

John P. Sibilia

Chemical Research Center
Allied Chemical Corporation
Morristown, New Jersey

I. INTRODUCTION

Infrared spectroscopy has been proven a powerful tool for the analysis of materials. It has been effectively applied to substances in the liquid, gaseous, and solid states. It has been particularly valuable to the polymer scientist in the areas of analytical chemistry and molecular structural analysis.

However, there has always been a problem with obtaining good quality infrared spectra of fibers. Some of the more common techniques for obtaining infrared spectra of fibers have included [1]:

1. chopping or grinding the fibers and dispersing them in a matrix such as nujol or KBr;
2. aligning fibers crudely;
3. utilizing an infrared microscope or beam condenser;
4. dissolving fibers in a solvent;
5. casting or molding the fibers into films; and
6. pyrolysis and hydrolysis.

More recently a new technique has been developed [2] that involves winding the fiber in a uniform manner around a spacer between infrared transparent windows. All these techniques are limited in some respects, and they are not generally applicable to all fiber types. For the most part these techniques provide information about the bulk of the fiber and are of limited value to the scientist interested in information about the fiber surface.

About 1960, Fahrenfort [3,4] and Harrick [5] described a new infrared technique called attenuated total reflectance (ATR). With this technique they were able to demonstrate that the infrared spectra of surface layers of various materials could be readily obtained. This work opened up a whole new frontier for the surface chemist and in particular it offered a new and useful way of obtaining information about the surfaces of fibers and about fiber structures in general.

Since that time there have been a number of reviews on the subject of internal reflection spectroscopy [6-18]. The number of review articles relating strictly to the application of ATR to fibers, however, have been limited [19,19a] although there has been considerable effort in the field. In this chapter, the principles, instrumentation, and applications of the technique to a wide variety of fiber- and fabric-surface problems will be presented. A number of other studies, although not made directly on fibers, are also presented since they should be of interest to fiber-surface scientists.

II. PRINCIPLES OF THE ATR TECHNIQUE

A. Attenuated Total Reflection

The mathematics and details of the theory of ATR have been discussed elsewhere [3,6] and will not be covered here. However, a brief discussion of the principles and terminology involved will aid in understanding the broad scope of applications of the technique.

In Fig. 1a is shown a hemicylindrical prism of refractive index η_1 in contact with a sample of refractive index η_2, where $\eta_1 > \eta_2$. If the

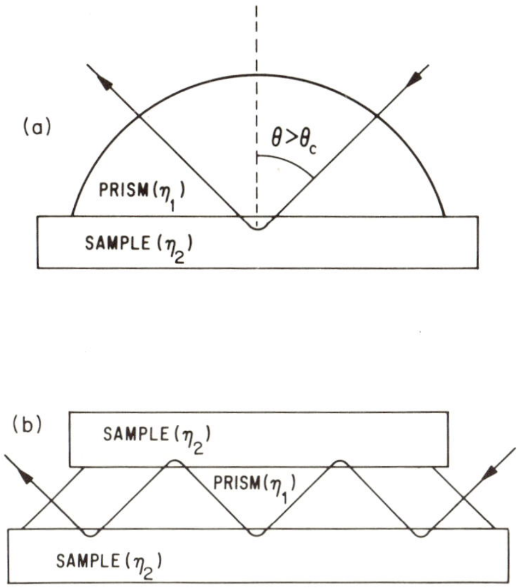

FIG. 1. Internal reflectance prisms. (a) Single reflection, hemicylindrical ATR (attenuated total reflectance) prism with sample, (b) multiple reflection, trapezoidal MIR (multiple internal reflectance) prism with sample.

prism and sample are transparent at the frequency of light used, then a light ray entering into the prism and impinging at the sample-prism interface at an angle greater than θ_c (the critical angle) will be totally reflected. The reflected wave, however, penetrates a small distance into the rarer medium. If the sample absorbs at the frequency of the impinging light, then the light leaving the prism will be attenuated, that is, suffer a loss in intensity. Therefore, a plot of the intensity of the reflected light as a function of infrared wavelength will result in a spectrum very similar to that obtained from a transmission infrared spectrum of the material.

The type of prism shown in Fig. 1a results in a single reflection at the prism-sample interface. The wavefront, therefore, makes a single interaction with the sample.

In Fig. 1b is shown another type of prism that is commonly used. This type, which has a trapezoidal shape, allows the light to make multiple reflections at the prism-sample interface. If a sample is placed against each side of the prism, then partial absorption will occur for each reflection of the incident light at the interface. A single wavefront, therefore, will have multiple interactions with the sample. The net effect is that the spectrum obtained is much more intense than in the case of the single reflection unit. This modification of the ATR technique has commonly been referred to as frustrated multiple internal reflectance (FMIR) or simply multiple internal reflection (MIR). The number of reflections will depend on the angle of incidence of the incoming light, the length of the prism, and the thickness of the prism.

For most purposes, trapezoidal-shaped, multiple reflection elements are used, which give about 20 reflections. However, with very thin trapezoidal elements, the number of reflections can be increased to 400 and thousands of reflections have been obtained from rod or light-pipe optical elements [8].

In Fig. 2 are shown the transmission spectrum and the MIR spectrum of a film of nylon-6. The film was prepared by casting from a solution of formic acid directly onto a KRS-5 reflection prism. The transmission spectrum was obtained by passing the light directly through the film and prism. The spectra are very similar; the same bands are present in both. However, the intensity of the longer-wavelength bands is relatively greater in the MIR spectrum. This is because the depth of penetration of the incident beam into the sample is greater at longer wavelengths.

The intensity differences are quite evident if bands at either end of the spectra are compared. For example the —NH stretching frequency at 3300 cm^{-1} is more intense in the transmission spectrum, while the —NH deformation frequency at 698 cm^{-1} is much stronger in the MIR spectrum.

B. Depth of Penetration

The depth of penetration (d_p) of a light wave into a sample is defined as the distance required for the electric field amplitude of the incident light wave to fall to 1/e of its value as it enters the sample from the prism of higher refractive index [6]. It is expressed as

$$d_p = \frac{\lambda}{2\pi\eta_1 (\sin^2 \theta - \eta_{21}^2)^{1/2}}$$

where λ is the wavelength, θ is the angle of incidence, η_1 the index of refraction of the denser medium, and η_{21} is the ratio of the index of refraction of the rarer medium to that of the denser medium.

Therefore, the depth of penetration is seen to increase linearly with increasing wavelength for a system where θ, η_1, and η_2 are constant.

Likewise the depth of penetration will decrease with increasing values of θ, that is increasing angles of incidence. Depths of penetration are of the order of fractions of a micrometer to several micrometers for polymeric materials.

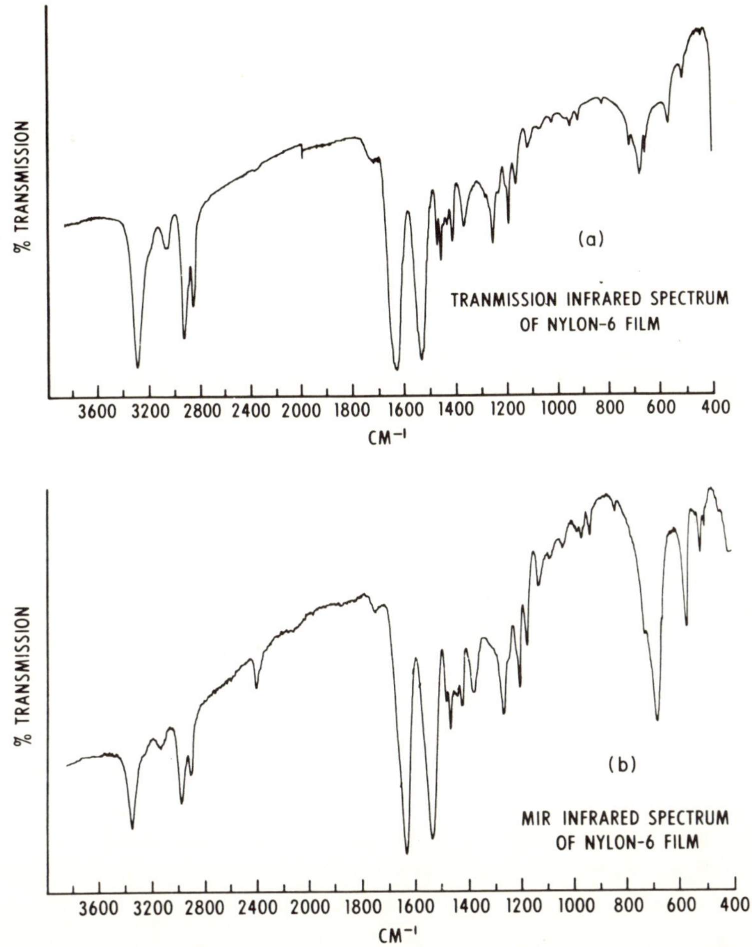

FIG. 2. Comparison of transmission and MIR spectra of nylon-6. (a) Transmission infrared spectrum of a thin nylon-6 film cast from formic acid onto a KRS-5 prism, (b) MIR spectrum of the film used in (a).

In Table 1 are given the indices of refraction of some materials commonly used as prisms for ATR work [20].

In Table 2 are shown the depths of penetration for these materials, expressed as d_p/λ, for various angles of incidence [21]. The depths correspond to penetration into a material of refractive index 1.5. This is approximately the value for many synthetic polymeric materials.

TABLE 1

Prism Materials Used for Attenuated Total Reflectance[a]

Material	Angle (θ_c)	Transmission range (μm)	Index of refraction
Ge	15	2 to 11.5	4
TlBr-TlI (KRS-5)	24	0.5 to 35	2.4
AgCl	30	0.4 to 23	2

[a]Reprinted in part from ref. 20 by courtesy of Beckman Instruments, Inc.

TABLE 2

Depth of Penetration (d_p/λ) for $\eta_2 = 1.5$[a]

	θ		
	30°	45°	60°
Ge	0.091	0.041	0.002
KRS-5	i[b]	0.290	0.113
AgCl	i	i	0.316

[a]Reprinted from ref. 21, p. 206, courtesy of Heyden and Son, Inc.
[b]i = All the light goes into the sample and reflection is minimal.

The depth of penetration is greatest near the critical angle, while at angles less than the critical angle, essentially all the light goes into the sample and reflection is minimal (indicated by i in Table 2).

C. Effective Sample Thickness

For cases of low absorption, the effective thickness (d_e) is the thickness of a material necessary to give a transmission spectrum of the same intensity as that obtained by a single reflectance prism [6]. It can be expressed as

$$d_e = \frac{\eta_{21} E_0^2 d_p}{2 \cos \theta}$$

where E_0 is the amplitude of the electric field in the rarer medium at the interface and η_{21}, d_p, and θ are as previously defined.

Effective thickness is a more complicated term than depth of penetration and depends on d_p as well as the indices of refraction, θ, and E_0.

III. INSTRUMENTAL TECHNIQUES

There are a number of suppliers of commercial ATR attachments which are adaptable to the infrared spectrometers produced today.

In Fig. 3 is shown a single reflection unit that contains a hemicylindrical prism. The optical diagram for this system is also shown. This equipment includes beam condenser optics and permits measurement on small samples (maximum sample area 4×9 mm). The angle of incidence can be varied from 25 to 68°.

In Fig. 4 is shown multiple reflection or MIR equipment mounted in the sample compartment of an infrared spectrometer. The optical path is also shown in Fig. 4. The angle of incidence can also be varied in units like this. The MIR units are usually the ones of choice for fiber and fabric studies because of the increased intensity over that of the single reflection units.

Mounting of the fibers and fabrics is critical in obtaining good MIR spectra. Good contact must be achieved between the sample and prism

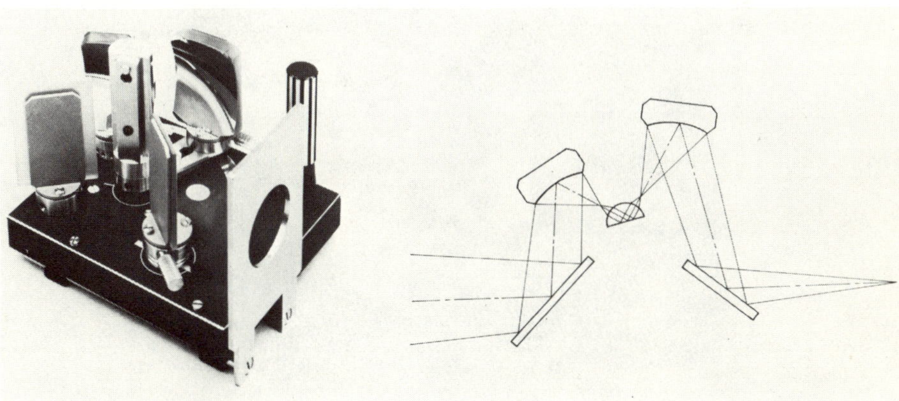

FIG. 3. Single reflectance micro-ATR accessory with capability of varying the angle of incidence. Reprinted from ref. 20, p. 14, courtesy of Beckman Instruments, Inc.

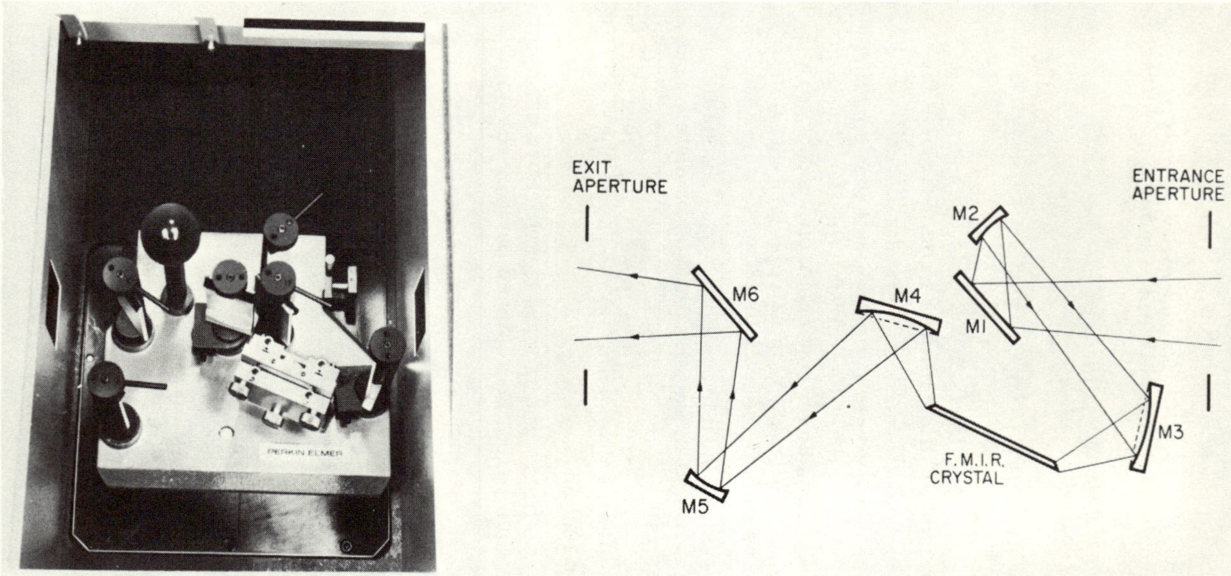

FIG. 4. Multiple reflection MIR unit and optical diagram. Reprinted courtesy of the Perkin Elmer Corporation.

surface. This is more difficult to do with fibers than with smooth solid films or liquids.

In Fig. 5 are shown ways in which three different types of fiber and fabric samples may be mounted for surface studies.

A fabric sample is shown in Fig. 5a just prior to clamping in the MIR holder. In this case, fabric is placed against both prism surfaces for enhanced intensity.

Fiber, yarn, or cord samples can also be examined by wrapping around the holder as shown in Fig. 5b. The MIR prism can then be clamped between the plates on which the fibers are mounted. Careful, parallel alignment will result in maximum packing, which will in turn result in the most intense spectrum.

Spectra may also be obtained on random fiber bundles as shown in Fig. 5c. This is particularly suitable for fibers in the form of staple padding. In this case the fiber pads may be attached to the mounting plates by means of adhesive tape.

The equipment and sample handling techniques described here will be discussed further, along with special techniques in the following section on applications.

IV. DIRECT APPLICATIONS TO FIBERS AND FABRICS

A. Fiber Identification

One of the most common applications of the ATR technique has been for the identification of fibers, fiber blends, and fabrics [15,22-25]. This has been demonstrated for both the qualitative identification and the quantitative measurement of individual as well as mixed fiber compositions. Siesler in particular has used this method for numerous investigations of fibers and fabrics including studies on high-temperature-stable fibers and textiles [25a-c]. It must be remembered, however, that the depth of penetration is of the order of several micrometers, so that in any identification of fibrous materials by the ATR technique, one must assume that the first several micrometers of depth of sample are representative of the bulk of the material.

Wilks and Iszard demonstrated the usefulness of the ATR technique for fiber identification by obtaining the spectra of 47 synthetic and natural fibers [22]. Some fiber blends were also included in this work. In Fig. 6 are shown the spectra of cotton (6a), Dacron (6b), and a 65% Dacron to 35% cotton blend (6c). Characteristic absorption is shown for both materials in the spectrum of the blend. For example, the carbonyl absorption at 1710 cm^{-1} (5.75 μm) is characteristic of the ester group in Dacron while the —OH stretching frequency at 3400 cm^{-1} (2.9 μm) is due to the cotton.

FIG. 5. MIR sampling techniques for fibers and fabrics. (a) Fabric sample ready for clamping to the internal reflector plate; (b) internal reflector plate ready to be mounted between two halves of the plate holder around which yarn is wrapped; (c) staple bundles taped to two halves of the plate holder with reflector plate ready for clamping. Reprinted from ref. 22, courtesy of Wilks Scientific Corporation.

FIG. 5 (continued)

Quantitative analysis of such blend samples has been carried out [21,25,25d] with some success. However, different materials vary in their degree of hardness, which will in turn effect the degree of contact of the fibers with the prism surface, and therefore effect the relative intensities of the ATR spectrum. One way to overcome this effect is to calibrate the technique by obtaining spectra on blends of known compositions.

Identification and analysis of fibers and fiber blends has also been applied to fabrics [19]. In Fig. 7a is shown the MIR spectrum of a wool fabric. A photomicrograph of the fabric is also included. The spectrum of a polyester taffeta and a photograph of the fabric from which the spectrum is obtained is given in Fig. 7b. In Fig. 7c is given the photograph of a fabric comprised of a blend of 50/50 wool and polyester and the MIR spectrum obtained from this fabric. Again, elements of each component can be identified. Wool shows its most characteristic bands, due to the amide group at 3300, 1650, and 1510 cm^{-1} while the remainder of the bands in the blend spectra are due for the most part to the polyester.

Identification and analysis of composite fibers have also been accomplished by use of the ATR technique. Carlsson et al. [19] point out that valuable information can be obtained on the identification of the constituent polymers in bicomponent fibers. Such fibers might be prepared by a side-by-side extrusion of two polymeric films, or by coat-

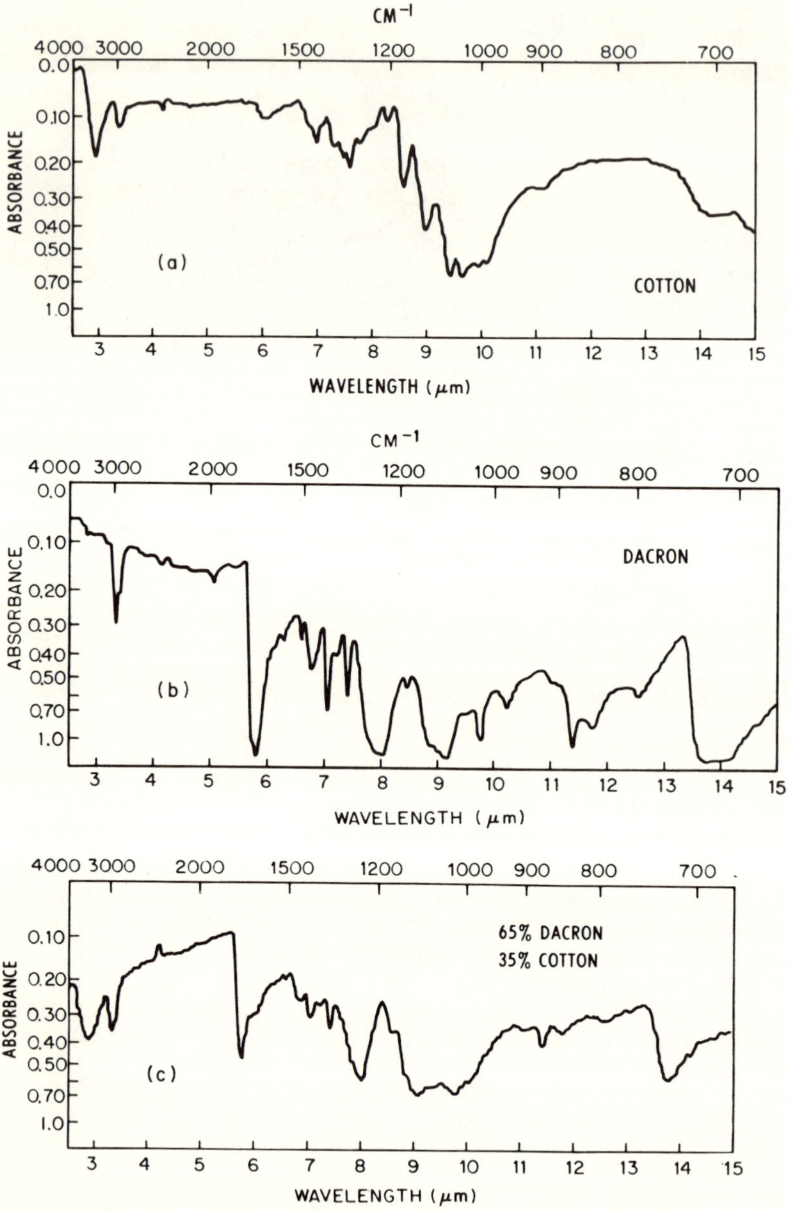

FIG. 6. MIR spectra of fibers and a fiber blend. (a) MIR spectrum of cotton, (b) MIR spectrum of Dacron, (c) MIR spectrum of a blend of 65% Dacron and 35% cotton. Reprinted from ref. 22, courtesy of Wilks Scientific Corporation.

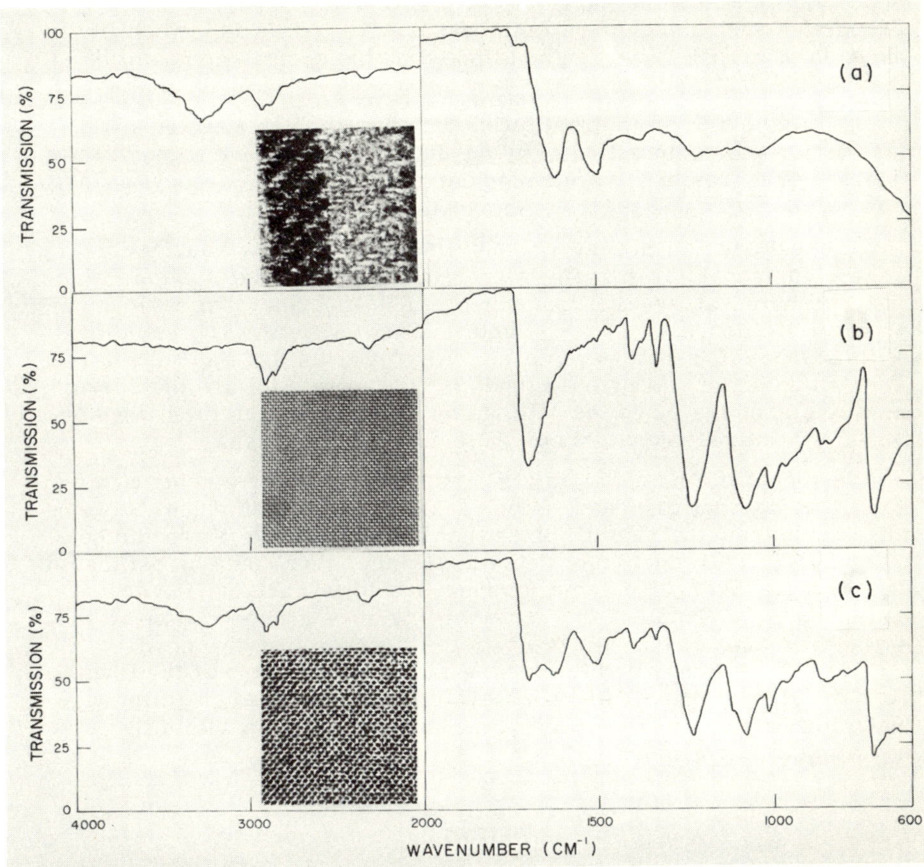

FIG. 7. MIR spectra of various fabrics: (a) wool, (b) polyester taffeta, (c) wool-polyester blend 50/50. Reprinted from ref. 19, courtesy of Canadian Textile Journal Publishing Co., Ltd.

ing an extruded film and subsequently slitting and/or fibrillating the bicomponent films into fibers.

Biconstituent fibers prepared from extrusion of a melt blend of nylon-6 and poly(ethylene terephthalate) have also been analyzed by ATR [26]. Fibers prepared in this manner can be made so that the major component will be the continuous phase and the minor component will be dispersed in the form of long, fine-diameter fibrils with their axes in the direction of the macro fiber axis.

Although nylon content and, therefore, the bulk composition of the fibers can be obtained by an elemental nitrogen analysis, it is also important to determine the fiber composition at or near the surface. Properties such as adhesion in the case of tire cords and dyeability in the case of carpet yarns can be affected by the surface composition of the fiber.

In Table 3 are given quantitative data based on an MIR technique for the surface composition of nylon-6 and poly(ethylene terephthalate) fibers. Measurements were based on the ratio of the carbonyl absorptions of the nylon and polyester components.

The data in the second column (number 1 below "Surface") were obtained at the natural penetration depth of the system, which was estimated at several μm since a KRS-5 prism was used at an angle of incidence of 45°. Samples A and B were yarn samples while samples C and D were cords prepared from 3 plies of the yarns. The data indicate that nylon-6 tends to concentrate toward the surface in all the samples since the MIR values are higher than the values obtained by a nitrogen analysis on the bulk of the samples.

The penetration depth was also reduced further by placing a thin, solid, transparent film such as poly(chloro trifluoroethylene) between the samples and the prism. Since the thickness of the film could be determined, the reduction of depth of surface layer could be estimated. These data, given in column number 2 below "Surface," indicate that there is a gradient of increasing nylon-6 concentration as the surface of the cord is approached and suggest that the cord-forming process may be altering the radial composition of the fibers relative to the yarns. Although there is an overall concentration of nylon toward the surface of the yarns, the composition within the first several μm is constant.

TABLE 3

Weight % Nylon-6 on the Surface of Biconstituent Fibers[a]

Sample	Surface 1[b]	Surface 2[c]	Whole sample[d]
A	91	91	86
B	45	45	27
C	77	85	66
D	63	69	48

[a]Reprinted from ref. 26, p. 2913, courtesy of Interscience Publishers (a division of John Wiley & Sons, Inc.).

[b]Estimated at several μm.

[c]Reduced by about 0.5 μm.

[d]From nitrogen analysis of whole sample.

TABLE 4

Comparison of Surface Compositions of Biconstituent Fibers and Films[a]

Sample	Weight % nylon-6		
	Film	Fiber	From nitrogen analysis of whole sample
A	97	91	86
B	42	45	27
C	81	77	66
D	69	63	48

[a]Reprinted from ref. 26, p. 2913, courtesy of Interscience Publishers (a Division of John Wiley & Sons, Inc.).

In Table 4 are shown MIR data on films prepared by melting and pressing the fiber samples. A change in surface composition from that of the fibers is noted, indicating that redistribution of the components can occur during the melting and cooling process used for preparing the films.

The method of reducing the depth of penetration by placing a transparent film between sample and prism can be applied to many of the applications to be discussed subsequently. It should be noted, however, that a reduction of penetration depth results in a decrease in the intensity of the spectrum. Therefore, a compromise must be reached between the "thinness" of the surface layer examined and the intensity of the infrared spectrum obtained.

B. Coatings and Reactions on Fibers and Fabric Surfaces

The ATR technique has been proven a valuable tool in the compositional analysis of film laminates. It has likewise been useful for the analysis of resin-coated fabrics. In Fig. 8 are shown examples of the ATR technique applied to a coating study on both a film and a fabric [19]. In Fig. 8a are shown the spectra of the uncoated side of a nylon fabric and the side that is coated with a polyester-type polyurethane coating. Since the nylon absorption at about 1650 cm^{-1} is not present in the spectrum of the coated side, it can be concluded that the coating is thicker than the degree of penetration, that is of the order of 1 μm or thicker. Spectra of nylon-coated polypropylene film are given in Fig. 8b to show how the ATR technique enhances the spectrum of the coating material as compared to the transmission spectrum.

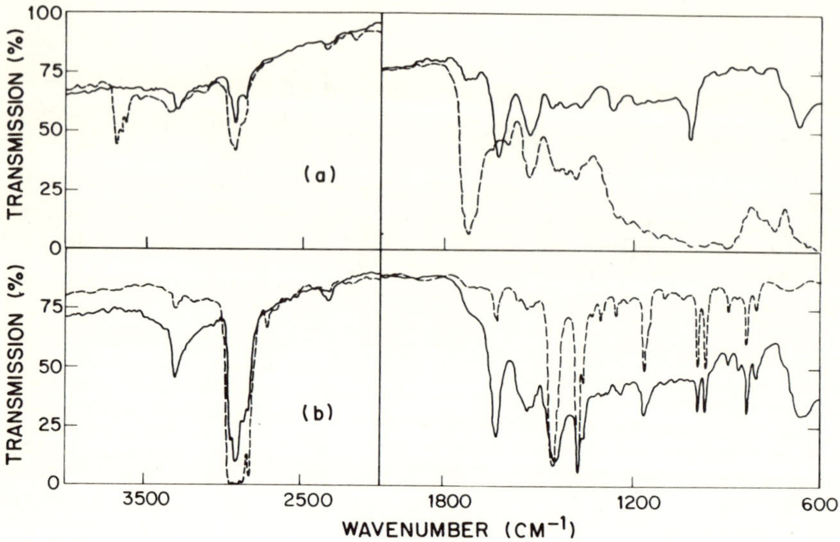

FIG. 8. ATR spectra of coated materials. (a) Polyurethane (polyester type) coated nylon fabric. Solid line: ATR spectrum of uncoated side of fabric. Dashed line: ATR spectrum of coated side of fabric. (b) Nylon-coated polypropylene film. Solid line: ATR spectrum of coated surface. Dashed line: transmission spectrum. Reprinted from ref. 19, courtesy of Canadian Textile Journal Publishing Co., Ltd.

Several papers [23,27-29] have dealt with identification of resin-coated cotton fabrics and chemically modified cottons. Identification of the various resins as well as variations in the composition of the resin over the surface was pointed out.

McCall et al. [27] showed the advantages of the ATR technique over prior techniques for the determination of surface coatings on cotton surfaces and for the determination of the effects on surfaces that have been chemically modified. They compared the spectra of nine samples of chemically modified cotton fabrics by a KBr disk method (which requires grinding the sample and dispersing in an excess of KBr from which a disk can be prepared) and an MIR technique.

The MIR spectra showed the nature of the coating or additive at least as well as the KBr spectra and in most cases the MIR spectra were more informative. In addition, the MIR technique did not destroy the sample, was much more rapid, and could indicate the presence of additive on one side of a treated fabric and the absence of additive on the untreated side. In Fig. 9a is shown the MIR spectrum of the side of a cotton fabric that was treated with a 5% solution of methylolmelamine; Fig. 9b shows the MIR spectrum of the untreated side. Absorp-

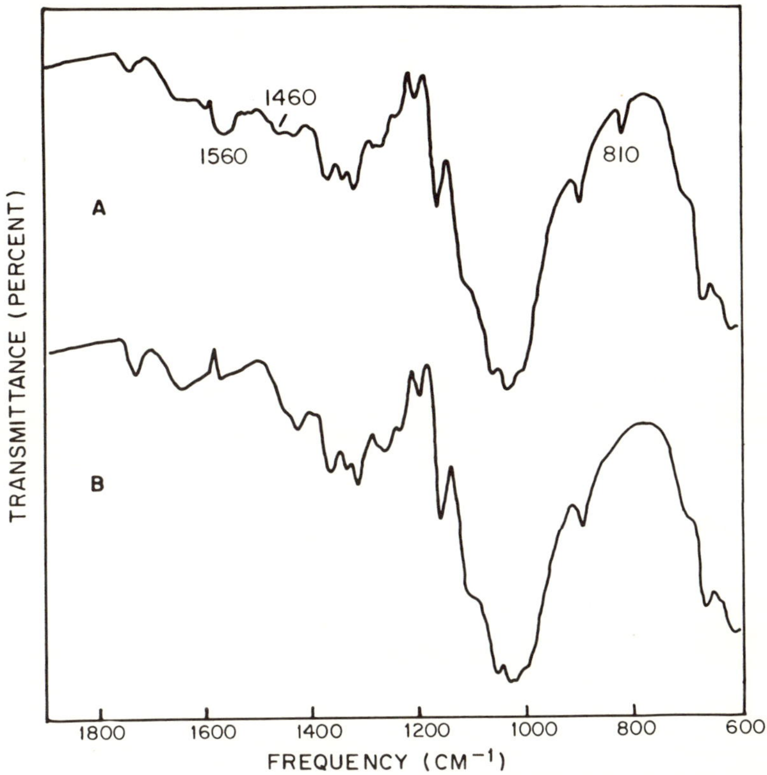

FIG. 9. MIR spectra of a cotton fabric treated on one side with a 5% solution of methylolmelamine. (A) Treated side, (B) untreated side. Reprinted from ref. 27, courtesy of American Dyestuff Reporter, published by SAF International, Inc.

tion bands are evident at 1560 cm^{-1} and 1460 cm^{-1} in the spectrum of the treated side. Most obvious in the spectrum of the treated side is the band at 810 cm^{-1} due to the melamine ring, which is missing in the spectrum of the untreated side.

In another example, the same authors [27] demonstrated the difference between the back and face of a cotton-filling-faced sateen. In Fig. 10 are shown the MIR spectra of the back and face of a sateen fabric. The fabric was woven with mercerized filling threads and unmercerized warp threads. Evidence for mercerized cellulose is shown in Fig. 10b by the presence of bands at 3480 and 3440 cm^{-1}. An increase in intensity of the 890 cm^{-1} band and a decrease of the 1110 cm^{-1} band is also pointed out by the authors as being indicative of the mercerized cellulose.

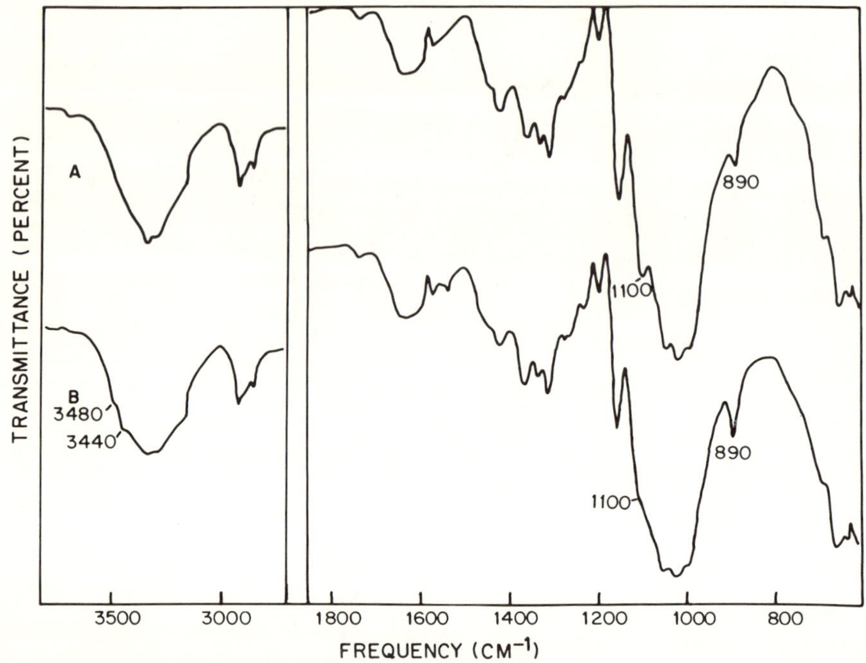

FIG. 10. MIR spectra of a cotton-filling-faced sateen. (A) Back, (B) face. Reprinted from ref. 27, courtesy of American Dyestuff Reporter published by SAF International, Inc.

Hermann [28] utilized a differential MIR technique to obtain information about a phenylcarbethoxyvinyl coating on cotton fibers. The differential spectrum was obtained by placing a pad of the treated cotton against a KRS-5 plate in the sample beam of a double-beam infrared spectrometer, while an untreated cotton pad was placed against a KRS-5 plate in the reference beam. The compensation spectrum cancelled out the absorption due to cotton and resulted in a spectrum essentially the same as the coating compound itself.

The sorption of pyridine on the surface of glass fibers was clearly demonstrated by the MIR technique [28]. In fact, based on the position of the ring modes of pyridine, it was concluded that the pyridine was chemisorbed at both Lewis and Bronsted sites on the glass surface.

Although it has been pointed out [19,29] that obtaining information about surface finishes on fibers by ATR is difficult, some success has been obtained. The technique was successful in demonstrating the existence of a 1% lubricant on the surface of nylon-66 fibers [8]. Spectra of the coated and uncoated fibers are given in Fig. 11. The presence of the lubricant is evident by the absorption at 5.8 μm (1725 cm^{-1}).

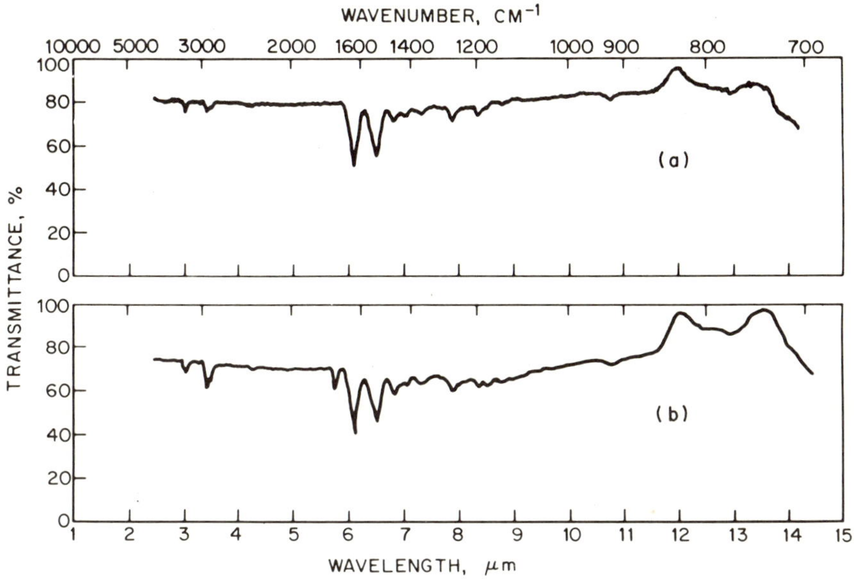

FIG. 11. Internal reflection spectra of coated and uncoated nylon-66 fibers using a 21-reflection germanium element. (a) Uncoated, (b) coated with 1% lubricant. Reprinted from ref. 8.

Some evidence for the presence of urea-type coatings on cotton surfaces, an acrylate-type coating on polyamide fibers, and mineral oil on polyamide fabrics was also obtained [29]. However, in most cases only the major band of the coatings was observed. In the same reference [29] a study of a heat-treated urea-coated cotton fabric was made. The distribution of the urea resin (imparted for wrinkle resistance) on the cotton surface could be followed by observing the intensity of the amide-I band in the ATR spectrum.

The ATR technique has been used with mixed success in determining surface oxidation or photo-oxidation effects. Krentz [29] observed only a weak carbonyl after oxidative treatment of cotton. However, Carlsson et al. [19] successfully used the ATR technique to obtain information about the photo-oxidation of polypropylene fibers. Their approach was to work on films instead of fibers. In one study they exposed polypropylene films of varying thickness to photo-oxidative degradation. They employed two reflection elements (Ge or KRS-5) each at two angles of incidence in order to obtain relative concentrations of ketone and hydroperoxide photo-oxidation products as a function of depth from the film surface. It was found that, even up to the point of brittle fracture of the films, essentially all of the photo-oxidation occurred within the first few micrometers of film surface and the bulk of the polymer remained unoxidized.

Other studies on polyolefins were also made although not directly on fibers. ATR studies on oxidized films of polyethylene and isotactic polypropylene were made to determine the distribution of oxidized groups as a function of thickness [29a]. It was also demonstrated [29b] by ATR techniques that oxidative aging of polyethylene increased as a function of the irradiation dose.

Andries and Diem [29c] measured the degree of ozone attack on elastomers by a study of their ATR spectra. They concluded that ozone attack on natural rubber was extensive and complex and could be related to the acetone extractables. The extent of reaction of ozone on synthetic rubbers such as cis-polybutadiene and butadiene-styrene rubbers however was much less severe.

The effect of corona discharge on the surface of polypropylene films was also investigated [30]. The use of electric-field plasma (normally a corona discharge) is known to improve such surface properties as printability, adhesion, and dyeability.

In this study the authors were able to alter the surface properties of substrates. This made it possible to form bicomponent fibers and film laminates from normally incompatible copolymers. In particular, they improved the adhesion between nylon-610 coatings and polypropylene films. They showed by use of the ATR technique that the formation of carbonyl and unsaturated groups in polypropylene film were related to the length of duration of the corona-discharge treatment.

It has already been pointed out that the ATR technique is useful in studying wrinkle-resistance treatments. It should also be possible to detect changes resulting from the grafting of vinyl polymers onto fibers and fabrics [19]. Information about the uniformity and depth of penetration of such treatment should also be obtainable by ATR.

C. Pyrolysis-ATR Techniques

The techniques of pyrolyzing a polymeric substance and identifying the pyrolysis fragments have been used by many workers [6,21,31-35] in order to identify structures and to gain insight into both polymerization and pyrolysis or combustion mechanisms. Cassels [35] used a combination of the ATR and pyrolysis techniques to qualitatively classify and to obtain information on the characteristics of the thermal degradation of 15 different synthetic fibers. He used a Wilks Model 40 Pyro-Chem accessory for his work. All fibers were fired at 800°C in the absence of oxygen. He obtained the infrared transmission spectrum of the gas phase; the MIR spectrum of the condensed phase resulting from pyrolysis; and the MIR spectrum of the fibers before pyrolysis.

The study included polyamides, polyolefins, cellulosics, halogenated fibers, polyesters, acrylics, and polyurethanes. In many cases the condensed phases were similar to the original fiber spectra although significant differences were noted for some fiber types. Most of the gas spectra showed low molecular-weight compounds characteristic of the original fibers. For example, four different acrylic types showed hydrogen cyanide in the gas phase while ammonia and hydrogen chloride were found, respectively, in the gas phase of pyrolyzed fibers containing polyamides and halogen.

Although this work was aimed at obtaining information about the bulk of the polymer, it should be of general interest to a fiber surface chemist. It may be feasible, by such a technique, to identify pyrolysis products characteristic of chemical modification on the surface of fibers. These products, however, would be expected to be quite small relative to the amount of products obtained from the bulk of the material.

The pyrolysis-ATR technique was applied with some success to identification of coatings or treatment on glass fibers and surfaces [36-38]. In many cases the direct ATR spectrum of the coated glass fibers does not show the presence of the coating due to the low concentration and the strong absorbing background of the glass. However, the pyrolysis-ATR technique applied to glass fibers for information about organic surface coatings is much more informative.

In Fig. 12a is shown the internal reflectance spectrum of a silane-treated glass fabric [36]. Little evidence of the treatment is visible in the spectrum. In Fig. 12b the nature of the surface finish is shown. Products from the pyrolysis of the treated glass fabric were condensed on an MIR plate and the MIR infrared spectrum obtained. Similar information was also obtained from phenolic-impregnated glass fiber. Weiner [37,38] points out that although the pyrolysis-ATR technique was useful for "fingerprint" identification of coatings from glass which were made from organic film formers, it was not particularly successful in the case of coatings comprised of functional silanes. He attributed the difficulty to a lack of condensable material forming from the pyrolysis of the silane materials.

D. Structure and Morphology

Structure and morphology of fibers are important aspects of fiber technology. Physical properties such as tensile strength, dimensional stability, and fatigue life are strongly dependent on the molecular arrangements of molecules in a fiber. Likewise, diffusion-controlled processes such as dyeability, gas permeability, moisture penetration, and organic liquid penetration are all dependent, to some extent, on molecular orientation and crystallinity. X-ray diffraction, birefringence, and transmission infrared [2] techniques on fibers have all been helpful in obtaining information about the bulk orientation and crystallinity of fibers. However, depending on the chemical nature of a fiber and the way in which it is processed, it is possible to obtain

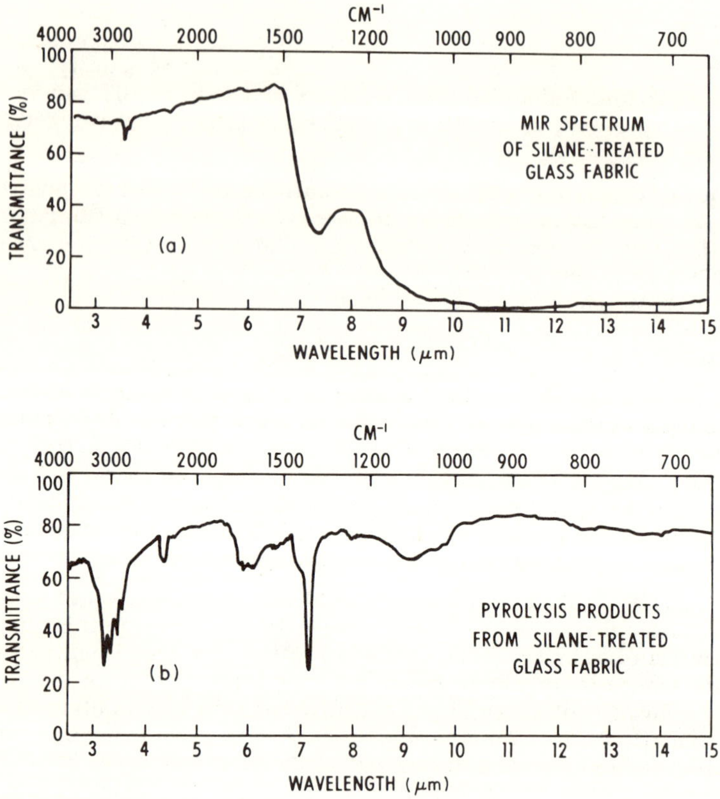

FIG. 12. Application of ATR pyrolysis for analysis of surface-treated glass fabric. (a) Internal reflection spectrum of silane-treated glass fabric; (b) MIR spectrum of condensed-phase pyrolysis products of sample in (a). Reprinted from ref. 36, courtesy of Wilks Scientific Corp.

structural features at the surface which are different from those in the interior or in the bulk of the fiber.

The process of drawing can impart many useful properties to a synthetic fiber. When this process occurs, molecular chains can orient in the fiber direction. The infrared dichroism of a particular vibration of known origin has been used to give valuable information about molecular orientation.

There is a relationship between the observed dichroic ratio and the transition moment direction of a particular infrared vibrational mode.

The dichroic ratio can be expressed as [8]

$$\frac{k_x}{k_y} = \frac{f \cos^2 \omega + (1/3)(1 - f)}{(1/2) f \sin^2 \omega + (1/3)(1 - f)}$$

where ω is the angle of inclination of the transition moment with the molecular axis and f the fraction of molecules that are perfectly oriented.

It is quite difficult to obtain quantitative transmission polarization data on fibers themselves [2] or on microtomed sections of fibers [39,40]. Barr [41], however, was quite successful in applying the ATR technique to a study of orientation on a series of drawn nylon-66 fibers. He determined that the surface dichroic ratios obtained by internal reflection spectroscopy are directly related to the bulk birefringence for fibers having a draw ratio of more than 1.4. However, the surface of the fibers was noted to have a greater orientation than the interior for fibers of draw ratio less than 1.4. Therefore, it was pointed out that the polarization data obtained by the ATR technique was very sensitive to small changes in surface orientation and complimented the X-ray diffraction technique.

In Fig. 13 are shown the polarized ATR spectra of oriented nylon-66 yarn. The intensity of the 1640 cm^{-1} band (amide I) is greater when the electric vector of the incident light is perpendicular to the fiber axis. The 1538 cm^{-1} band (amide II), on the other hand, is stronger when parallel incident light is used. These, as well as other absorption bands, were used by Barr in the analysis of nylon-66 fibers [41].

Heidemann [42] discusses the skin-core effect in more detail. He explains, how, based on microscopic dyeing and swelling experiments, the skin of fibers is considered very oriented. When filaments have different core-surface structures, streaking of dyes on the surface can occur. Therefore, the absorption of the dye depends, at least in part, on the diffusion rate of the dye across the skin of the fiber. Therefore, it is very important to define the orientation of the surface and core of the fiber.

Heidemann used the 1430 cm^{-1} band in nylon-66 as a measure of the α-methylene groups in the gauche conformation and the 925 cm^{-1} band as a measure of the crystalline content. Measurements were made on both these bands relative to the 2910 cm^{-1} band as an internal standard. ATR spectra were obtained on nylon-66 fibers as a function of varying angles of incidence. The depth of penetration was, therefore, varied by varying the angle of incidence.

In Table 5 are given the angles of incidence and absorbance ratios for the 1430 and 925 cm^{-1} bands of nylon-66. These data were also compared to values obtained from an isotropic film. Heidemann

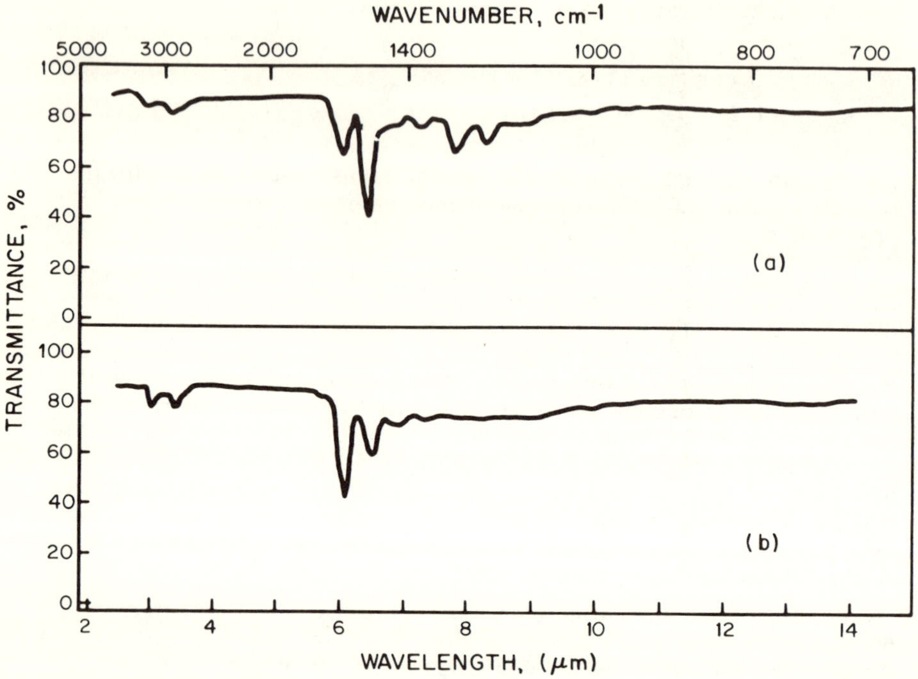

FIG. 13. Polarized internal reflection spectra of nylon-66 yarn. (a) Electric vector parallel to fiber axis, (b) electric vector perpendicular to fiber axis. Reprinted from ref. 8.

TABLE 5

Absorbance Ratios from the ATR Spectra (KRS-5) for a Drawn Nylon-66 Filament[a]

Angle of incidence	A_{1430}/A_{2910}	A_{925}/A_{2910}
51°	0.25	0.09
45°	0.35	0.09
39°	0.33	0.06

[a]Reprinted in part from ref. 42, p. 25, courtesy of n.v. Publi—Synthese and R. L. Dupuy, s.a.

concluded that the gauche concentration, which represents the non-crystalline phase, increases as the penetration depth increases (angle of incidence decreases) while the crystalline phase decreases toward the center of the fiber. This technique was instrumental in showing the reasons for two different shades of two nylon-66 filaments that were dyed superficially under identical conditions. The lighter shade was attributed to a higher level of orientation in the skin, which in turn inhibited the diffusion of the dye toward the center of the fiber. The surface-core differences between the fibers were in turn related to temperature variations that occurred in the spinning process.

In Fig. 14 are shown spectra of a cotton fabric obtained by a KBr disk technique and by an MIR technique using a 1-mm thick KRS-5 prism [27]. The (A) curve shows the relatively clear background obtained when the MIR spectrum is taken without sample and with a KRS-5 prism in both the sample and reference beams. The KBr disk spectrum (Fig. 14B) is given to show the lack of definition that occurs in the —OH stretching region when this technique is used. The MIR spectrum (Fig. 14C), however, shows definition in this region with peaks observable at 3330 and 3280 cm^{-1}.

The authors point out that these bands could be indicative of crystallinity or polymorphic forms and had not been observed clearly prior to their work without the benefit of deuteration, polarization, or hydro-

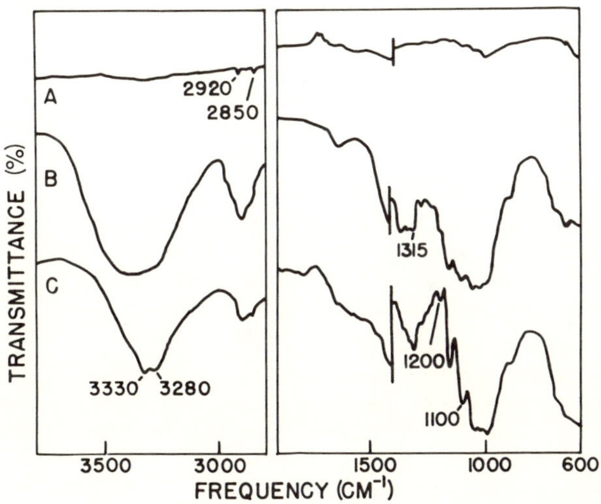

FIG. 14. Comparison of KBr transmission and MIR spectra on cotton fabrics. (A) KRS-5 versus KRS-5, (B) untreated cotton fabric (KBr disk), (C) untreated cotton fabric (MIR). Reprinted from ref. 27, courtesy of American Dyestuff Reporter published by SAF International, Inc.

cellulose preparation. Differences are also pointed out in regions of the spectra between 600 and 1500 cm^{-1}.

V. APPLICATION OF ATR TO FIBER-RELATED PROBLEMS

There are reports in the literature of many applications of the ATR technique which, although not directly concerned with fibers, are related to fiber-forming materials or materials that could be utilized in fiber and fabric products and processing. Although no attempt will be made to cite all of the related works, it is appropriate that at least some of these areas of study should be noted in this chapter.

A. Diffusion and Absorption

One area of direct interest to fiber-surface chemists would be diffusion-related processes. Such processes might include materials diffusing into fibers as well as substances exuding out to the surface.

The process of dyeing fibers is of obvious importance to textile scientists. Unfortunately, it has been difficult to apply the ATR technique directly to this problem. Carlsson et al. [19] have studied the problem of dye uptake and penetration by examining fiber-forming polymer in film form. In Fig. 15 are shown the ATR spectra of dyed and undyed films of a copolymer of polyacrylonitrile and an acrylic ester. The dyed spectrum shows a number of differences from the undyed spectrum, the most obvious being a loss of the absorption band at 1650 cm^{-1} in the dyed spectrum. This change is evidence that the dyeing process has altered the chemical nature of the substrate.

Deley et al. [43] obtained good ATR spectra of four dye-coated paper samples by varying the angle of incidence and using either KRS-5 or AgCl reflectance prisms. The papers had surface coatings of dyes that exceeded 5 μm in thickness. Resolution of good enough quality to provide identification of the dyes was obtained.

The ATR technique can lend itself to study of the diffusion rates of materials other than dyes. In many cases a coating or adhesive will diffuse into a polymeric material. For example, the diffusion of a silicone grease into a sample of natural rubber was studied as a function of time [8]. The 1269 cm^{-1} absorption of a silicone grease was monitored by ATR as it diffused into a rubber sample. An initial coating of 1-μm thick was found to diffuse completely into the rubber after 4 hr.

The degree of effusion can be measured in a similar manner. In Fig. 16a is shown the poorly defined spectrum of a carbon-filled elastomer. In Fig. 16b is shown the ATR spectrum of surface bloom [10].

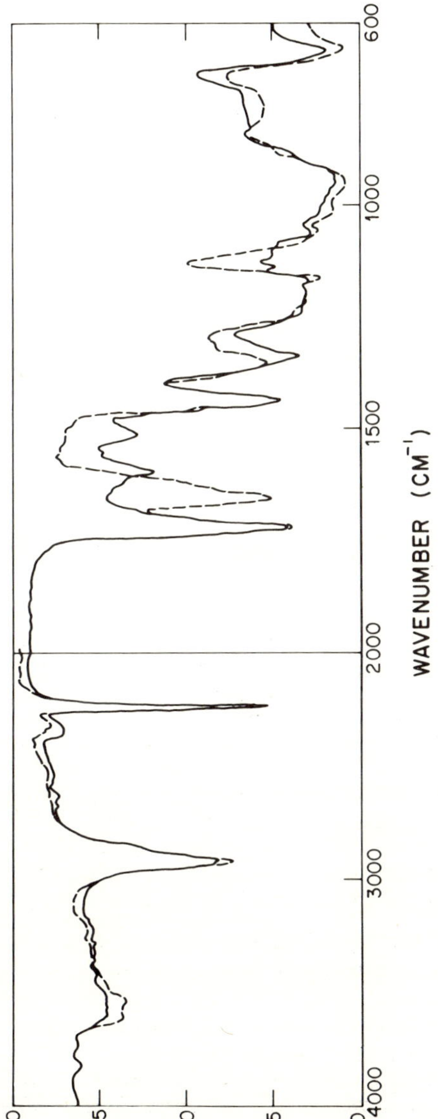

FIG. 15. Effect of dyeing on an acrylonitrile/acrylic ester copolymer film. Dashed line: undyed film. Solid line: dyed film. Reprinted from ref. 19, courtesy of Canadian Textile Journal Publishing Co., Ltd.

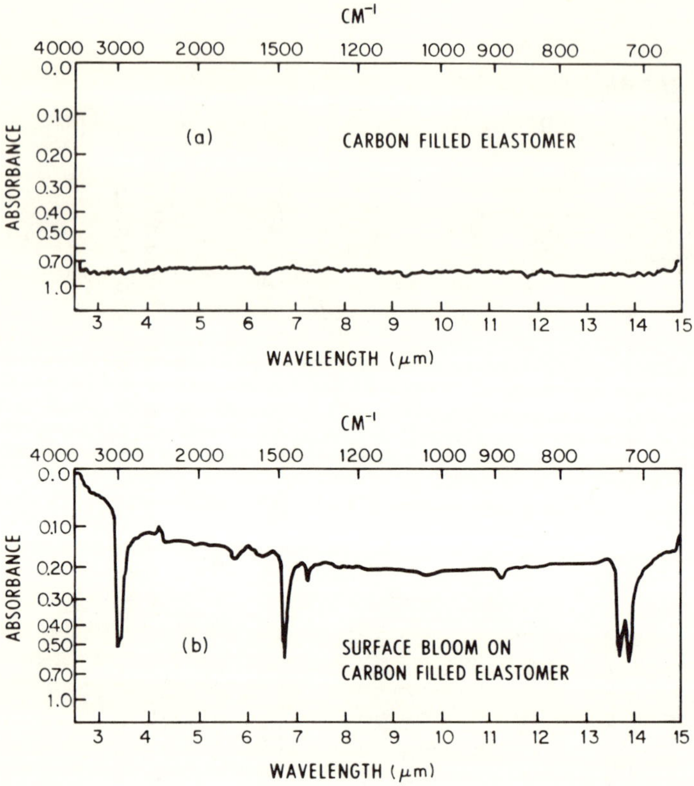

FIG. 16. Identification of surface bloom from carbon-filled elastomer. (a) Spectrum of carbon-filled elastomer, (b) ATR spectrum of surface bloom on carbon-filled elastomer. Reprinted from ref. 10, courtesy of Wilks Scientific Corp.

The spectrum indicates that a polyethylene-like species has migrated to the surface of the elastomer.

In Fig. 17 is shown a similar experiment in which the rate of diffusion of a silicone oil through polyethylene films was monitored [8]. In this case the polyethylene films were thoroughly impregnated with silicone oil. The surface of the film was initially cleaned by extracting for 12 hr in acetic acid. No detectable silicone was found on the surface of this material, as evidenced by lack of absorption at 1050 cm^{-1} as shown in Fig. 17a. However, after heating for 5 min at 90°C and a subsequent examination of the surface by means of ATR, the silicone was found on the surface as evidenced by the presence of the 1050 cm^{-1} band in Fig. 17b. Rates of diffusion of the silicone through the polyethylene could be determined in this manner.

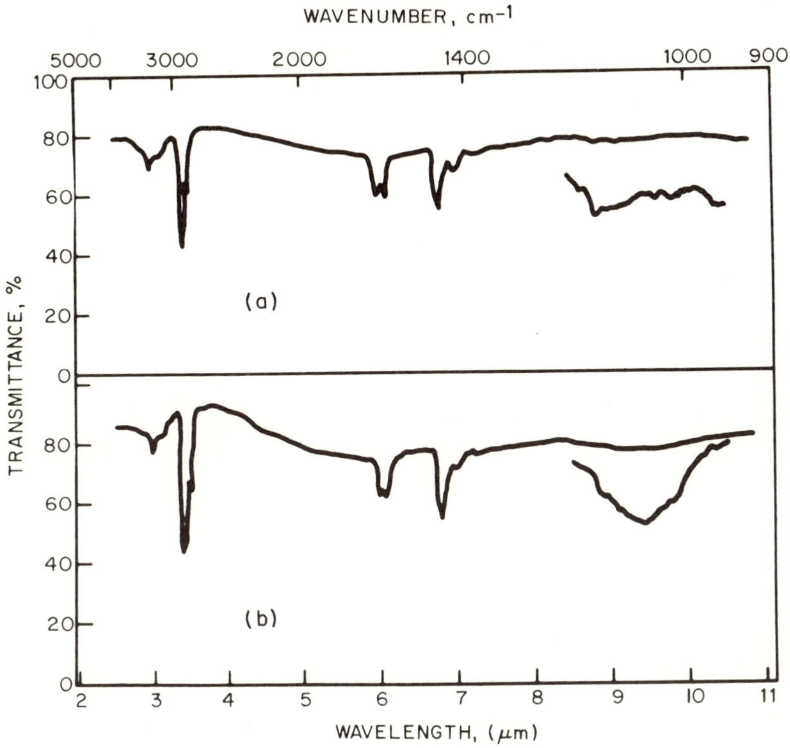

FIG. 17. Effusion of silicone-oil impregnated polyethylene film using 5× scale expansion at 9.5 μm. (a) After 12 hr extraction to remove surface oil, (b) same sample heated for 5 min at 90°C to show reappearance of oil at surface. Reprinted from ref. 8.

Examples of utilizing the ATR technique for the determination of absorbed hydrogen on silicon and absorbed gases on an active carbon surface have also been given [8].

B. Polymerization and Crosslinking Studies

The application of the ATR technique to surface reactions such as those involved in forming wrinkle-resistant cottons and grafting of vinyl polymers to fibers and fabrics has already been mentioned. An interesting extension of this type of work has been demonstrated by Fujimoto and Wataya [44]. They constructed a high-temperature ATR (HATR) apparatus that enabled them to study changes occurring during chemical reactions such as polymerization, isomerization, and crosslinking. Although the work was done primarily to obtain information about the bulk reactions, the technique was particularly applicable to surface studies of these reactions. They studied polymers prepared from ethylene and propylene with termonomers of methylene-norbornene, dicyclopentadiene, and 1,4-hexadiene. Each termonomer had a characteristic band due to unsaturation. The terpolymers were then formulated with such crosslinking agents as sulfur and dicumylperoxide and the crosslinking reactions followed by the disappearance of unsaturation as a function of temperature and time. Reaction temperatures of 140 and 150°C were used and reaction times monitored up to 100 min. The greatest crosslinking effect was observed in the case of the methylene-norbornene-type terpolymers.

The infrared polymerization index (IRPI) has been used to determine degree of resin polymerization. An ATR approach to this determination has been taken, which has a number of advantages [45] over the conventional IRPI methods. The conventional IRPI method involves an extraction followed by an infrared-transmission-type determination of the degree of reaction. The ATR technique is simpler because direct measurement of the resin can be obtained. Various phenolic formulations, impregnated in glass fabrics, were studied in this way. Absorption at 10 μm was used to monitor the unreacted methylol phenol and absorption at 12.2 μm (820 cm^{-1}) was characteristic of the reacted monomer. Data on partially reacted phenolics obtained by the ATR technique were compared with data obtained by more conventional techniques. It was concluded that the ATR measurements correlated with fabrication performance. The method was also applied qualitatively to study of the degree of reaction of polyester, diallyl phthalate, and epoxy formulations.

C. Adhesion

The ATR technique should be helpful in understanding adhesion since adhesion for the most part is dependent on surface characteristics. Failure of an adhesive bond between two polymeric materials may occur in a number of ways. The area that fails to adhere may be either of the two interfaces. Therefore, examination of the failure surfaces by ATR should provide information as to which surfaces had the weaker bond to the adhesive. Cohesive failure of either the adhesive or the polymer should also be obvious from the ATR examination.

Wettability of surfaces bears an important relationship to adhesion. Baier and Zisman [46] point out the significance of wettability and adhesion in biological systems. They studied the wettability of a series of solvent-cast polyamide films by ATR techniques in order to gain insight into the behavior of polypeptides and proteins in general.

Films of nylon-11, nylon-6, nylon-66, and nylon-2 (polyglycine) were cast from solutions of dichloroacetic acid or formic acid. "Wettability" data were obtained on the films by measuring the contact angle with both hydrogen-bonding and non-hydrogen-bonding liquids. Measurements were made on the MIR spectra by comparing the ratio of intensities of the 3300 cm^{-1} (—NH band) to the 2900 cm^{-1} band (characteristic of the $-CH_2-$ group). Critical surface tension and wettability of hydrogen-bonded solvents was shown to be greater for nylon-6 than for nylon-11 due to the higher concentration of amide groups on the surface.

Adhesive tapes are multicomponent, consisting of various layers of single polymers, blends of polymers, and formulated polymers. Internal infrared reflection spectroscopy has been shown [47] to be helpful in identifying and characterizing these various layers.

It has also been used to give information about curing, contaminants, and migration of components with respect to adhesion properties. In the same reference, it was pointed out that insensitive regions of the MIR crystals can be a problem, particularly when small samples are examined.

D. Surface Coatings

Iszard and Wilks [10] used a differential technique to study the surface coating on cellophane. In Fig. 18a is shown the transmission spectrum of a cellophane film suspected of having a surface coating. No evidence of a coating is present in this spectrum. However, in Fig. 18b, the MIR spectrum shows that a material other than cellophane is present on the surface as evidenced by new absorptions at 4.48 μm (2230 cm^{-1}) and 5.8 μm (1725 cm^{-1}). A sample of uncoated cellophane was then put in the reference beam and the differential spectrum obtained with the coated material in the sample beam. Figure 18c clearly shows the nature of the coating material with interferences from the cellophane removed.

Leukroth [48] evaluated the ATR technique for quality control and other analysis on polyurethanes, polyacrylates, poly(vinylchloride), natural and styrene-butadiene rubbers, and adhesive resins such as ethylenevinylacetate copolymers. The technique was found to be advantageous, particularly for the analysis of crosslinked and insoluble polymers.

The technique has also been used to identify coatings such as silicones, acrylics, polystyrene, and paraffin waxes on paper [43]. This normally difficult analysis is solved readily and nondestructively with the ATR method.

Michell [49] extended the analysis of coatings on paper and also applied it to coated boards. He demonstrated how the ATR technique could provide information about adhesives, pigments, and binders. The quality of the spectra of inks was found to be strongly dependent on the film thickness and therefore on the printing process. One of

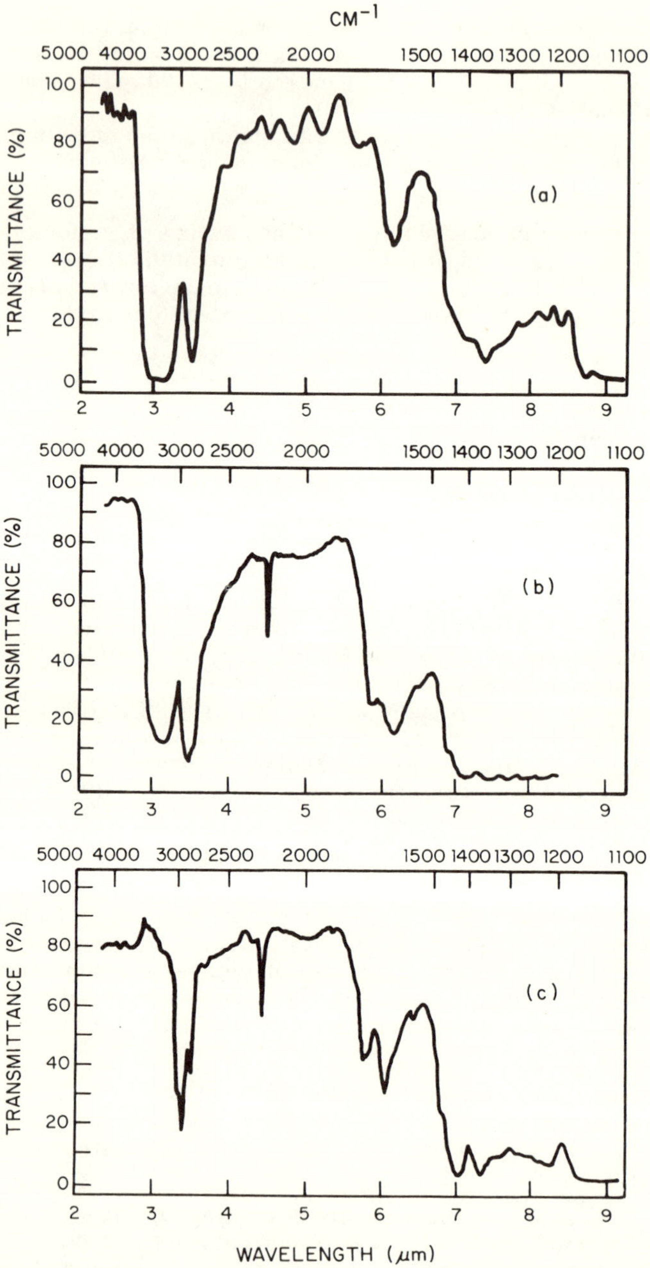

FIG. 18. Coating on cellophane film. (a) Transmission spectrum of cellophane suspected of having a thin polymer coating; (b) MIR spectrum of surface of same film showing absorption from cellophane and coating; (c) differential MIR spectrum of polymer-coated cellophane vs. uncoated cellophane. Reprinted from ref. 10, courtesy of Wilks Scientific Corp.

the main advantages of the technique is that surface layers can be studied in situ on the paper or board.

Other studies along these lines were carried out by Jayme and Traser [50,51]. They applied the FMIR technique to the quantitative determination of kaolin-casein coatings on paper.

REFERENCES

1. B. C. Stace, in Laboratory Methods in Infrared Spectroscopy (R. G. J. Miller and B. C. Stace, eds.), Heyden and Son, New York, 1972, Chap. 17.
2. G. A. Tirpak and J. P. Sibilia, J. Appl. Polymer Sci., 17, 643 (1973).
3. J. Fahrenfort, Spectrochim. Acta, 17, 698 (1961).
4. J. Fahrenfort, in Molecular Spectroscopy Proceedings-IV. International Meeting, Bologna, 1959 (A. Mangini ed.), Vol. 2, Pergamon, London, 1962, p. 701.
5. N. J. Harrick, Phys. Rev. Letters, 4, 224 (1960).
6. N. J. Harrick, Internal Reflection Spectroscopy, Wiley, New York, 1967.
7. W. W. Wendlandt and H. G. Hecht, Reflectance Spectroscopy, Wiley-Interscience, New York, 1966.
8. J. K. Barr and P. A. Flournoy, in Physical Methods in Macromolecular Chemistry (B. Carroll, ed.), Marcel Dekker, New York, 1969, Chap. 2.
9. P. A. Wilks, Am. Laboratory, 4, 42 (1972).
10. M. R. Iszard and P. A. Wilks Jr., 148th National American Chemical Society Meeting, Chicago, Ill., Sept. 3, 1964.
11. D. Kunath and A. Reklat, Z. Chem., 11, 361 (1971).
12. G. Kortuem, Reflectance Spectroscopy Principles, Methods, Applications, Springer-Verlag, New York, 1969.
13. J. Fahrenfort, in Molecular Spectroscopy Proceedings of Fourth Conference (P. Hepple, ed.), American Elsevier, Brighton, England, 1968, pp. 111-130.
14. J. A. Pawlah, G. Fricke, and H. Szymanski, Progr. Infrared Spectrosc., 3, 39-54 (1967).
15. P. A. Wilks Jr. and T. Hirschfeld, Appl. Spectrosc. Rev., 1(1), 99-130 (1968).
16. T. Hirschfeld, Appl. Spectrosc., 20(5), 336-8 (1966).
17. W. J. Potts, Am. Chem. Soc. Org. Coatings Plastics Chem., Preprints, 25(2), 364-373 (1965).
18. J. Fahrenfort, Proceedings Colloquium Spectroscopicum Internationale, 10th, University of Maryland, June 18-22, 1962, Spartan Books Inc., pp. 437-460.

18a. S. Siewierska and W. Kozlowski, Polimery, 18(5), 221 (1973).

19. D. J. Carlsson, T. Suprunchuk, and D. M. Wiles, Can. Textile J., 87, 73 (1970).

19a. W. Kozlowski, Polimery, 18, 221 (1973).
20. R. J. Manning, Beckman Instruments Applications Research Technical Report #565.
21. P. A. Wilks Jr., in Laboratory Methods in Infrared Spectroscopy (R. G. J. Miller and B. C. Stace, eds.), Heyden, New York, 1972, Chap. 14.
22. P. A. Wilks and M. R. Iszard, Fifteenth Mid-America Spectroscopy Symposium, Chicago, Illinois, June 1964.
23. E. Loeser and M. Schuetz, Melliand Textilber., 12, 1412, 1966.
24. J. A. Knight, M. P. Smoak, R. A. Porter, and W. E. Kirkland, Textile Res. J., 37, 924 (1967).
25. P. A. Wilks Jr. and J. W. Cassels, Develop. Appl. Spectrosc., 7B, 280 (1970).
25a. H. Siesler, Symp. Mol. Spectros. [Proc.], 4, Counc. Sci. Ind. Res.; Pretoria, S. Afr., 1973 (South African Council for Scientific and Industrial Research).
25b. H. Siesler and D. O. Hummel, Melliand Textilber. Int., 54(10), 1093 (1973).
25c. D. O. Hummel, H. Siesler, E. Zoschke, I. Vierling, U. Morlock, and T. Stadtlaener, Melliand Textilber. Int., 54(12), 1340 (1973).
25d. A. Basch and E. Tepper, Appl. Spectrosc., 27, 268 (1973).
26. J. P. Sibilia, J. Appl. Polymer Sci., 17, 2911 (1973).
27. E. R. McCall, S. H. Miles, and R. T. O'Connor, Am. Dyest. Rep., p. 31, May 1966.
28. T. S. Hermann, J. Appl. Polymer Sci., 9, 3953 (1965).
29. V. Krentz, Melliand Textilber., 5, 557 (1969).
29a. Y. A. Pentin, B. N. Tarasevich, and B. S. Eltesfon, Vestn. Mosk. Univ. Khim., 14(1), 13 (1973).
29b. P. Svoboda, Anal. Fys. Metody Vyzk, Plastic. Pryskyric [Proc. Conf.], 1, 230 (1971), Vyzk, Vstav. Syn. Pryskyric Laku., Pardubice, Czech.
29c. J. C. Andries and H. E. Diem, J. Polym. Sci., Polym. Lett. Ed., 12(5), 281 (1974).
30. P. Blais, D. J. Carlsson, and D. M. Wiles, J. Appl. Polymer Sci., 15(1), 129 (1971).
31. D. L. Harms, Anal. Chem., 25, 1140 (1953).
32. P. F. Kruse and W. B. Wallace, Anal. Chem., 25, 1156 (1953).
33. B. Cleverly and R. Hermann, J. Appl. Chem., 192, May (1960).
34. D. N. Kendall, in Applied Infrared Spectroscopy (D. N. Kendall, ed.), Reinhold, New York, 1966, Chap. 4, p. 148.
35. J. W. Cassels, 18th Pittsburgh Conference on Analytical Chemistry and Applied Spectroscopy, March, 1967.
36. J. W. Cassels, 19th Southeastern Regional Meeting, American Chemical Society, Atlanta, Georgia.
37. L. R. Weiner, Picatinny Arsenal, Dover, N. J. Technical Memorandum 1849, Dec., 1968.
38. L. R. Weiner, Picatinny Arsenal, Dover, N. J. Technical Memorandum 1893, Sept., 1969.
39. R. G. Quynn and R. Steele, Nature, 173, 1240 (1954).

40. G. Caroti and J. H. Dusenbury, J. Polymer Sci., 22, 399 (1956).
41. J. K. Barr, Nature, 215, 844 (1967).
42. G. Heidemann, Text. Chimiques, 27, 19 (1971).
43. J. P. Deley, R. J. Gigi, and A. J. Liotti, Tappi, 46 No. 2, 188A (1963).
44. K. Fujimoto and K. Wataya, J. Appl. Polymer Sci., 13, 2573 (1969).
45. G. Brown and C. A. Murray, Plastics Technology, p. 45, Sept. (1965).
46. R. E. Baier and W. A. Zisman, Macromolecules, 3, 462 (1970).
47. C. M. Paralusz, 165th National American Chemical Society Meeting, Dallas, Texas, April 12, 1973.
48. G. Leukroth, Gummi. Asbest. Kunsts., 28, 1118 (1970).
49. A. J. Michell, Appita, 26, 25 (1972).
50. G. Jayme and G. Traser, Papier (Darmstadt), 26, (7), 345; (10), 506 (1972).
51. G. Jayme and G. Traser, Angew. Makromol. Chem., 21, 87 (1972).

Chapter 9

EFFECT OF PHOTOCHEMICAL AND ENVIRONMENTAL DEGRADATION ON THE SURFACE PROPERTIES OF TEXTILE FIBERS

S. H. Zeronian

Division of Textiles and Clothing
University of California
Davis, California

I. INTRODUCTION

Weather resistance is an important consideration for several applications of textile materials, as prolonged outdoor exposure results in a gradual deterioration in the properties of textile fibers. This degradation can be caused by many factors, such as sunlight, atmospheric contaminants, and mildew. Good light resistance is required also for many textile products that have indoor use, such as carpets, curtains, and upholstery. Although these materials may be partially or fully protected from sunlight, they may be damaged by radiation from other light sources.

The deterioration of textile materials by weathering or by exposure to artificial light sources is reflected in both the chemical and physical properties of the fibers. In this chapter, the effects of weathering and of radiation from artificial light sources on the surface structure of textile fibers will be considered. First, however, factors that control the rate of deterioration will be discussed.

II. FACTORS AFFECTING THE RATE OF DETERIORATION OF TEXTILE FIBERS BY WEATHERING OR BY EXPOSURE TO ARTIFICIAL LIGHT SOURCES

A. Exposure Conditions

It is generally accepted that, normally, the primary cause of the photodegradation of fibers by exposure either to solar radiation or to artificial light sources is the uv portion of the radiation. Studies have indicated that different types of fibers lose their strength at different rates during outdoor exposure [1-4]. For a given type of fiber, the rate of loss of strength can be affected by the exposure site [1]. The geographical location determines the amount and spectral distribution of radiant energy from the sun that reaches the exposure site. The location will, in addition, determine the general climatic conditions, which can also affect the rate of degradation of textile or polymeric materials [5]. In relation to this, the effect of temperature on the photochemical degradation of rayon, acetate, and nylon-66 fibers has been studied, and it has been reported that degradation in moist air is particularly marked at high temperatures [6]. The loss in strength of a fiber for a given time of exposure can also be influenced by whether the sample is exposed directly to sunlight or is exposed under window glass [4], as window glass will filter out shorter wavelengths of uv radiation. Strength retention of a fiber is affected also by the season of the year in which it is exposed, as solar radiation and weather conditions at a particular site vary with the time of year. However, it has been claimed that discrepancies caused by seasonal changes can be eliminated by relating fiber strength loss to uv radiation rather than to total radiation [7,8]. The results of both outdoor studies [9-13] and laboratory studies [14-16] indicate that atmospheric contaminants can influence the rate of degradation of fibers.

When correlations are being made between outdoor exposures and indoor photodecomposition of textiles, consideration must be given to the fact that indoor sources of uv radiation do not duplicate natural sunlight, and differ greatly among themselves in spectral energy distribution [17]. The marked effect that the wavelength of the radiation has on the photodegradation of light-exposed fibers has been considered by several workers [18-22].

B. Fiber Characteristics

Many characteristics of fibers can influence their resistance to light or to the environment. For example, the chemical composition has an effect. The photochemical degradation of both natural fibers [23,24] and of polymers used in manufacturing man-made fibers [25] has been reviewed.

The physical characteristics of the fiber will also affect its light resistance. For example, light resistance will increase with the denier of the fiber, probably because less radiation penetrates into the interior [4]. In addition, the cross-sectional shape of a fiber can influence the reflection, refraction, and transmission of radiation striking the fiber. Thus, yarns of different cross-sectional shapes of the same fiber may have different degrees of light resistance [4].

Different types of fiber from the same generic class may differ in their resistance to light. Such differences could result from modifications of the fiber polymer or structure, or from the presence of additives. Pigments also affect the rate of degradation of fibers. The light resistance of polyester and nylon-66 fibers greatly improves when black pigment is dispersed in the polymer during fiber manufacture [4]. TiO_2, which is used as a delusterant during the manufacture of fibers, decreases the light resistance of man-made fibers [1,4,6,19, 21,26,53]. The presence of TiO_2 in nylon fibers enhances the photodegradation only above 3,000 Å in the presence of oxygen, and therefore is connected with photo-oxidation rather than photolysis [19]. The mechanism by which anatase TiO_2 increases the photodegradation of nylon-66 has been discussed [6,26].

Such agents as dyes, finishes, and coatings can affect the rate of deterioration of fibers on exposure to light. Dyes can either adversely influence the light resistance of fibers [4,27] or can increase the resistance [4]. Resin finishes applied to cellulosic fabrics can affect the deterioration of the light-exposed fabric [14,28-35]. Loss of yarn strength is not a sensitive measure of the degradation of resin-treated fabric exposed to light [35]. Uv absorbers [4,18] applied to the fiber product also influence light and weather resistance.

III. SURFACE STRUCTURE OF TEXTILE FIBERS EXPOSED TO LIGHT OR TO WEATHERING

A. Nylon Fiber

There is evidence that the surface structure is the first part of the nylon fiber to be affected by exposure to light. Kato [36] studied photodegradation of nylon-6 filaments by microscopy, using a staining technique. White taffeta cloth of TiO_2-delustered nylon-6 filament was irradiated in a Fade-Ometer for long periods of time. Details of the irradiation procedure are not given. A staining solution was prepared from Procion Black HGS, urea, and soda ash. Unexposed nylon-6 filaments were not stained by this solution. Exposure of the nylon-6 fabric in the Fade-Ometer for increasing lengths of time showed that staining by the Procion Black HGS occurred at the peripheral layer of the filaments and then progressively proceeded into the core of the filaments. Kato [36] also cut nylon-6 filaments exposed in the Fade-Ometer into short lengths, and then shredded the filaments in water with a mixer and examined the product with the transmission electron microscope. Kato observed that the photodegraded nylon-6 had a microfibrillar structure that was not present in the unexposed starting filaments. Changes in the supramolecular structure of nylon materials resulting from light exposure have also been claimed by Achhamer et al. [37]. Films were prepared from a copolymer of nylon-66 salt and ϵ-caprolactam, and from a copolymer of nylon-66 salt, nylon-610 salt, and ϵ-caprolactam. The films, which did not contain TiO_2, were exposed to radiation from a sunlamp. The authors stated that comparison of electron micrographs of the shadowed surfaces of untreated samples with those of the irradiated samples indicated that the crystallinity of the nylon-66/ϵ-caprolactam sample appeared to be increased, whereas that of the nylon-66/nylon-610/ϵ-caprolactam sample was not affected as a result of the treatment.

Bigler [38] has cited examples of the use of microscopy in the examination of polyamide fibers exposed to light. Examination of nylon-66 fabric exposed in a showcase indicated that portions of the nylon-66 that had been exposed directly to light were distinguished by an apparent magnification of the TiO_2 delustering agent. Bigler refers to this as an apparent grain coarsening. This effect was reproduced in the laboratory by exposures in a Xenotest. Bigler suggested that the TiO_2 acts as a photosensitizer. Consequently, photodegradation occurs in the periphery of the TiO_2 particles leading to an enclosure of disintegrated material about the pigment particles. This results in an apparent magnification of the pigment particles when the fiber is examined by microscope.

Zeronian, Tomioka, and Pangborn [39,40] exposed samples of bright, semidull, and dull nylon-66 filament yarn in air simultaneously to light from a borosilicate-filtered xenon arc lamp and to intermittent water spray. These exposures were made in a Weather-Ometer. Each type of yarn contained the same number of filaments, and the filaments in each yarn had the same denier. The anatase-TiO_2 content of the bright, semidull, and dull nylon-66 filament yarn was 0.02, 0.30, and 2.0%, respectively. The surface structure of the filaments was examined with a scanning electron microscope (SEM). The surfaces of the unexposed nylon filaments at all TiO_2 levels were roughly similar, and free from pits. A typical example of surface structure is shown in Fig. 1. The surface of the semidull and dull nylon-66 filaments became pitted after the exposures. The number of pits on the surface of exposed semidull and dull nylon increased with the time of exposure

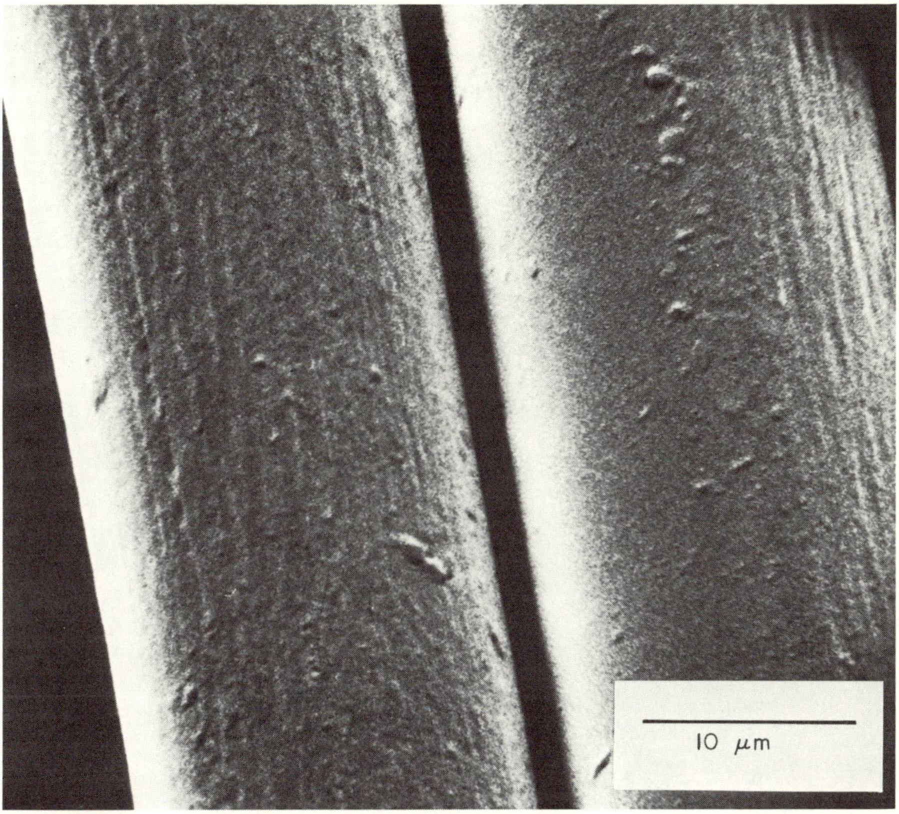

FIG. 1. Surface of unexposed semidull nylon-66 filaments.

(compare Figs. 2a and 2b and Figs. 3a and 3b). For a given exposure time, surface damage on the semidull nylon (Fig. 2b) was less than on the dull fibers (Fig. 3b). Also, for a given loss in breaking load, surface pitting of the semidull filament (Fig. 2a) was less than that of the dull filaments (Fig. 3a). Only a minimal number of pits were observed on the bright filaments whether comparison with the delustered filaments was made for equal time of exposure (Fig. 4) or for equal loss in strength. Thus the concentration of anatase-TiO_2 particles appears to be an important factor in inducing surface pitting in nylon-66 fibers exposed to borosilicate-filtered radiation and water spray. To determine if TiO_2 particles were located in the vicinity of the pits, Zeronian and co-workers [39,40] compared the surface of unexposed dull nylon-66 filament yarn to dull nylon-66 filament yarn exposed to borosilicate-filtered light and to intermittent water spray for 267 hr with the SEM operating in the X-ray mode. Ti was found to be localized at discrete

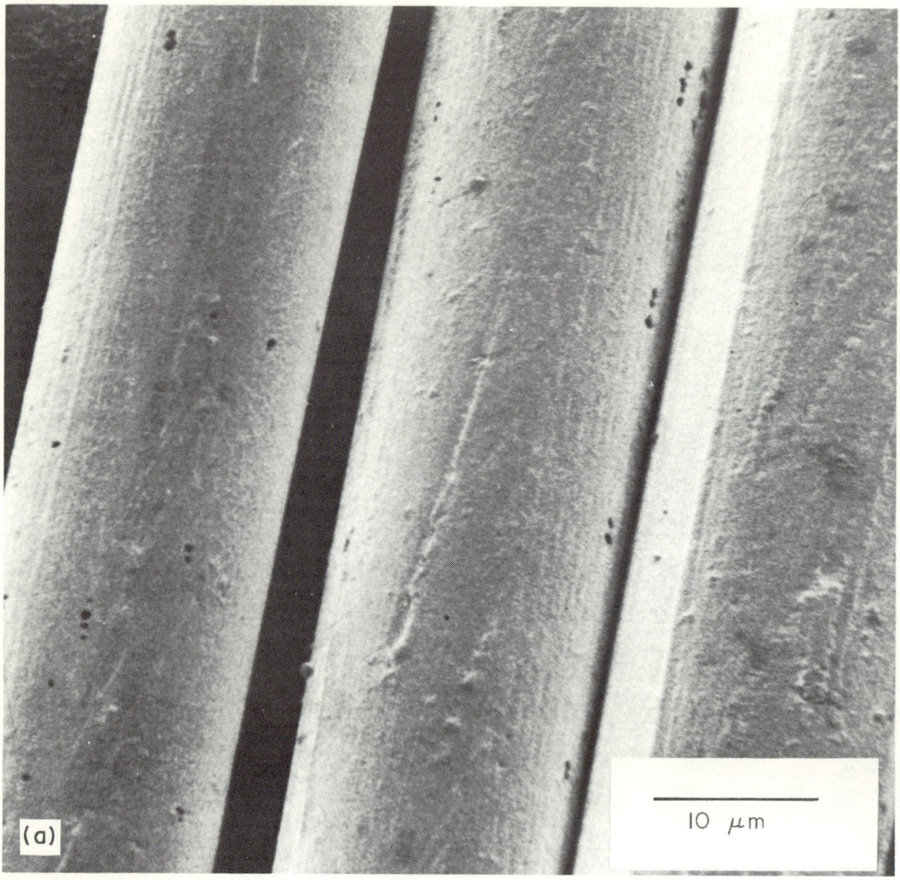

FIG. 2. Surface of semidull nylon-66 filaments after exposure to borosilicate-filtered light and to intermittent water spraying. In (a), exposure time 267 hr and yarn strength retention 67%; in (b), exposure time 672 hr and yarn strength retention 11%. Reprinted from ref. 39, p. 59, courtesy of Claitor's Publishing Division.

sites on the filaments. It was found close to several holes, but could not be detected in the vicinity of other holes. However, the nylon filaments were damaged by exposure to the large electron beam required for the operation of the SEM in the X-ray mode. Consequently, the correlation of the Ti sites with the positions of the pits could only be approximate. It is also possible that the TiO_2 particles may have been washed out of some of the pits during the water spraying.

Zeronian and co-workers [40] also studied the effect of the spectral distribution of the radiation on the surface structure of light-

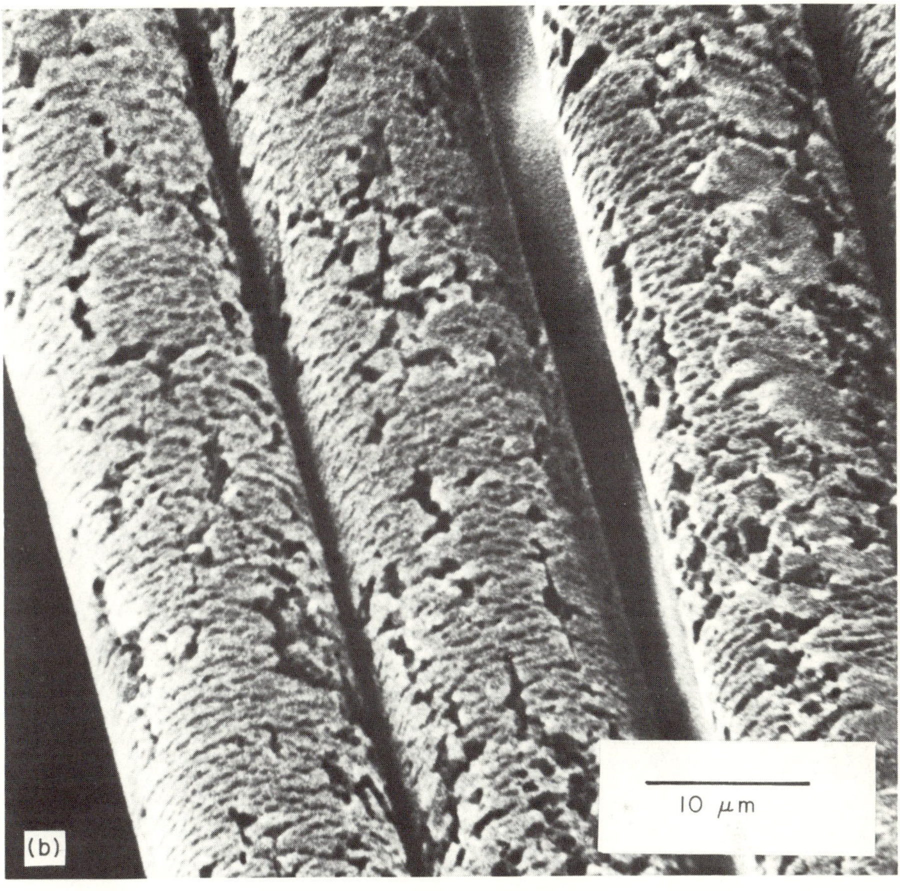

FIG. 2 (continued)

exposed, semidull, nylon-66 staple fiber. Exposures were made in a Weather-Ometer in air simultaneously to light from a xenon arc lamp and to intermittent water spray. The radiation was filtered either by borosilicate glass or by quartz. For samples showing approximately similar losses in yarn strength, the surface structure of nylon-66 fibers exposed to radiation from the quartz-filtered xenon arc lamp was less affected than fibers exposed with borosilicate filters on the lamp. In general, the former samples contained fewer and smaller pits. Also, fissures were not present with the pits. Compared with borosilicate filters, the quartz filters yielded radiation from the xenon arc lamp having lower-wavelength and greater-intensity uv. Thus, it was suggested that part of the damage in fibers exposed to quartz-filtered radiation was due to photolytic degradation rather than photo-

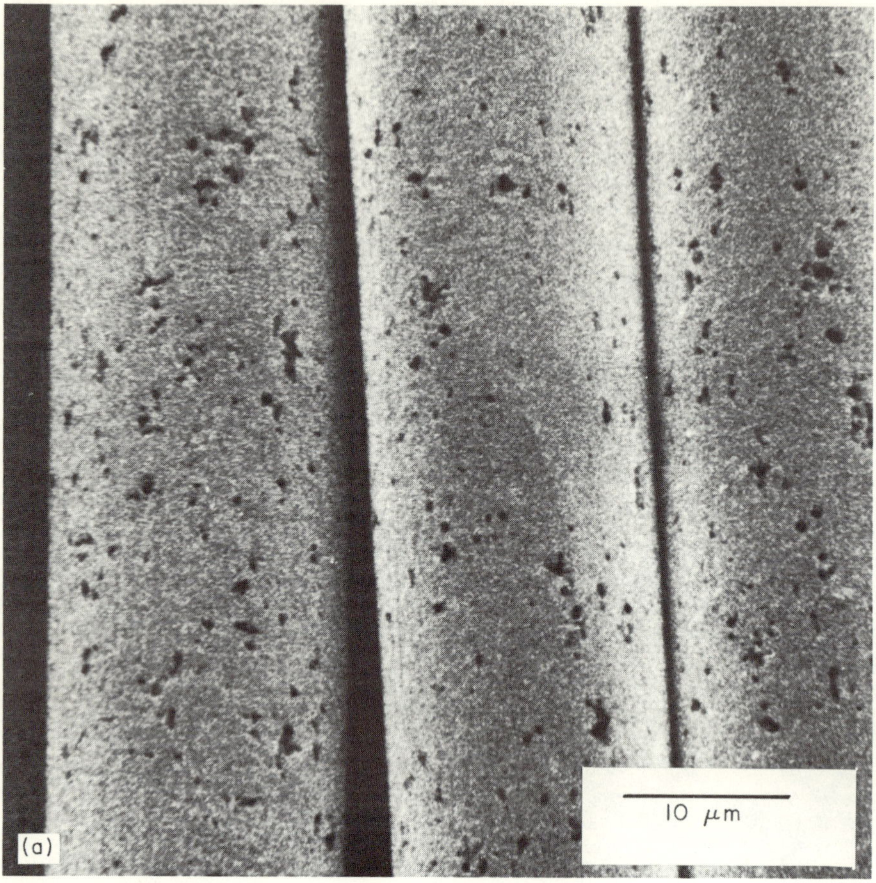

FIG. 3. Surface of dull nylon-66 filaments after exposure to borosilicate-filtered light and to intermittent water spraying. In (a), exposure time 168 hr and yarn strength retention 68%; in (b), exposure time 672 hr. Yarn was too fragile for strength measurements. Reprinted from ref. 39, p. 59, courtesy of Claitor's Publishing Division.

oxidation. Unlike photo-oxidation induced by TiO_2, photolytic degradation need not be localized in specific areas, and thus would not produce degradation leading to the formation of pits. The authors noted that other explanations of the difference in surface damage caused by different radiations were possible, i.e., they stated that the differences may have been caused by differences in the times of exposure. For an equal loss in strength of nylon-66, less exposure to the quartz-filtered radiation than to the borosilicate-filtered radiation was required. Consequently, the nylon exposed to the quartz-filtered radiation was sub-

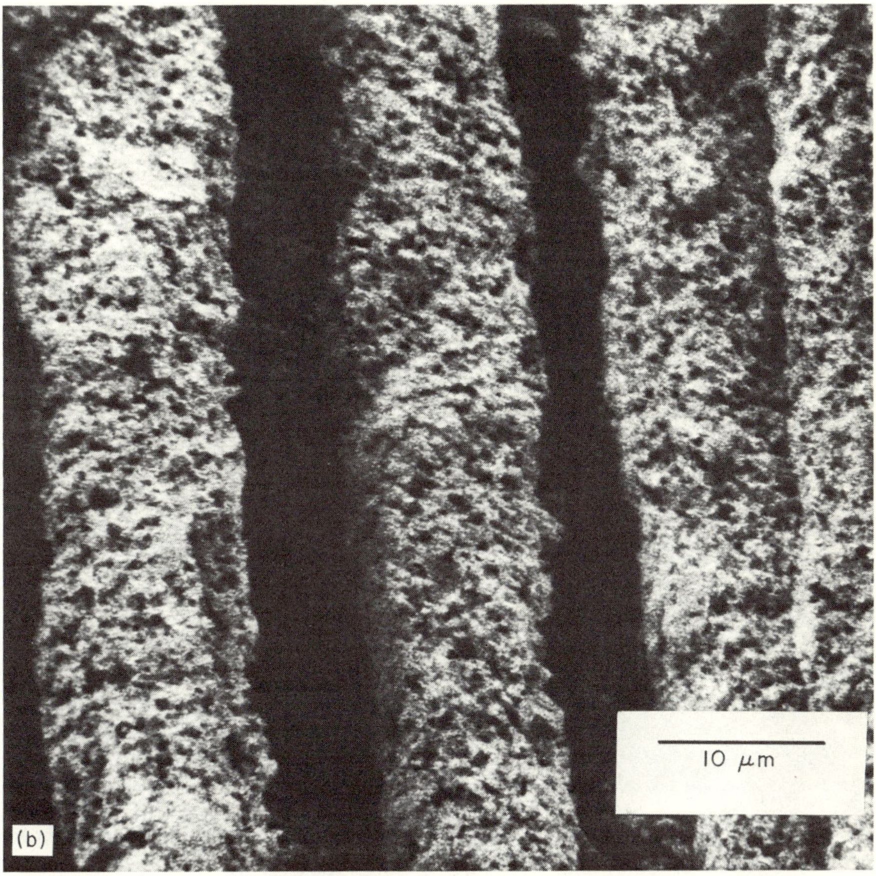

FIG. 3 (continued)

jected to less water spray. The authors suggested that exposure to less water spray may result in less damage, irrespective of other factors. This statement is supported in the finding of Zeronian and coworkers [40] that for equal exposure times, samples of delustered nylon-66 degraded in the Weather-Ometer by exposure to light without water spray exhibit no surface damage, whereas samples degraded by exposure to light and water spray show surface pitting. However tensile testing indicates that the former samples are degraded more than the latter. The authors give two possible explanations. The first is that in the exposures with water spray a higher concentration of peroxides is produced, leading to extensive degradation in the vicinity of the TiO_2 particles. The degraded material is washed out during water spraying and pits are formed. The second is that similar damage to

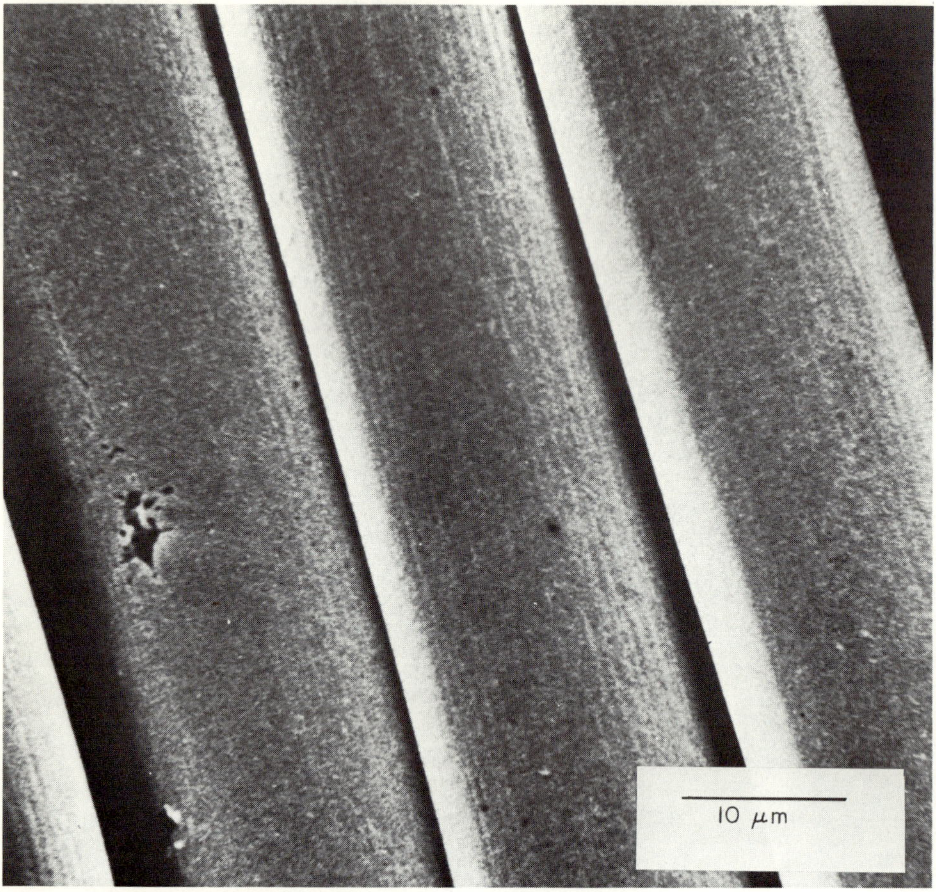

FIG. 4. Surface of bright nylon-66 filaments after exposure to borosilicate-filtered light and to intermittent water spraying for 672 hr.

that in the exposures with water spray is also present in the exposures without water spray, but the damage is latent. The damage in the exposures with water spray develops as pits due to the contraction and expansion of the fibers as the result of periodic wetting and drying. The contractions and expansions form small cracks that develop eventually into pits and cavities. This process may be assisted by water-spray removal of low molecular-weight material formed as a result of the phototendering. Evidence to support the suggestion that the damage is latent in unsprayed light-exposed samples is that large fissures appear in such samples if they are immersed in boiling water (Fig. 5) [40].

FIG. 5. Surface of semidull nylon-66 fibers exposed to quartz-filtered light without water spraying for 120 hr and then immersed in boiling water for 18 hr.

This treatment does not cause damage to the surface structure of unexposed nylon.

Zeronian [41] has studied the effect of SO_2 on the surface structure of semidull nylon-66 fibers. Nylon was exposed in a Weather-Ometer to air contaminated with 0.2 ppm SO_2 and simultaneously to light from a borosilicate-filtered xenon arc lamp and to intermittent water spray. The surface structure of the exposed nylon contained large cavities. When comparison of the damage was made with similarly exposed nylon in air uncontaminated with SO_2, it was found that SO_2 accelerated the degradation of nylon-66 by light, but had not changed the mode of degradation of the fiber surface.

There is other evidence that under specific conditions exposure to light can cause pitting of nylon-66 fibers. Zeronian and co-workers [16] exposed semidull nylon-66 staple fiber immersed in 20% sulfuric acid to radiation from a sunlamp. After exposure, the fiber surface was roughened and contained small pits. If a sample of the same fiber is highly degraded by hydrolysis with sulfuric acid without light exposure, the only damage evident is a roughening of the surface of some fibers [41].

Evidence that voids can be created in delustered nylon-66 on exposure to light has been given by Hearle [42,43], who reported briefly on studies made on the fracture of light-exposed nylon-66 filament. Hearle found that unexposed nylon subjected to a tensile force fractures by crack growth followed by a rougher zone of catastrophic failure running across the fiber. However, when the nylon had first been exposed outdoors for 24 weeks in Manchester, England during the summer months, the break surface contained a complex collection of tunnels and cavities. Hearle suggested that the light causes material around the TiO_2 particles to oxidize and leave voids from which breaks can start. It is not clear from this brief note [42] when the voids appear. One possibility is that the voids appear after the application of the tensile force, as the degraded material in the vicinity of the TiO_2 particles is not able to sustain any load. Hearle also reported that the fracture of delustered nylon-66 monofilament degraded by exposure in a Xenotest for 200 hr is similar to that of nylon-66 exposed for 3 weeks outdoors during the summer months. He noted the existence of a thick layer of elongated crystal-like structure on the surface of the filaments exposed in the Xenotest. Such a structure was not present in the outdoor-exposed nylon. No explanation was given as to what caused this change in surface structure in Xenotest samples.

B. Polyester Fiber

Kuligina et al. investigated the properties of photodegraded Lavsan fabric [44]. Samples were degraded by outdoor exposures and by exposure to an indoor light source. The structures of exposed and unexposed samples were examined by electron microscopy. From a study of ultrasonically dispersed material, the authors claim that the structure of the polyester fiber is predominantly fibrillar, and is degraded by light exposure. For fiber that had received three months of weathering, there was a weakening of the intermolecular bonds, primarily in the amorphous region of the polymer. For fiber exposed to artificial light for 400 hr, the long, thick fibrillar bundles had been broken into individual short, thin fibrils, indicating that the bonds between the elements in the crystalline region of the polymer had been weakened. Kuligina also made replicas of the fiber surfaces and examined them by electron microscopy. The unexposed fabric appeared to lack a clearly defined surface structure. The surface is described as dense and slightly undulating. The authors stated that after weathering, or exposure to indoor artificial light exposure for 400 hr, the hillocks and cracks on the fibers became much more clearly defined.

Weber [45] investigated the properties of polyester fibers by both light and electron microscopy. TiO_2-delustered polyester fiber exposed to light through windows exhibited an increased delustering effect. Studies of polyester fibers from which the surface had been removed by an alkali treatment indicated that the fiber material itself was damaged by light. An examination of light-exposed polyester fibers with a SEM indicated that transverse cracks and fiber deformation had been caused by light exposure.

Day and Wiles [22] have stated that prolonged uv irradiation of poly (ethylene terephthalate) film results in the formation of surface cracks. This film did not contain uv stabilizers. Irradiations were made with a Hanovia high-pressure mercury arc as the light source. These authors state that photochemical reactions give rise to chain scission, and consequently to smaller polymer molecules. They suggest that these fragments may then regroup and result in restructuring of the surface. In a later communication, Blais, Day, and Wiles [46] irradiated Mylar polyester film in a xenon arc Weather-Ometer. The film was reported to contain no light stabilizers. It was observed from transmission electron micrographs of direct replicas of irradiated material that the initially featureless surface of the film remained featureless, even after prolonged uv exposure, with the exception of small amounts of particulate debris that appeared on some areas of irradiated films. A test was devised to determine if volatilization or volume contraction occurred during irradiation of the film. It was concluded that some volatilization appeared to be taking place from the irradiated regions, and that volume contraction seemed unlikely. The volatilization was thought to be of low molecular-weight fragments resulting from chain scission in the uppermost surface layers. It was suggested that the chain scissions occurred in the amorphous regions of the polymer or in the molecules linking the crystalline regions. The authors also stated that the possibility exists that oligomers undergo secondary reactions leading to some gaseous products. Scanning electron micrographs were taken of the surface revealed by peeling a layer of film from the Mylar, portions of which were shielded from uv irradiation. The texture, as well as the thickness of the stripped layer, varied with the degree of uv exposure. Thus a different mode of cohesive failure appeared to characterize the shielded Mylar as compared to the exposed material. Scanning electron micrographs of irradiated polyester film subjected to tensile failure revealed that the side exposed to the irradiation was granular in appearance, whereas that facing the opposite direction appeared to have the liquid-like features of a plastic deformation. It was suggested that the application of tension to an irradiated film resulted in crack formation at the film-front surface. The crack then propagated at right angles to the major stress direction through the surface-degraded layer until high molecular-weight polymer was encountered, whereupon tearing occurred. In additional experiments, it was observed that when irradiated polyester film was flexed in a convex direction to the irradiated surface, cracks appeared on the uv-exposed side of the film.

It should be noted that comparisons between results of light exposure on films and those on fibers have to be made with caution, as these materials differ in morphological and fine structures, and there may be differences in the polymer itself.

Zeronian, Tomioka, and Pangborn [40] exposed samples of bright, semidull, and dull Dacron polyester filament yarn in air simultaneously to light from a borosilicate-filtered xenon arc lamp and to intermittent water spray. The number of filaments, and their denier, were the same in each yarn. The anatase-TiO_2 content of the bright, semidull, and dull polyester filament yarns was 0.10, 0.30, and 2.0%, respectively. The surface structure of the filaments was examined by SEM. The surface structure of unexposed polyester filaments was similar irrespective of TiO_2 content, and contained irregularly shaped protuberances of a type normally seen on the surface of polyester fibers. A typical example of surface structure is shown in Fig. 6. The surface of dull polyester filaments contained a few pits after 267 hr of exposure. After 672 hr of exposure, a considerable number of pits were present (Fig. 7). In contrast, the semidull polyester filaments contained no pits after 267 hr of exposure and only a small number after 672 hr (Fig. 8). No pits were observed on the surface of the bright polyester filaments even after 672 hr of exposure (Fig. 9). Thus, the greater the concentration of anatase TiO_2 in Dacron, the greater was the damage to the filament surface on exposure to light and intermittent water spray.

Zeronian and co-workers subjected nylon-66 filament yarn of similar denier and containing the same number of filaments as the Dacron polyester yarn, to identical exposures of light and intermittent water spray. Then, they compared the surface damage of the polyester and nylon-66 filament yarns. They reported that for the same length of exposure, nylon-66 in the semidull or dull state suffered greater surface damage than did the polyester counterparts. Thus, delustered polyester fibers may be intrinsically more resistant to surface damage than delustered nylon-66 fibers for a given exposure to light and water spray. The authors also stated that comparisons became more complicated when similar losses in strength were considered, as the relation between the rate of change of surface structure and rate of loss of strength was complex.

C. Orlon Acrylic Fiber

Zeronian and Tomioka [47] have studied the effect of light on Orlon acrylic fibers. Normally, the rate of loss of strength of Orlon yarn is less than that of nylon-66 or polyester yarn on exposure to sunlight

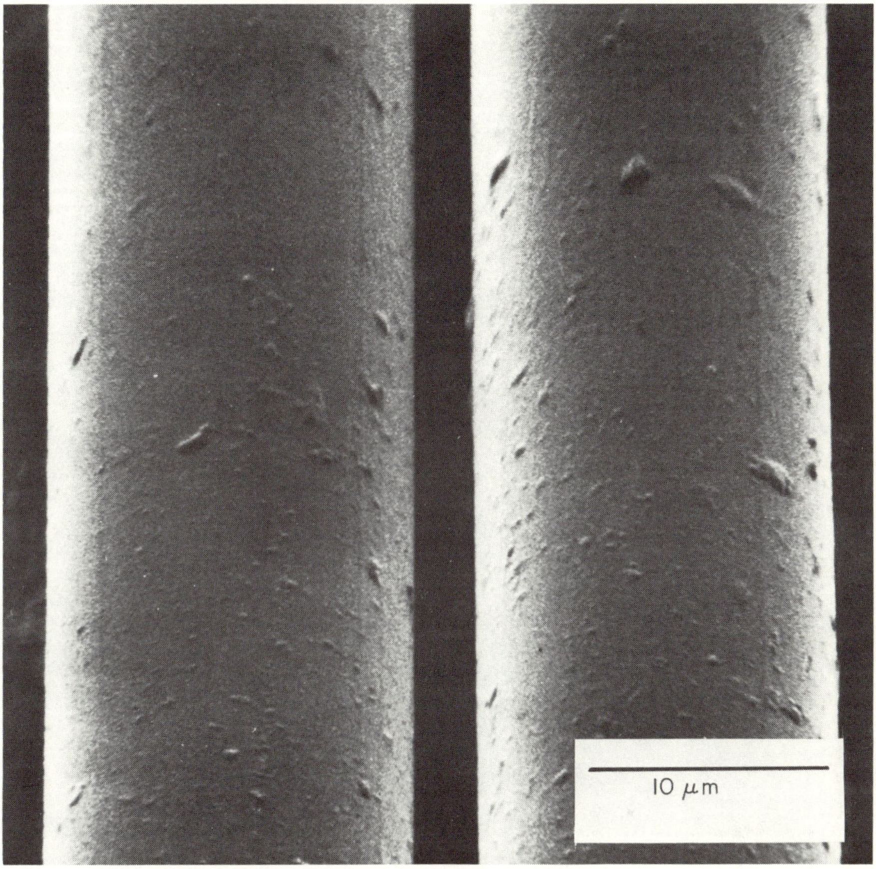

FIG. 6. Surface of unexposed bright polyester filaments.

[4]. However, when semidull Orlon acrylic fiber was exposed in air to light from a borosilicate-filtered xenon arc lamp and to intermittent water spray, pitting of the fiber surface occurred (compare Fig. 10 and Fig. 11). Thus, the damage was of the type observed on Dacron polyester or delustered nylon-66. The surface damage on semidull Orlon acrylic fiber was less extensive than that on semidull nylon-66 fiber exposed for a comparable period of time. However, the degradation of the Orlon fiber was markedly less than that of the nylon fiber, as indicated by breaking-strength measurements.

D. Other Fibers

There have been reports on the effect of light on the fine structure of other fibers, which are relevant to this review.

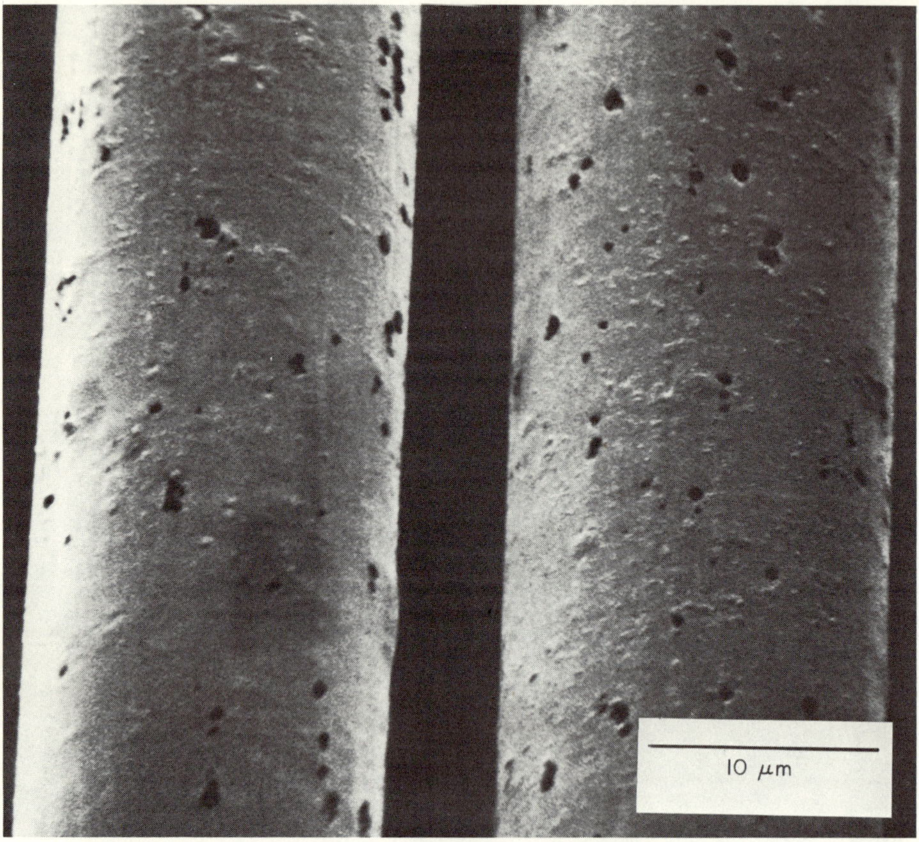

FIG. 7. Surface of dull polyester filaments after exposure to borosilicate-filtered light and to intermittent water spraying for 672 hr.

Müller [48] has reported that the action of light on delustered acetate rayon causes haziness in the fiber, which he attributes to the production of small gas bubbles, probably of CO_2. For light-damaged delustered viscose rayon, Müller, as well as Reumuth [49], concluded that photodegraded material surrounded the TiO_2 particles. These centers of degraded material would change the refractive properties within the fiber mass.

Wiles and Blais [50] studied the effect of uv-initiated oxidation on (crystalline) polypropylene fiber and found that the unexposed fiber consisted of a bundle of small filaments enclosed in a smooth sheath-like envelope. They stated that after mild uv oxidation (full details of the type of exposure are not given), it became possible to strip off

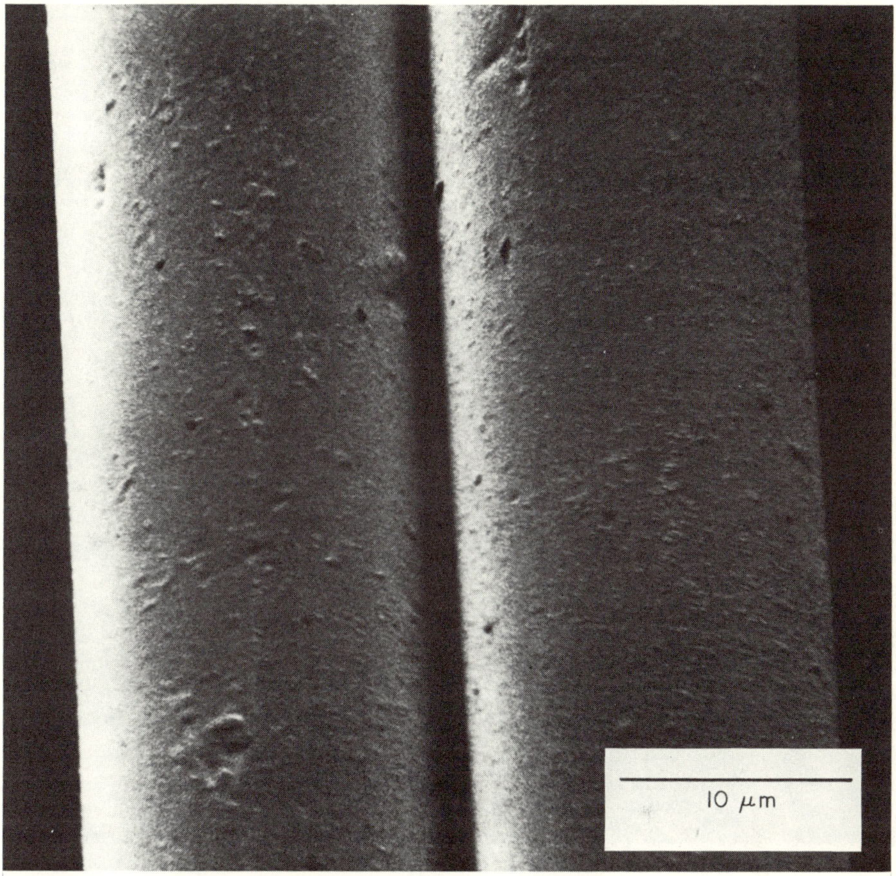

FIG. 8. Surface of semidull polyester filaments after exposure to borosilicate-filtered light and to intermittent water spraying for 672 hr.

this outer sheath with various adhesives. Finally, after extensive oxidation, large portions of the outer envelope were eroded, leaving the core filaments relatively unaltered. Thus, the outer envelope of the fibers appeared to be made up of uv-vulnerable disordered material in which the deterioration of the fiber was initiated.

It is interesting also that Blais and co-workers [51] studied the surface changes of uv-irradiated polypropylene film. The film was predominantly isotactic, and it was exposed in air to the radiation of a Hanovia medium-pressure mercury lamp. A filter transmitting wavelengths greater than 3,200 Å was used with the lamp. It was found

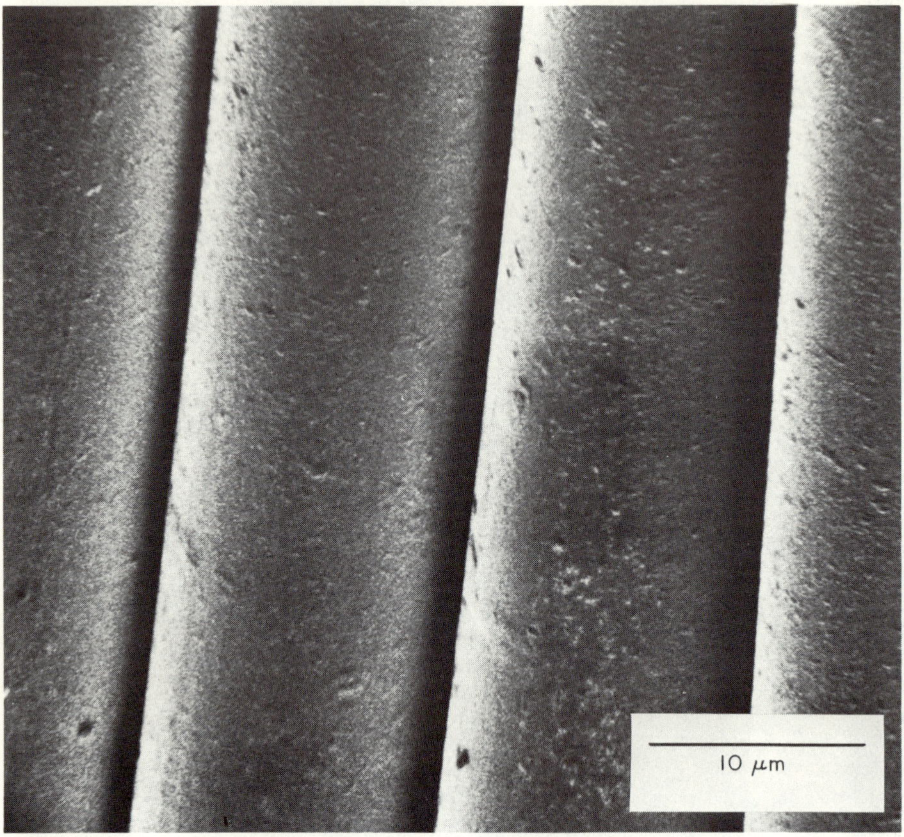

FIG. 9. Surface of bright polyester filaments after exposure to borosilicate-filtered light and to intermittent water spraying for 672 hr.

that photo-oxidation of the film resulted in extensive surface restructuring, namely roughening and parallel crack formation, probably resulting from large local density increases brought about by the photo-oxidation. The cracks were preferentially aligned along the direction of uniaxial orientation (roughly the machine direction). Lamellar crystallites were also formed on the photo-oxidized surface. The difference between the results on film and on fiber could be caused by such factors as differences in morphological structure, in exposure conditions, or in the polymer itself.

Blaga [52] studied the effect of outdoor weathering in Saskatoon, Canada, on glass-fiber-reinforced polyester sheets. These sheets

FIG. 10. Surface of unexposed Orlon acrylic fibers.

consisted of 25% treated reinforcing glass fiber (E-glass) (in the form of chopped strand mat) and resin, fillers, and pigments. Breakdown of the sheet was reported, but no damage to the fibers was indicated.

IV. CONCLUSIONS

It is difficult to compare the experimental results of different workers who have studied the surface structure of light-irradiated fibers. In many instances full details of the exposure are not given, nor is the amount of degradation measured by such means as loss in

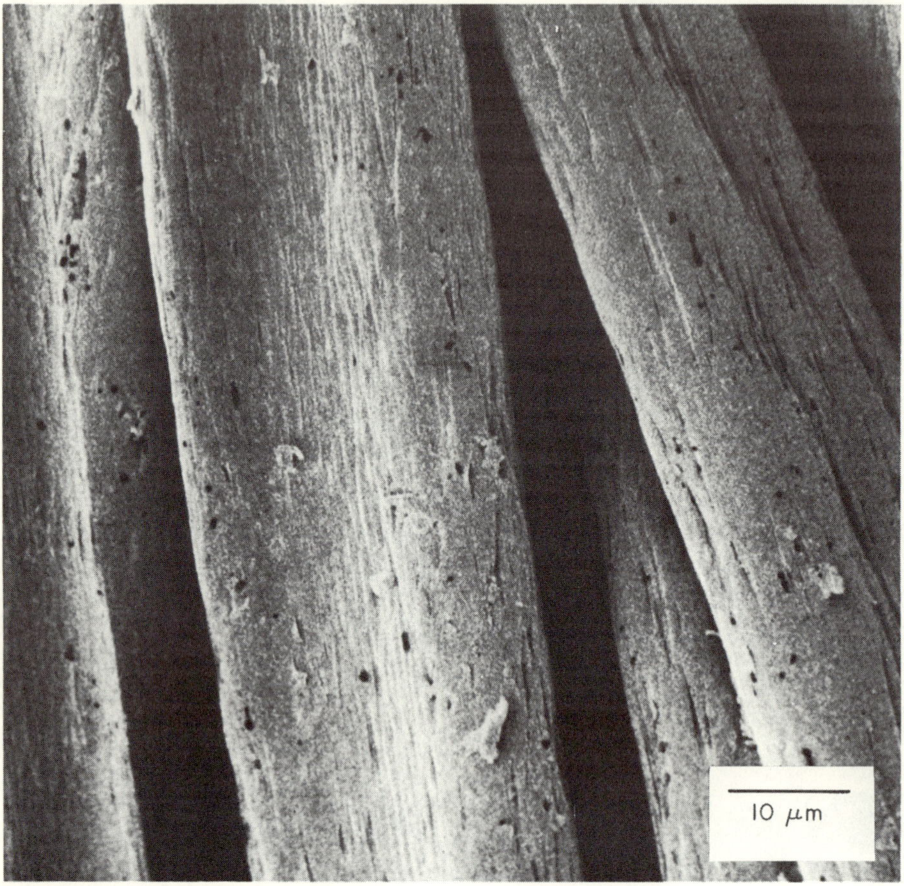

FIG. 11. Surface of semidull Orlon acrylic fibers after exposure to borosilicate-filtered light and to intermittent water spraying for 315 hr.

strength or change in chemical properties. However, it can be concluded that the surfaces of many man-made fibers are modified by uv irradiation, and that the type and amount of damage depend on such factors as exposure conditions, the polymer, and the presence of anatase TiO_2 as delusterant.

With the increasing availability of the scanning electron microscope, further studies may be anticipated on the surface structure of uv-irradiated fibers, as researchers attempt to characterize the damage caused to fibers by exposure to light.

REFERENCES

1. M. Fels, J. Textile Inst., 51, P648 (1960).
2. Du Pont Bulletin X-189, 1964.
3. Chemstrand Bulletin TT-17, 1965.
4. Du Pont Bulletin X-203, 1966.
5. R. E. Eshenaur, Plastics Polymers, 37, 401 (1969).
6. G. S. Egerton and K. M. Shah, Textile Res. J., 38, 130 (1968).
7. R. W. Singleton, R. K. Kunkel, and B. S. Sprague, Textile Res. J., 35, 228 (1965).
8. R. W. Singleton and P. A. C. Cook, Textile Res. J., 39, 43 (1969).
9. E. Race, J. Soc. Dyers Colourists, 65, 56 (1949).
10. M. A. Morris, M. A. Young, and T. A. Molvig, Textile Res. J., 34, 563 (1964).
11. R. J. Brysson, B. J. Trask, J. B. Upham, and S. G. Booras, J. Air Poll. Control Assoc., 17, 294 (1967).
12. A. H. Little and H. L. Parsons, J. Textile Inst., 58, 449 (1967).
13. D. A. Clibbens, Textile Inst. Ind., 6, 20 (1968).
14. S. H. Zeronian, Textile Res. J., 40, 695 (1970).
15. S. H. Zeronian, K. W. Alger, and S. T. Omaye, Proceedings of Second International Clean Air Congress (H. M. Englund and W. T. Beery, eds.), Academic Press, New York, 1971, pp. 468-476.
16. S. H. Zeronian, K. W. Alger, and S. T. Omaye, Textile Res. J., 43, 228 (1973).
17. R. C. Hirt, R. G. Schmitt, N. D. Searle, and A. P. Sullivan, J. Opt. Soc. Am., 50, 706 (1960).
18. P. T. Standring and G. W. Penmore, J. Textile Inst., 51, T336 (1960).
19. R. F. Moore, Polymer, 4, 493 (1963).
20. M. J. Wall and G. C. Frank, Textile Res. J., 41, 32 (1971).
21. M. J. Wall, G. C. Frank, and J. R. Stevens, Textile Res. J., 41, 38 (1971).
22. M. Day and D. M. Wiles, J. Appl. Polymer Sci., 16, 191-202 (1972).
23. P. J. Baugh and G. O. Phillips, in Cellulose and Cellulose Derivatives (N. M. Bikales and L. Segal, eds.), Part V, Vol. 5, Wiley-Interscience, New York, 1971, pp. 1047-1078.
24. A. Bendak, Am. Dyest. Rep., 62, (1) 46 (1973).
25. R. B. Fox, in Progress in Polymer Science (A. D. Jenkins, ed.), Vol. I, Permagon Press, New York, 1967, pp. 47-89.
26. H. A. Taylor, W. C. Tincher, and W. F. Hamner, J. Appl. Polymer Sci., 14, 141-146 (1970).
27. G. S. Egerton and N. E. N. Assaad, J. Soc. Dyers Colourists, 86, 203 (1970).
28. W. N. Berard, G. A. Gautreaux, and W. A. Reeves, Textile Res. J., 29, 126 (1959).
29. A. G. Kempton, M. R. Rogers, and A. M. Kaplan, Am. Dyest. Rep., 52, (20) 745 (1963).

30. H. H. St. Mard, C. Hamalainen, and A. S. Cooper Jr., Am. Dyest. Rep., 55, (25) 1046 (1966).
31. H. H. St. Mard, C. Hamalainen, and A. S. Cooper Jr., Textile Chem. Colorist, 2, (8) 136 (1970).
32. J. D. Dean, P. J. Fynn, W. B. Strickland, J. E. Sands, R. O. Simmons, H. M. Waddle, and M. A. Grimes, Am. Dyest. Rep., 36, (24) P705 (1947).
33. P. J. Fynn, C. N. Daly, C. M. Fleming, and J. D. Dean, Textile Res. J., 21, 116 (1951).
34. F. C. Wood, J. Textile Inst., 30, P142 (1939).
35. S. H. Zeronian and B. A. E. Miller, Textile Chem. Colorist, 5, (5) 89 (1973).
36. K. Kato, Textile Res. J., 32, 181 (1962).
37. B. G. Achhammer, F. W. Reinhart, and G. M. Kline, J. Res. Nat. Bur. Stand., 46, 391 (1951).
38. N. Bigler, SVF-Fachorgan, 18, (2) 82 (1963).
39. S. H. Zeronian, M. K. Tomioka, and J. Pangborn, Proceedings 31st Annual Meeting Electron Microscopy Society America (C. J. Arceneaux, ed.), Claitor's Publishing Division, Baton Rouge, 1973, pp. 58-59.
40. S. H. Zeronian, M. K. Tomioka, and J. Pangborn, to be published.
41. S. H. Zeronian, Textile Res. J., 41, 184 (1971).
42. J. W. S. Hearle, Textile Mfr., 99, (1174) 36 (1972).
43. J. W. S. Hearle, Textile Mfr., 100, (1177) 24 (1973).
44. M. S. Kuligina, L. P. Kolov, and A. F. Tumanova, Tech. of the Textile Ind., U.S.S.R. No. 5, 90 (1970); and Izv. Vyssh. Ucheb. Zaved. Tekhnol. Tekst. Prom. No. 5, 88 (1970).
45. R. Weber, Textilveredlung, 5, 703 (1970).
46. P. Blais, M. Day, and D. M. Wiles, J. Appl. Polymer Sci., 17, 1895 (1973).
47. S. H. Zeronian and M. K. Tomioka, to be published.
48. J. Müller, Melliand Textil Ber., 33, 742 (1952).
49. H. Reumuth, Melliand Textil Ber., 33, 743 (1952).
50. D. M. Wiles and P. Blais, Proceedings Fourth Annual Scanning Electron Microscope Symposium, Part I., 185 (1971), IIT Research Institute, Chicago Illinois.
51. P. Blais, D. J. Carlsson, and D. M. Wiles, J. Polymer Sci. Part A-1, 10, 1077 (1972).
52. A. Blaga, Polymer Eng. Sci., 12, 53 (1972).
53. L. M. Lock and G. C. Frank, Textile Res. J., 43, 502 (1973).

Chapter 10

COLOR OF FIBERS AND FABRICS

George Goldfinger, Kwok Ching Lau,* and Ralph McGregor

Department of Textile Chemistry
School of Textiles
North Carolina State University at Raleigh
Raleigh, North Carolina

I. INTRODUCTION

The most important property of a textile material is often its appearance. In this chapter we concentrate attention on the color of

---

*Present address: Color Science Laboratory, Lehigh University, Bethlehem, Pennsylvania.

textile materials, regarded as assemblies of fibers. As will be seen, the surface properties of fibers have an influence on their color and, thereby, the subject of color is appropriate for a text on the surface characteristics of fibers and textiles.

We begin by considering the psychological and physical aspects of color, its scientific definition, and its measurement. This discussion is followed by a brief review of existing mathematical approaches to the scientific description of the color of scattering-absorbing substrates, and leads to the presentation of a simple theoretical model for the prediction of the color of assemblies of cylindrical fibers. This model is used to illustrate the dependence of the color of a fiber assembly on fundamental parameters such as fiber denier and refractive index, the average dye concentration and dye distribution in the fibers, and the extinction coefficient of the dye-fiber system. The dependence on the average dye content in the fibers, and on the distribution of this dye within these fibers, provides a link with the physical chemistry and technology of dyeing and necessitates a discussion of these aspects of the total color problem.

II. THE PHENOMENON OF COLOR

When we talk about the appearance of a textile material we are attempting to summarize, in verbal terms, certain visual information which we have received. The essence of this visual information is contained in the distribution, in space and in time, of wavelength and energy in the electromagnetic radiation emanating from the object under consideration. Visual phenomena are associated with electromagnetic radiation of wavelengths ranging from 0.4 to 0.7 μm.

Color is that aspect of the visual experience which is independent of the spatial or temporal aspect [1]. The sensation of color is the human organism's interpretation of nerve impulses transmitted from the color-sensing nerve ends, the cones, in the retina of the observer's eye. These nerve impulses result from the response of the cones to electromagnetic radiation which is impinging on them. This radiation emanates from the object which is viewed, and often originates in an independent light source. The radiation from the light source may or may not be modified by processes of absorption or fluorescence in the object which reflects its radiation.

In the case of textiles we deal, for all practical purposes, with an independent light source whose radiation distribution is to a greater or lesser extent modified by absorption or fluorescence in the textile material. In the following discussion, only the absorption processes are considered to modify the radiation from the light source.

In order to permit a textile substrate to generate the sensation of color, dyes or pigments are added to the substrate. "White" materials are exceptions, though even here substances may be added to modify the "whiteness" through fluorescence or the addition of "blueing." The dyes or pigments cause the substrate to absorb some of the incident radiation, and to permit only a selected portion to be reflected. If we dye a fabric "green," for example, we use dyes which absorb both the short-wavelength and the long-wavelength portion of the visible spectrum, but permit the fiber to reflect the intermediate wavelengths.

III. COLOR THEORY AND DYEING THEORY

The nature of the phenomena which give rise to the sensation of color is described by the theories of color and of color vision.

The physical conditions which enable the textile material to absorb or to reflect in specific wavelength regions are dependent on the presence in the fibers of the complex chemical entities known as dyes, and are determined by the physical-chemical processes of dyeing. These two aspects are completely independent in the theoretical sense. However, the dyer's primary interest is in the relationships between the nature and the quantity of the dye used, and the color of the finished product. For this reason these two independent problems have to be handled together.

The assumptions on which the theory of color is based have by now acquired the nature of axioms: deductions based on them are definitive. The uncertainties in the theory of the color of textile materials arise from two distinct sources. The first problem relates to the very complex psycho-physiological process of color perception, which by necessity is based on the visual experiences of individuals and requires the arbitrarily defined concept of the standard observer. The second problem originates in the optical and geometrical complexity of the textile substrate itself, which makes simplifying assumptions unavoidable.

In spite of these limitations, we can now make reliable predictions of the color of textile substrates from information on the optical properties of the system (i.e., refractive indexes of fiber, coefficient of absorption of dye, dye concentration and distribution) and on its geometry and dimensions.

The general nature of the thermodynamic and kinetic processes involved in dyeing is also well known. However, dyeing systems are so complex that the formulation of quantitative mathematical models for the description or prediction of their behavior has so far proven impossible, except perhaps in the simplest of cases.

IV. THE DESCRIPTION AND MEASUREMENT OF COLOR

The basic concepts of colorimetry were expressed by Grassman [2] in 1853, and have been stated by Judd and Wyszecki [3] in the following form:

1. (As far as color is concerned) the eye can distinguish only three kinds of variations.
2. If, in a two-component color mixture, one component is steadily changed, the color of the mixture steadily changes.
3. Lights of the same color produce identical effects in mixtures, regardless of their spectral composition.

On this basis, when we attempt to reproduce a given color by mixing other colors, it does not matter what mixtures, if any, these other colors were to begin with. Consequently, it is not in general necessary to use exactly the same dyes which were used to dye the original sample, when we wish to reproduce the color of that sample.

Thus a variety of absorption patterns, modifying the radiation from a given light source, can create the same color sensation. However this same group of different absorption patterns, when modifying the radiation from a different light source, will as a rule produce different color sensations. This phenomenon, known as metamerism, can lead to significant mismatches.

Any color can be duplicated by an infinity of sets of other colors in proper combination. However, there does not exist any one set of a finite number of colors from which all other colors can be generated.

Several systems have been developed by which colors can be systematically described, or arranged. These systems fall into two principal groups: (1) color appearance systems; and (2) colorimetric systems.

A. Color Appearance Systems

In color appearance systems, colored samples are systematically arranged so that the color differences between adjacent samples follow some simple regularity: the color differences may, for example, be equal. Since the number of different colors is estimated to be of the order of 10^7, it is manifestly impractical to prepare such a one-dimensional array of all possible colors. In very special circumstances, as in the grading of particular commodities, it may be both possible and useful to develop sets of closely spaced color samples involving a limited and small number of colors.

The practical systems, which cover that very sizeable fraction of the color gamut realizable with colored solids, are three-dimensional.

One possible (polar) coordinate, representing hue, encompasses the colors of the spectrum from red to blue (spectral colors), with the purples, or mixtures of blue and red (nonspectral colors) linking those two component colors into a closed loop. A second coordinate may represent saturation, a measure of purity: the color sensation produced by narrow-band monochromatic radiation is one of high purity, or high saturation. The progressive addition of gray to a saturated color produces a progressive decrease in saturation. The remaining coordinate is then that of lightness.

The most impressive of these three-dimensional color systems is the Munsell system, as represented in Munsell's Book of Colors [4]. The DIN-Color Chart is conceptually similar, but differs in detail from the Munsell system: it is widely accepted as a standard in Germany [5]. The well-known system developed by Ostwald is of this general type, but is rarely used in the textile industry [6].

The "shade cards" distributed by dye manufacturers as guides to the use of their products are of great practical value to the textile processor, and represent a similar type of color appearance system, which has reached a high point of development in Imperial Chemical Industries recent monumental color book [7].

B. Colorimetric Systems

The colorimetric systems are conceptually different from the color appearance systems, though they also make use of three coordinates to define a three-dimensional color space. The most widely used system of this type is the so-called CIE system (Commission Internationale d'Éclairage), which has international acceptance [8].

The three vectors, X, Y, Z, of the CIE system are defined as the coordinates of the "color space." These three vectors are noncolors, but are often referred to inappropriately as nonreal colors. The vectors that define colors (real colors) originate at the origin $X = Y = Z = 0$ and always have positive values. The vector defining a color may be defined by its components, or projections, on the X, Y, Z axes.

The coordinate vector Y provides a measure of lightness: the Y coordinate vector has been defined in terms of the spectral sensitivity of the color sensors of the eye of a "standard observer," at levels of illumination high enough to render the grey sensors inoperative.

It has become customary to represent color in terms of the two-dimensional chromaticity diagram of Fig. 1, rather than in terms of the three-dimensional X, Y, Z space. The following transformation of coordinates is made, defining new chromaticity coordinates x, y, z

$$x = \frac{X}{X + Y + Z} \qquad (1)$$

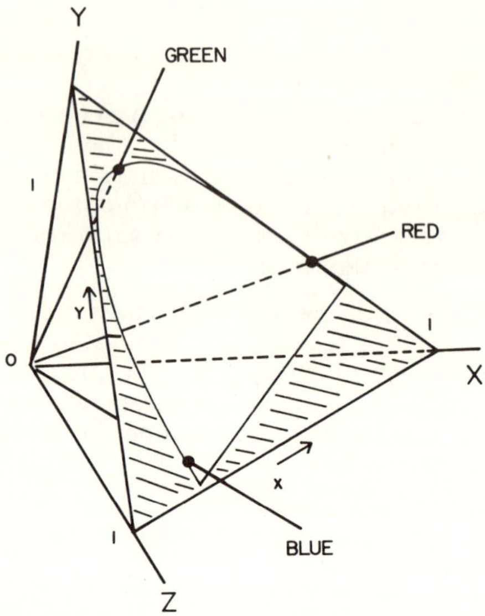

FIG. 1. Projection of the CIE color space and perspective view of the chromaticity diagram. The three vectors labeled BLUE, GREEN, and RED are the approximate location of the vectors representing the color of narrow-band (monochromatic) radiation of 0.45, 0.515, and 0.65 μm, respectively.

$$y = \frac{Y}{X + Y + Z} \tag{2}$$

$$z = \frac{Z}{X + Y + Z} \tag{3}$$

Since $x + y + z = 1$, two independent chromaticity coordinates suffice to define the coordinate z. The third independent coordinate necessary for the definition of a color is conveniently the coordinate Y, the chromaticity-diagram of Fig. 1 utilizes the x - y plane for $X + Y + Z = 1$.

In the strictest sense, those points in color space which have the same x and y coordinates but different Y values define different colors. Conceptually however, within fairly wide limits on the coordinate Y, we perceive as the same color all those colors with the same chromaticity coordinates x and y. In this limited sense, the area bounded by the horseshoe shape of the chromaticity diagram represents or includes all colors.

V. THE COLOR OF SCATTERING-ABSORBING SUBSTRATES

The first thorough treatment of the color of a system simultaneously absorbing and scattering light was proposed by Paul Kubelka and Franz Munk in 1931, and was meant to relate the color of paint films to the concentration of pigment and to the equally important effect of the background on the appearance of the system (hiding power) [9,10].

The model (Fig. 2) assumes a film of infinite width, thus neglecting the effect of the edges. The thickness X of this film is large compared to the dimensions of the absorbing and scattering pigment particles. It is subdivided into elementary layers dx, which themselves are thick compared to the dispersed particles and to the wavelength of light, but which are thin compared to the thickness X. Thus from an optical point of view the entire system is homogeneous. Placing the observer and the light source "above" the layer, two diffuse light fluxes are considered proceeding, respectively, "downward" and "upward" through the elementary layer dx.

The effects of absorption (K) and back-scattering (S) on the downward light flux i are to diminish it by Ki dx and Si dx, respectively. Similarly the upward flux j is diminished by Kj dx and Sj dx. Thus the total "upward" flux through one layer is

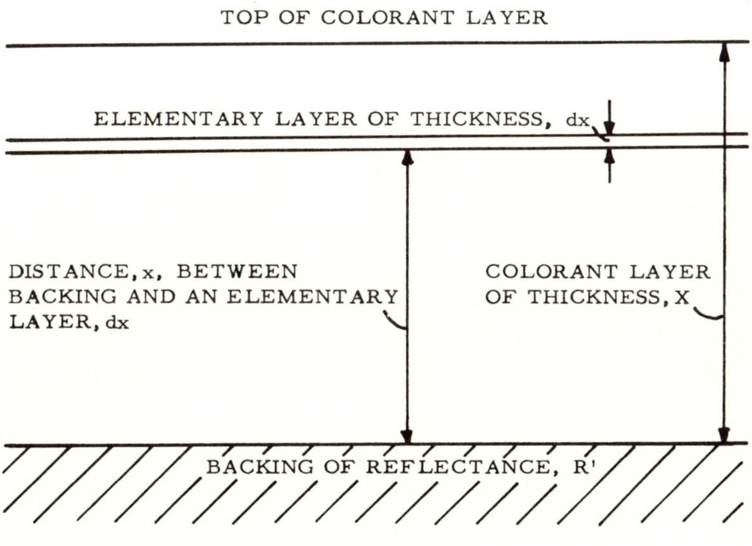

FIG. 2. The Kubelka and Munk model. The light source and the observer are located above the layer.

$$dj = -(K + S)j\,dx + Si\,dx \tag{4}$$

and the total "downward" flux is

$$-di = -(K + S)i\,dx + Sj\,dx \tag{5}$$

from which it follows that

$$d\ln(j/i) = 2(K + S)\,dx + S(j/i + i/j)\,dx \tag{6}$$

If the intensity of the light incident on the film from "above" is I_0, and if the intensity of the light reflected "upward" is I, we may define the fraction R reflected upward by $R = I/I_0$, and for any one of the infinitesimal layers the corresponding fraction r is $r = j/i$.

Since $d\ln r = dr/r$, Eq. (6) can be written as

$$dr = [-2(K + S)r + S + Sr^2]\,dx \tag{7}$$

setting $(K + S)/S = a$, we find that

$$\int_{R'}^{R} \frac{dr}{r^2 - [2(K + S)r/S] + 1} = S\int_0^X dX \tag{8}$$

where R' is the reflectance of the background, R is the reflectance of the sample, and X is its thickness. After substitution one finds

$$2SX\sqrt{a^2 - 1} = \ln\left[\frac{(R - a - \sqrt{a^2 - 1})\,(R' - a + \sqrt{a^2 - 1})}{(R' - a - \sqrt{a^2 - 1})\,(R - a + \sqrt{a^2 - 1})}\right] \tag{9}$$

The effect of the background vanishes for a layer of infinite thickness X, and then $R = R_\infty = a - \sqrt{a^2 - 1}$, or

$$R_\infty = 1 + \frac{K}{S} - \sqrt{\left(\frac{K}{S}\right)^2 + 2\left(\frac{K}{S}\right)} \tag{10}$$

so that

$$\frac{K}{S} = \frac{(1 - R_\infty)^2}{2R_\infty} \tag{11}$$

If the effect of the background is not negligible,

$$R = \frac{[(R' - R_\infty)/R_\infty] - R_\infty(R' - 1/R_\infty) \exp SX(1/R_\infty - R_\infty)}{R' - R_\infty - (R' - 1/R_\infty) \exp SX(1/R_\infty - R_\infty)} \tag{12}$$

This treatment has been widely used in predicting the color of paint films, plastics, paper, and textiles [11], particularly in predicting the colorant concentration. The model is adequate for paint films and plastics, which indeed are homogeneous within the meaning of the Kubelka and Munk treatment. Paper, and to an even greater extent, fabrics are poorly represented by the model. Among the attempts at improvement of the treatment, that by Atherton in 1955 is noteworthy [12]. While Kubelka and Munk assumed a diffuse incident flux, Atherton assumed illumination by a collimated beam of light. The argument in favor of this approach is that, in optical measuring instruments, illumination is frequently by a more or less collimated beam, and even if the illumination is diffuse, observation is made on a collimated portion of the reflected light.

The treatment is similar in all mathematical essentials to that of Kubelka and Munk. A restrictive simplifying assumption is that the light passing through each infinitesimal layer remains collimated in the direction of incidence and suffers scattering only in the opposite, "upward" direction.

Atherton [12] found

$$R_\infty = \frac{1}{(K/S + 1)(\ell - 1)} \tag{13}$$

and

$$\frac{K}{S} = \left[\frac{1}{R_\infty(\ell - 1)} - 1\right] \tag{14}$$

where ℓ is a diffusing factor, approximately equal to 1.5.

Preston and Tsien [13] proposed a model in better accord with the geometry of a fabric. They considered a pile of parallel plates in which each plate represented a layer of fabric and had the thickness of one fiber diameter. The novel features of their treatment were that they considered the surface reflectance of the fibers ρ, and expressed the transmittance (g) through each layer in terms of the exponential Beer function $e^{\eta ct}$, where η is the extinction coefficient of the dye fiber-polymer system, c is the concentration of the dye, and t is the thickness of the plate. Neglecting all multiple reflections within the pile they obtained the result

$$R_r = \frac{e^{2\eta ct} - 1 + 4\rho}{\rho(e^{2\eta ct} + 1 - 2\rho)} \tag{15}$$

where R_r is the ratio of the light reflected by the dyed fiber to that by the undyed fiber. Predictions based on this treatment are unfortunately not good.

In 1972 Allen and Goldfinger [14] attempted to predict the color of a fabric based on a model which recognized the discreteness of the individual fibers (Fig. 3). The fabric was represented by a parallel bundle of cylindrical fibers which were optically homogeneous and large in diameter compared to the wavelength of light. These fibers were sufficiently widely separated so that the light paths of their neighbors were not influenced. In the published version of this study it was assumed that the incident light flux was collimated and normal to the plane of the sample and that, beyond the first layer, collimation was

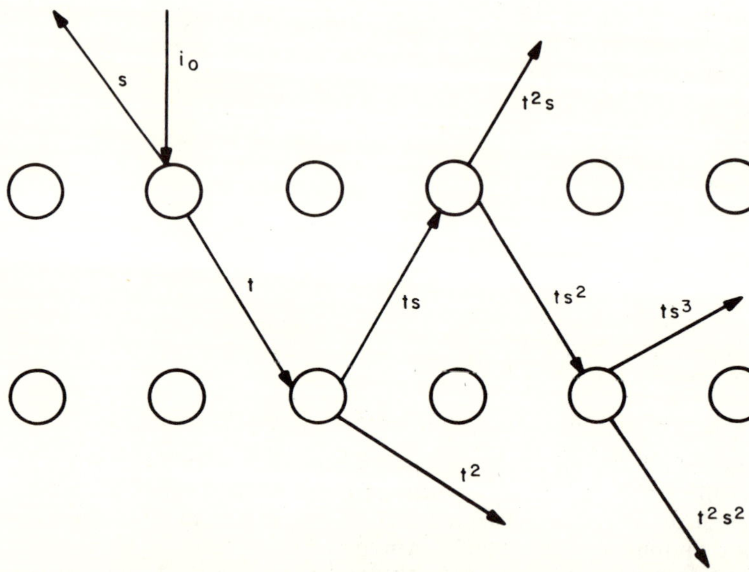

FIG. 3. The Allen and Goldfinger model. The relative positions of the fibers shown in cross section have been distorted. The model assumes a spacing sufficient to render negligible the interference of adjacent fibers in each other's reflection and refraction patterns. The symbols associated with the arrows representing light paths show the magnitude of the modification of the light path by the different optical events occurring in and at the different fibers.

preserved on incidence. A fraction t of the incident flux, on encountering the first layer, will be permitted to continue in the general direction of incidence, a fraction s of the incident flux will have its direction reversed, and a fraction a will be absorbed within the layer Thus

$$t + s + a = 1 \qquad (16)$$

Without, at this point, examining whether the fractions t and s of the incident flux represented light that had been reflected from the fiber surfaces or refracted within the body of the fibers, new variables, τ and σ, were introduced to define the events in the first layer:

$$\tau_1 = \frac{t^2}{1 - s^2} = \frac{k^2(1 - a)^2}{(1 + k)^2 - (1 - a)^2} \qquad (17)$$

and

$$\sigma_1 = 1 + \frac{t^2}{1 - s^2} = \frac{(1 - a)(1 + \tau_1)}{(1 + k)} \qquad (18)$$

where k = t/s. The events within a pile of such cylindrical layers can be defined in terms of the variables

$$\tau_n = \frac{\tau_{n-1}^2}{1 - \sigma_{n-1}^2} \qquad (19)$$

and

$$\sigma_n = \sigma_{n-1}(1 + \tau_n) \qquad (20)$$

where the subscript n refers to the 2^nth layer of fabric.

The quantity a is obtained by calculating the length of the light path within the fiber between points of subsequent refractions, as a function of the refractive indices of the fiber and continuous medium, and of the angle of incidence of the incident light flux. Some of the incident light is reflected simultaneously with the first refraction, according to Fresnel's law. From there, until the second refraction, light is lost by absorption. Then some of it is again refracted to emerge from the fiber and the rest is reflected to continue its path inside of the fiber, and so on. In each step between successive refractions the

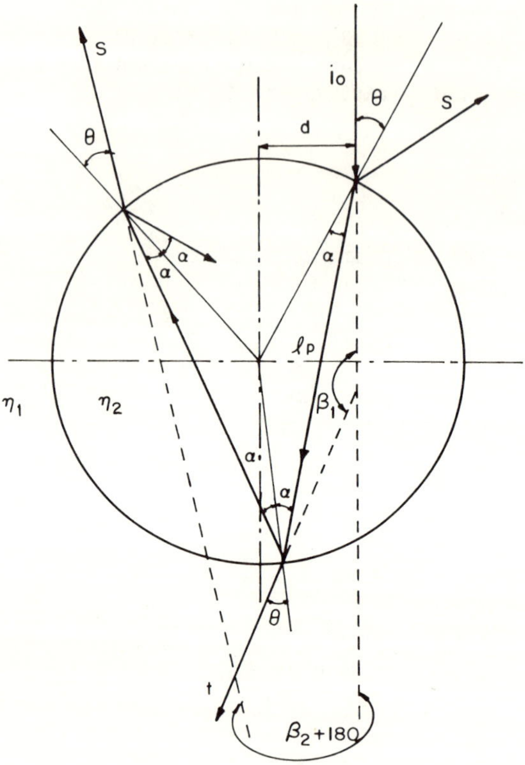

FIG. 4. Cross section through a fiber showing the different light reflections and refractions that occur.

length of the light path (Fig. 4) ℓp is given by

$$\ell p = \frac{2}{m}\sqrt{m^2 - d^2} \qquad (21)$$

where m is the ratio of the refractive index of the fiber to that of the continuous medium, and d is the sine of the angle of incidence θ of the light. From this one can define a quantity α from Beer-Lambert's law:

$$\alpha = 10^{-(2c\eta r/m)\sqrt{m^2 - d^2}} \qquad (22)$$

where c is the concentration of the dye, η the coefficient of extinction of the dye in the fiber-polymer, and r the radius of the fiber. From this

$$a = \frac{(1 - \rho)(1 - \alpha)}{1 - \rho\alpha} \tag{23}$$

where ρ is the Fresnel reflection at the fiber surface.

The quantity k is obtained by determining for each angle of incidence and for each step of refraction whether the general direction of the resulting light flux is "downward" or "upward":

$$k = \frac{t}{s} = \frac{\rho_{\cos 45° < d \leq 1} + (1 - \rho)^2 \sum_{n=1}^{\infty} \alpha^n \rho^{n-1}_{\cos \beta_n(-)}}{\rho_{0 \leq d < \cos 45°} + (1 - \rho)^2 \sum_{n=1}^{\infty} \alpha^n \rho^{n-1}_{\cos \beta_n(+)}} \tag{24}$$

where the first terms of the numerator and the denominator refer to the first reflection, and the second terms to the refraction as the light winds its way within the fiber [note: α is the quantity defined in Eq. (22)]. The quantities β_1 and β_n are given by

$$\beta_1 = 180 - 2\left[\arctan\left(\frac{d}{\sqrt{1 - a^2}}\right) - \arctan\left(\frac{d/m}{\sqrt{1 - d^2/m^2}}\right)\right] \tag{25}$$

and

$$\beta_n = \beta_{n-1} + \left[180 - 2 \arctan\left(\frac{d/m}{\sqrt{1 - d^2/m^2}}\right)\right] \tag{26}$$

Substituting a and k back into Eqs. (17) and (18) one can then solve numerically for σ_n. By setting t_n sufficiently small, say 10^{-3} or 10^{-4}, the color predicted is that of a pile of "infinite" thickness.

This approach permits the prediction of the color of fabrics (within its limitations) from the quantities η, c, r, and m, all determined independently from the sample itself. In Fig. 5 the reflectance R of a sample is shown as a function of the product CK (= $c\eta r$) of the concentration c of the dye, the coefficient of extinction η of the dye in the fiber-polymer, and the radius r of the fiber. Also shown are the parallel and perpendicular polarized components of this reflected light flux. It will be noted that significant polarization occurs. The effect of the ratio m of the refractive index of the fiber to that of the viewing medium is shown in Fig. 6. If the viewing medium is air then, of

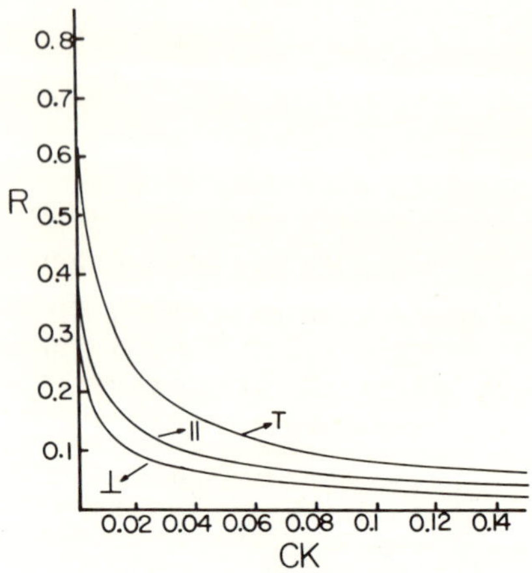

FIG. 5. The reflectance R of an array of fibers of refractive index 1.6 in air plotted against CK. The top curve gives the total reflectance, the middle curve the parallel polarized component, and the bottom curve the perpendicular polarized component of the reflectance.

course, for all intents and purposes, m is equal to the refractive index of the fiber. The prediction of the effect of the refractive index of the continuous medium on the color of fabrics makes it possible, for example, to predict the color of fabrics immersed in water or in other media. The effect of fiber diameter, represented in this instance by the denier for a cylindrical fiber of density 1.4, is shown in Fig. 7. In Fig. 8 the reflectances of fibers of different deniers are plotted against the reflectance of a 10-denier fiber, all fibers having the same value of CK, for any given reflectance of the 10-denier fiber.

The question of the limits of the color gamut which can be realized on textile samples has been treated by Vickerstaff [14a] and Atherton and Peters [14b] based on the limits of visual efficiency as proposed by Mac Adam [14c] and essentially arbitrary correcting factors. On the basis of the Allen and Goldfinger model such limits can be calculated without arbitrary assumptions, based on the shape of the spectral transmittance curve of the dye-fiber polymer system. The results of such a calculation have been shown [14d] (Fig. 8a), assuming a transmittance band 0.01 μm wide. Broadening of this band would constrict the gamut. In Fig. 8a the actual gamut achieved with a variety of dyes by Atherton et al. [14e] in 1954 is shown by the shaded area.

The effect of one type of nonuniformity of dye distribution, ring-dyeing, has also been calculated. Garrett and Peters [15] tackled this problem first, by extending Atherton's treatment. Allen and Goldfinger [16] modified the value of ℓp to distinguish between the distance traveled by the light in the dyed sheath, and in the undyed core of the fiber

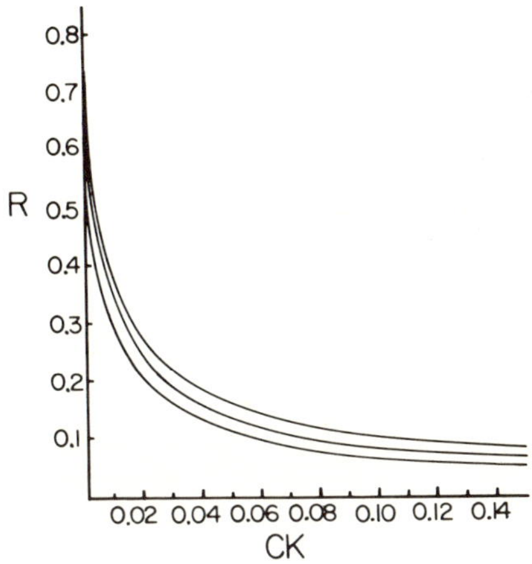

FIG. 6. The reflectances R of arrays of fibers plotted against CK. The top curve refers to an array of fibers of refractive index 1.7, the middle curve to fibers of refractive index 1.6, and the bottom to fibers of refractive index 1.5, all viewed in air.

(Fig. 9). It was also necessary to relate c, the concentration of the dye in the ring, to the average concentration of the dye in the fiber. Thus α of Eq. (22) becomes

$$\alpha = 10^{-[c\eta r/th(2 - th)][(2/m)(\sqrt{m^2 - d^2} - \sqrt{m^2(1 - th)^2 - d^2})]} \quad (27)$$

The effect of ring dyeing is shown in Figs. 10 and 11. In Fig. 10 the reflectances of ring-dyed and homogeneously dyed fibers, and the ratios of these two values, are plotted against CK. In Fig. 11, the reflectance of a ring-dyed fiber is plotted against that of a homogeneously dyed fiber, and the same reflectance ratio as in Fig. 10 is also shown. The change of the scale of the abscissa changes the shape of the curve representing the reflectance ratios.

Assuming that the surface reflectance is influenced by surface roughness, this effect on reflectance can be predicted by increasing the contribution of the first terms of the numerator and denominator in Eq. (24). Those terms express the Fresnel reflection of the fibers. It appears that factors F between 1 and 1.5 do in effect change the reflectance curves in the direction observed, as shown in Fig. 12.

Figure 13 shows a comparison of the classical (Kubelka and Munk) and the fundamental (Allen and Goldfinger) predictions. At reflectances down to about 0.07 the two predictions are identical for all intents and purposes. At lower reflectances the new treatment permits correct

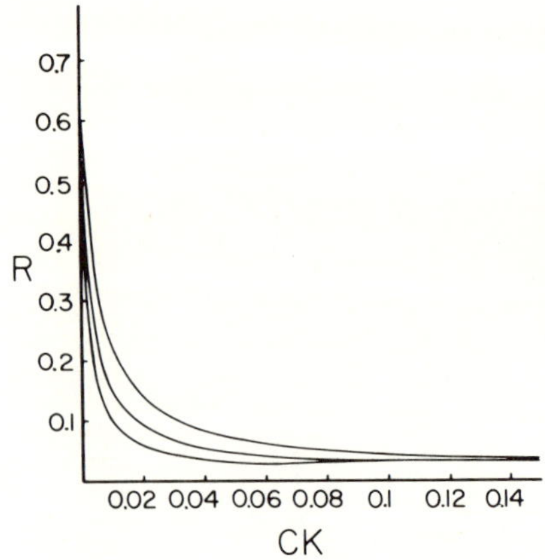

FIG. 7. The reflectances R of arrays of fibers of different denier, of fiber density 1.4 and of refractive index 1.6 viewed in air, plotted against CK. The top curve represents the reflectance of 3-denier fibers, the middle curve the reflectance of 10-denier fibers, and the bottom curve that of 30-denier fibers.

prediction while the classical approach breaks down: its predictions are too low because it does not consider Fresnel reflection on the fiber surface.

VI. THE COLOR OF COMMERCIAL FABRICS

The calculations presented earlier for a regular, parallel array of cylindrical fibers have shown the reflectance of the array at any given wavelength to be dependent on:

1. the thickness of the array;
2. the fiber radius;
3. the dye concentration;
4. the extinction coefficient of the dye-fiber polymer system;
5. the distribution of dye concentration in the fibers; and
6. the refractive index ratio of the fibers and the medium in which the array is immersed.

The behavior of even this simple model is dependent on many variables. The behavior of commercial fabrics cannot at this time be predicted from first principles, by using detailed information on the fiber, yarn, and fabric properties and dimensions together with data on the dye concentrations and characteristics, for example. However, the

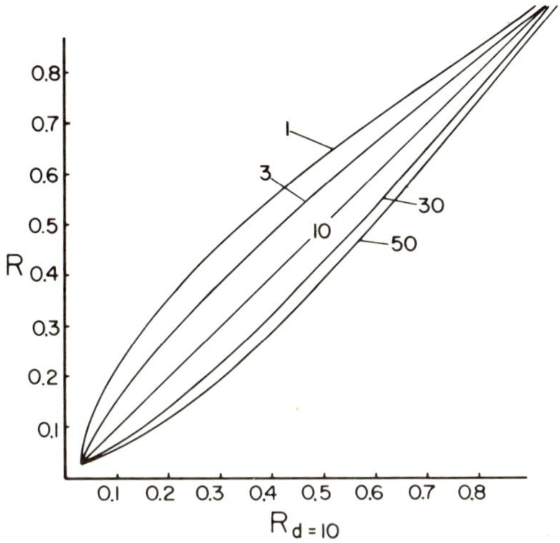

FIG. 8. The reflectances R of arrays of fibers of different deniers of fiber density 1.4 and of refractive index 1.6, viewed in air, plotted against the reflectance $R_{d=10}$ of an array composed of fibers of 10 denier. The numbers identifying the curves indicate the denier of the fibers.

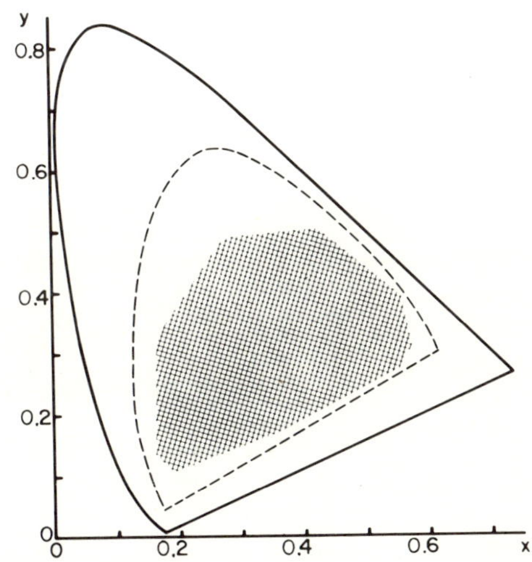

FIG. 8a. The limits of the color gamut (- - -) which can be obtained on a textile material of refractive index 1.6 with a dye whose transmittance curve consists of a band 0.01 μm wide. The shaded area shows the location of actual colors of textiles [14e].

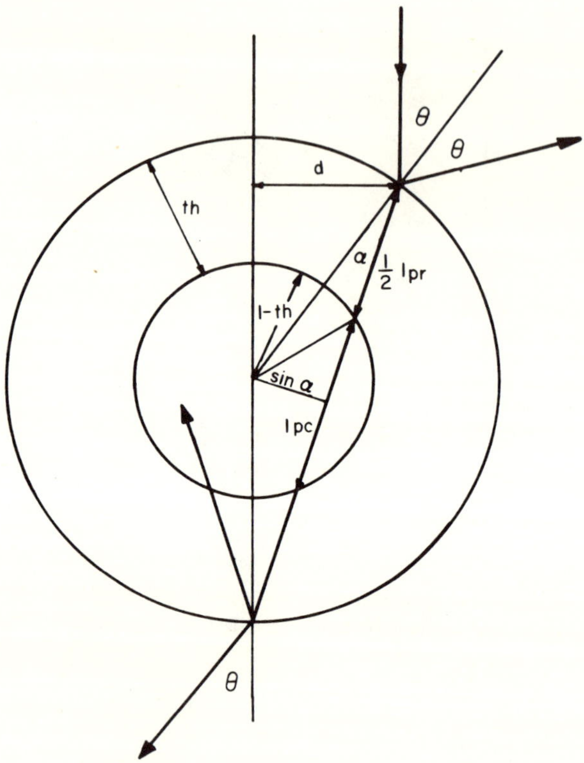

FIG. 9. Cross section through a ring-dyed fiber showing the calculation of the fraction of the light path in the dyed ring and that in the undyed core.

variables of importance in the Allen and Goldfinger model prove to be important for commercial fabrics also, and the general characteristics of the behavior of commercial fabrics resemble those of the Allen and Goldfinger model.

Textile colorists have been long familiar with the dependence of the appearance of a dyed fabric on factors such as

1. fiber denier;
2. fiber cross-sectional shape;
3. the presence of delusterant or of voids in fibers;
4. fiber roughness or smoothness;
5. fiber crimp;
6. yarn construction, twist, tension, and texture; and
7. fabric construction.

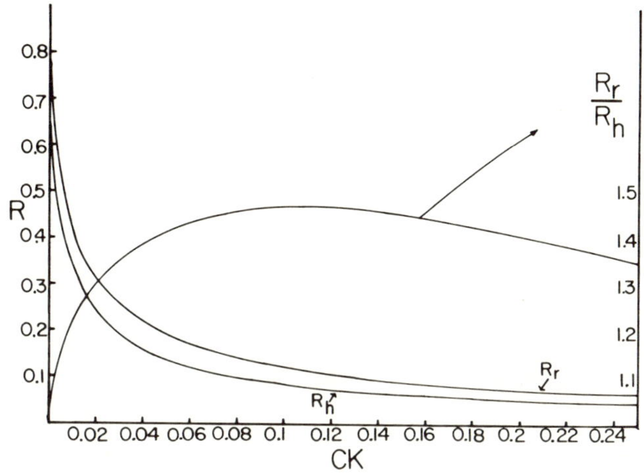

FIG. 10. The reflectances R of two arrays of fibers of refractive index 1.6 viewed in air and containing the same amount of dye, for each CK value, plotted against CK. The middle curve gives the reflectance of a ring-dyed sample R_r and the bottom curve that of a homogeneously dyed sample R_h. The top curve gives the ratio R_r/R_h against CK.

The effects of these variables can be large [17]. A fiber of modified cross section, for example, may require as much as 50% more dye to give the same depth of shade as a cylindrical fiber of the same denier. The presence of delusterant can cause equally large changes in appearance, the delustered yarns requiring more dye than the bright yarns for a comparable depth of shade. Those familiar with the dyeing procedures used for quality control in fiber manufacture are aware that mere differences in crimp level between two yarns can give rise to significant differences in the depths of shade of otherwise identical dyeings. An accidental product mix leading to the presence of a yarn with fibers of lower denier than those in the bulk of a fabric will lead to local differences in appearance, even if the dyer inadvertently succeeds in dyeing all the fibers to the same dye content: the thinner fibers are paler in color.

Opinions are divided as to the origins of the increase in depth of shade of vat dyeings on cotton when the dyed fibers are mercerized. The changes in fiber roughness and in fiber shape on mercerization, with their implications for the scattering and reflection of light by the fibers, presumably contribute to the marked increase in depth of shade in mercerized cottons [18]. This effect may be enhanced by an accompanying increase in the extent of penetration of the fibers by the dye, though this has not yet been established. An effect which almost certainly is

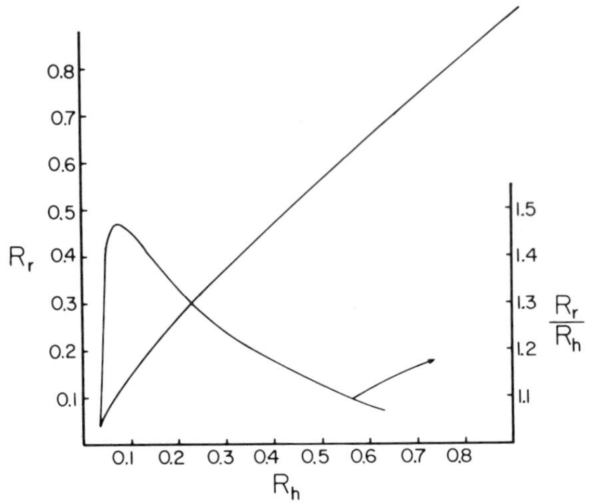

FIG. 11. The reflectance of an array of ring-dyed fibers R_r plotted against the reflectance R_h of the corresponding homogeneously dyed array, both having a refractive index 1.6 and viewed in air. The curve with the sharp maximum is a plot of R_r/R_h against R_h.

related to an increase in the extent of penetration of fibers by dye is the increase in depth of shade which occurs when certain dyed materials are heat-set [19]. The increasing importance of rapid-dyeing procedures can be expected to lead to situations in which definite differences in dye penetration may exist between different batches of dyed material, with inevitable consequences for the reproducibility of color.

This latter observation stems from the physical-chemical nature of dyeing processes, and leads to an examination of some troublesome aspects of the general problem of color control in dyeing.

VII. COLOR CONTROL IN DYEING

As we have seen, the color of a dyed fiber, yarn, or fabric is influenced by many variables in addition to dye content. The dependence of the color on the nature of the fabric, regarded as an assembly of fibers, poses special problems of color measurement, especially for pile fabrics or for materials with reflectance characteristics that are very dependent on the orientation of the material with respect to the light source and with respect to the observer. With appropriate

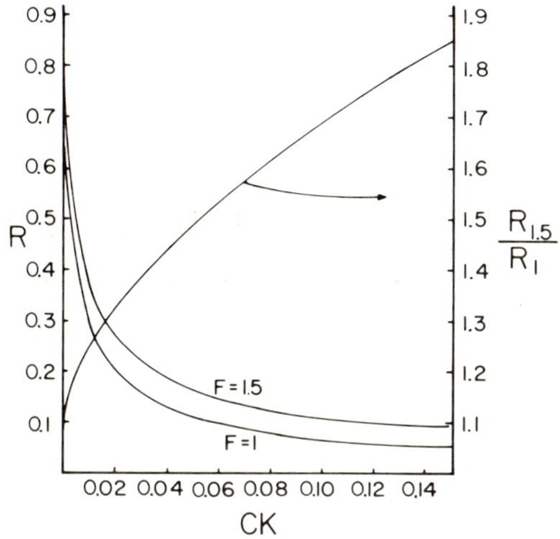

FIG. 12. The reflectances R of two arrays of fibers of refractive index 1.6 viewed in air, plotted against CK. The bottom curve represents the reflectance of an array of smooth cylinders while the middle curve shows the reflectance of cylinders with rough surfaces, showing the maximum possible effect. The top curve gives the ratio of the reflectances of arrays of rough and smooth fibers.

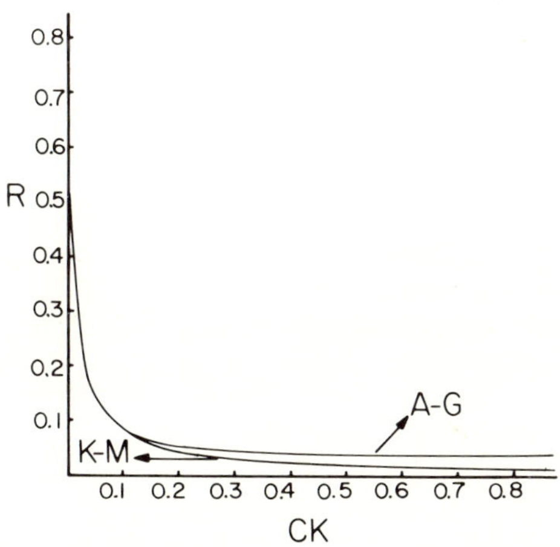

FIG. 13. The reflectances R of an array of fibers of refractive index 1.6 viewed in air, plotted against CK. The top curve is calculated from the Allen and Goldfinger treatment while the bottom curve is a Kubelka and Munk prediction based on the conditions of 50% reflectance.

precautions, some of these difficulties may be overcome, and reproducible measurements of color differences become possible [20]. The reproducibility of measurements of color differences on a given instrument, or between different instruments, appears to be a lesser problem than that of the reproducibility of measurements of color, i.e., of X, Y, Z values, from sample to sample and from instrument to instrument [20].

In the establishment of instrumental match prediction (IMP) procedures, one is currently advised to ensure that the same instruments are used both in establishing the original data banks and in the practical applications of these data to problems of color matching or of dye formulation [21]. These instrumental problems will not be pursued here; instead, attention will be concentrated on certain limitations inherent in the dyeing process itself.

A. Standardization of Dyes

Commercial dyes invariably contain diluents and rarely consist of only a single colored species. The manufacturing processes generally yield a mixture of colored isomers, or at best a product containing predominantly the desired colored species together with small amounts of other colored products. The diluents may serve specific functions in stabilizing the product, as marketed or in the dyebath, but their natures and proportions by weight in the commercial product are rarely specified. Some companies have begun to specify the contents of pure dyes in their products, at least indirectly, but only when this information is necessary for the proper control of a recommended dyeing procedure for these products. The Color Index nomenclature is helpful, although significant differences may exist between products with the same Color Index description but from different manufacturers.

The standardization of commercial dyes is usually based on the appearances of standard dyeings. Once a standard has been established, the dye content of a commercial product will be kept within definite tolerances, (for example within $\pm 2.5\%$), either through standard dyeing procedures or through colorimetric measurements on solutions of the dye [22]. The diluent content is adjusted to maintain a product of the correct dye content, and occasionally small quantities of other dyes will be added to adjust the hue of the product.

There appears to be little unanimity among the different dye manufacturers in the matter of uniform standardization procedures, and in view of the dependence of the color of a fabric on all the parameters specified earlier, it is clear that there is some ambiguity about the results to be expected from a given dye on an unfamiliar fabric. The existing products would probably be more diverse than they are if it were not for the commercial pressures generated through the comparison of products by users on a "money value" basis. Reputable dye companies take pains to ensure that their products are within specified tolerances on shipment, but problems may arise through mechanical segregation of solid particles on shipment and storage, or through variations in moisture content from improper handling and storage. It appears that the starting product, the dye itself, is seldom characterized to better than ±2.5%, and the possibility of occasional variations greater than this leads some users to check the dye contents of these products on receipt from the manufacturer, and even again, after storage. The drug rooms of some modern dyehouses have temperature and humidity controls to eliminate variability from changes in moisture content during storage, transfer, and weighing of dyes and textile chemicals.

In general, each commercial product must be treated as a distinct entity, and specific color-control procedures must generally be tied to specific commercial products.

B. Uniformity of Fibers, Yarns, and Fabrics

Let us suppose that the dye standardization problems have been overcome. The variability of fabrics, yarns, and fibers introduces a new set of problems. Some of these problems are optical problems, and some of the optical problems relate to pattern recognition by the observer.

Knitted rib structures have a geometrical regularity which makes small local differences in rib spacing very obvious to the eye. The juxtaposition in a fabric of areas with quite small differences in construction, such as may arise from small differences in knitting tensions or in the mechanical properties of yarns, may lead to serious differences in appearance even when the fibers are otherwise identical in dye content and dye penetration. These geometrical effects are important contributors to barré problems, for example, but will not be considered further.

As we shall see, dyeing processes are sensitive to differences between fibers, which are not made apparent through detectable changes in the physical or mechanical properties of the fibers. Minor differences in extrusion conditions, or in draw ratio, can give differences in fiber structure which become apparent only on dyeing, because of the great sensitivity of dye diffusion to the structure of a fiber. For this reason, commercial synthetic fibers are segregated into batches, or "merges," of similar dyeing characteristics: uniformity of dyeing behavior is guaranteed only within yarns of the same merge number.

This guarantee is not absolute, because commercial merge control cannot economically be based on the behavior of the least-favorable dyeing system, and therefore there may be some restrictions on the dyes or dyeing procedures to be employed.

Related problems occur of course with the natural fibers, because of the natural accidents or variations of growth and because of differences in age, in origin, in storage, or in handling.

From this aspect, each different batch of textile material represents a different dyeing problem, and a different problem of color control.

C. Control of Dyeing Processes

Most commercial dyeings use mixtures of dyes and, since each commercial dye is itself often a mixture of products, the dyeing process is often one in which three or more different colored species are sorbed by the fiber. Each different colored species will have a different chemical structure and its molecules will have a different size and geometry, together with a different pattern of detailed chemical interactions with the fiber and with all other constituents of the system.

Each distinct colored species will tend to diffuse into the fiber with its own characteristic diffusion coefficient D, at a given temperature and concentration, and at equilibrium the distribution of that species between dyebath and fiber will be determined by its own characteristic distribution coefficient K. As though this diversity of properties alone were not sufficient, it remains to be emphasized that the magnitudes of D and K for any given dye will be influenced by local variations in temperature or composition of the dyeing system, and will differ from fiber to fiber, though the sensitivity of D to differences in fiber structure will generally much exceed that of the distribution coefficient K.

Color measurement and instrumental match prediction are based on the premise that the material is uniformly dyed, though opportunities for nonuniformity are to be found at every stage in dyeing.

If the dyebath does not properly wet the fibers there will be imperfect penetration of the fiber assembly: air will be trapped within yarns and will prove difficult to remove. Modern rapid dyeing processes use a preliminary evacuation of the dyeing vessel, before the dye solution is admitted, to obviate this problem and facilitate uniformity of dyeing.

Even when the material is properly wetted-out and completely penetrated by the dyebath, the ease with which the dyebath can flow through the material may show considerable local variations: the fluid flow will tend to follow the larger "pores" in the assembly, to pass around yarns rather than through them [23], and to "channel" through the less-densely packed regions of the material. There will be considerable local variations in the efficiency of the supply of the dye solution to the fiber surfaces. Only if the rate of sorption of dye by the fibers is kept low throughout by comparison with the local rate of supply of fresh dye by the dyebath will the initial sorption of dye be uniform. This ideal is often not attained in practice: the consequence is an unacceptable dyeing, unless the dyes have very good migration and leveling properties, which tend in some systems to correlate with the low distribution coefficients K of these dyes [24].

Suppose that our best endeavors have resolved this problem, and that we have succeeded in bringing fresh dye solution to each fiber at an adequate rate. The dye molecules must then pass through the solution-fiber interface and penetrate the fibers by diffusion. The great sensitivity of diffusion in polymers to the molecular size, shape, and chemistry of penetrant molecules makes this a very favorable situation for the separation of the different dyes in the mixture, rather than one which favors uniformity of uptake of all species. There are strong similarities in principle between chromatographic separation procedures and the events which occur during dyeing.

The dye manufacturer uses trial and error to select "compatible" dyes for use in mixture dyeing: such dyes have similar dyeing properties but do not of course behave identically in all respects.

In general one would expect dyes to differ more in their diffusion characteristics (D) than in their equilibrium properties (K). Dyeing procedures which depend primarily on similarities in the diffusion, or rate of dyeing characteristics of dyes, will probably be less successful than those based on a close approach to equilibrium in the system; the shorter and the more rate-dependent the dyeing process is, the more troublesome is its control likely to be and the less feasible any predictive control of the dyeing process.

We have seen that in dyeing processes nonuniformity of the dye distribution is inherently more probable than the desired uniformity. The success of many dyeing procedures is critically dependent on the leveling and migration processes whereby dye is transferred from heavily dyed regions to less heavily dyed areas: the time required for leveling of the initially nonuniform dye distribution may constitute a major fraction of the total dyeing time, and very short dyeing times will tend to accentuate lack of levelness and give imperfectly penetrated fibers. A fabric containing such fibers may be subject to color change in subsequent processing, if that processing increases the penetration of dye.

Enough has been said to emphasize the difficulty of predicting, in any real quantitative sense, the detailed kinetic behavior of a dyeing system: control of color in dyeing cannot be based on this type of prediction.

Programmed control of the dyeing cycle is helpful, but does not correct for any variability in the dyes or fibers themselves.

Since the equilibrium distribution coefficients of dyes in general vary less from dye to dye or from fiber to fiber than do the diffusion properties of the dyes, prediction of the equilibrium behavior should be more feasible than prediction of the kinetic behavior. Even here the variability is too large to be ignored, and current theory is inadequate as a basis for prediction. How then can successful dyeings be obtained?

There seem to be two distinct approaches to the successful coloration of a fabric with known amounts of dyes.

In one type of procedure, such as pad-thermofix application, a calculated amount of dye solution or dispersion of known composition is applied uniformly to the fabric by mechanical means. The solution is dried uniformly onto the fabric and the penetration of the dye into the fibers occurs in situ, in subsequent thermal or steam treatments. In this type of procedure the characteristics of the individual dyes play a minimal role in influencing the uniformity of dye uptake and distribution.

In the second type of procedure, exhaustion dyeing, the desired quantities of dyes are added to the dyebath. The dyeing process is so controlled that the exhaustion of the dyes onto the material is slow enough to proceed uniformly from place to place, or is controlled in a manner which permits initial nonuniformities to level out. This latter requirement of migration and leveling is favored by low exhaustions of dye, i.e., by the retention of dye in the dyebath, but this very requirement enhances the importance of differences in equilibrium distribution characteristics between different dyes in the system.

Consider for example the equilibrium uptake of a dye of distribution coefficient K from a dyeing system of liquor ratio L; the fractional exhaustion E of the dyebath at equilibrium is

$$E = \frac{1}{1 + (L/K)} \tag{28}$$

and the behavior of E as a function of the ratio L/K is shown in Fig. 14. The exhaustion E is insensitive to the numerical value of L/K if L/K is sufficiently small and the exhaustion E is high. The sensitivity of E to changes in L/K is greater for larger numerical values of L/K, and at these lower exhaustions the behavior can become very sensitive to differences in the distribution coefficients K of different dyes. It is therefore necessary to ensure that the exhaustions of the different dyes in the dyebath not only increase gradually during dyeing, but also approach unity (or 100%) at the end of dyeing: the correct uptake of dye can be assured, strictly speaking, only for complete exhaustion of the dyebath.

This section of the discussion may be concluded by emphasizing that differences in the dyeing characteristics of the individual dyes in a dye mixture are inevitable. These differences will appear in the end result of the dyeing process to an extent determined by the chosen conditions of dyeing: inadequate dyebath penetration or circulation through the fabric, high dyeing rates, short times of dyeing, and low equilibrium

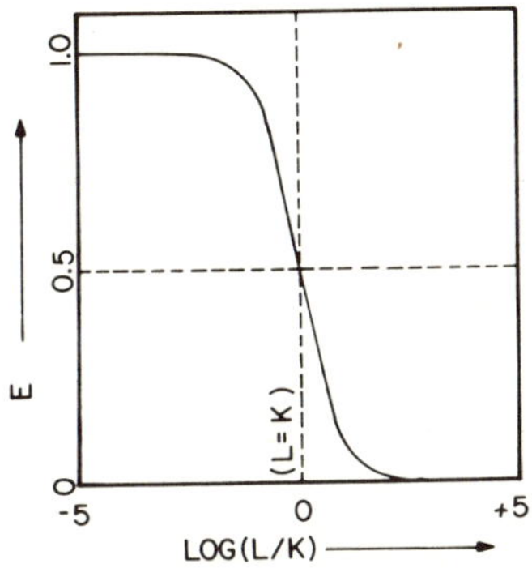

FIG. 14. The dependence of the equilibrium exhaustion E on the liquor ratio L and the dye distribution coefficient K. Kinetic differences between dyes, in their sorption by fibers, disappear at equilibrium, when only the equilibrium distribution coefficients K remain important. The sensitivity of the behavior of E to changes in K decreases as the parameter (L/K) decreases, and as the fractional exhaustion E approaches unity.

exhaustions are factors which can lead to unacceptable dyeings, and to serious incompatibility in the behavior of dyes in mixtures.

D. Instrumental Match Prediction and Color Control in Dyeing

The development of color theory has reached a point at which the relationships between the dye content of a textile material and its color are well understood. Admittedly there are deviations from simple theory, but modern techniques of instrumental match prediction can provide useful initial estimates of the quantities of dyes necessary to produce a given color on a given material: indeed a selection of possible dye formulations can be generated, with different levels of cost or with different metameric properties.

Unfortunately, only in chemically stable systems giving complete exhaustion is there equivalence between the amounts of dyes added to a dyebath and the amounts of dyes sorbed by the material being dyed. The variability of dyes and fibers, when combined with variations in dyeing procedures and human errors, generally results in initial dyeings which differ from the expected result. The existing procedures can give useful estimates of the amounts of dyes needed to bring the dyeing "on shade," but cannot guarantee that the addition of these amounts of dyes to the dyebath will result in their complete sorption by the material. The absence of this guarantee weakens the usefulness of the predictive computations.

As extreme examples, but examples which can occur in practice, we may cite the dyeing of polyamide fibers with mixtures of acid dyes, and of acrylic-polymer fibers with cationic dyes. Under normal dyeing conditions there is an effective limit to the total concentration of dye which may be sorbed by these fibers. Suppose that a red dye and a blue dye are used to obtain a purple dyeing on such a fiber, and that the initial dyeing is on-shade, but too pale. The natural response of a computer program, or of a dyer, would be to estimate the additional red and blue dye required, and these would be added to the dyebath. Suppose the total dye content of the system now exceeds the capacity of the fiber for dye. Then the dyebath will be incompletely exhausted and the characteristics of the individual dyes will become important: one may easily find that the red dye is sorbed in preference to the blue dye, and that a bright red dyeing results instead of a deeper purple one [25].

Effects such as these reveal the limitations placed on the instrumental measurement, control, and prediction of color by our lack of ability to control the properties of dyes, fibers, and dyeing processes in a consistent manner, within close tolerances, and by our inability to predict dyeing behavior on a quantitative level.

Dyeing processes are difficult to reproduce [26]; there may be greater differences from place to place within a given piece of fabric than there are, on average, between different pieces of that fabric [26, 27]. Successful use of instrumental match prediction requires that much more attention be paid to the reproducibility of dyeing, and to the quality of the original dyeings on which the data banks are based, than is generally recognized [21,27].

VIII. GENERAL CONCLUSIONS

Color theory has advanced to the point at which the color of a fiber assembly of simple geometry can be quantitatively related to fundamental parameters of the fibers and of the dyes used to color the fibers. At present the extension of these ideas to textile materials is possible only with the aid of empirical correction factors.

The primary obstacles to the success of existing procedures for instrumental match predictions, and for instrumental control of color in dyeing processes, appear to arise from the inherent variability of dyes, of fibers, and of dyeing procedures, and from our inability to control consistently, or to predict, the behavior of dyeing systems within close tolerances. These obstacles appear to be inherent in the complex physical-chemical factors in dyeing processes: dramatic new advances seem unlikely in the absence of a new and radically different approach to these problems.

REFERENCES

1. R. W. Burnham, R. M. Hanes, and C. J. Bartelson, Color: A Guide to Basic Facts and Concepts, Wiley, New York, 1963, p. 11.
2. H. Grassman, Poggendorffs Ann., 89, 69 (1853); Phil. Mag. (4), 7 254 (1853).
3. D. B. Judd and G. Wyszecki, Color in Business, Science and Industry, 2nd ed., Wiley, New York, 1963, p. 48.
4. Munsell Book of Color, Munsell Color Company, Inc., Baltimore, Maryland.
5. M. Richter, J. Opt. Soc. Am., 45, 223 (1955).
6. W. Ostwald, Die Farbenlehre, Unesma, Leipzig, 1918; Colour Science, Winsor & Newton, London, 1931.
7. The ICI Colour Atlas, Butterworths, London.
8. G. Wyszecki and W. S. Stiles, Color Science, Wiley, New York, 1967.
9. P. Kubelka and F. Munk, Z. Tech. Physik, 12, 593 (1931).
10. P. Kubelka, J. Opt. Soc. Am., 38, 448 (1948).
11. E. I. Stearns, U.S. Patent 2,540,798 (1951); Calco Tech. Bull. 820, Calco Chemical Division, American Cyanamid Co., Bound Brook, N. J., 1951.
12. E. Atherton, J. Soc. Dyers Colourists, 71, 389 (1955).
13. J. M. Preston and P. C. Tsien, J. Soc. Dyers Colourists, 62, 242 (1946).
14. E. Hope Allen and G. Goldfinger, J. Appl. Polymer Sci., 16, 2973 (1972).
14a. T. Vickerstaff, Proc. Phys. Soc., 57, 15 (1945).
14b. E. Atherton and R. H. Peters, Congress FATIPEC, III, 147 (1955).
14c. D. L. Mac Adam, J. Opt. Soc. Am., 25, 249, 361 (1935).
14d. G. Goldfinger, J. Polymer Sci., Polymer Letters, 13, 231 (1975).
14e. E. Atherton, D. A. Garrett, and T. Vickerstaff, Text. Inst. J. Proc., 45, 350 (1954).
15. D. A. Garrett and R. H. Peters, J. Textile Inst., 47, 166 (1956).
16. E. Hope Allen and G. Goldfinger, J. Appl. Polymer Sci., 17, 1627 (1972).
17. J. L. Rush and C. J. Miller, Nylon Dyeing, Its Relation to Physical and Chemical Properties of Yarns, Research Monograph M. 13, Allied Chemical Corporation Fibers Division, Petersburg, 1968.
18. J. T. Marsh, Mercerizing, Van Nostrand, New York, 1942, p. 378.
19. H. U. von der Eltz, Textilveredlung, 6, 195 (1971).
20. E. Coates, S. Day, J. Durrans, and B. Rigg, J. Soc. Dyers Colourists, 88, 151 (1972).
21. J. R. Aspland, Textile Chem. Colorist, 5, 222 (1973).
22. C. E. Garland, Textile Chem. Colorist, 5, 227 (1973).
23. R. McGregor, J. Soc. Dyers Colourists, 81, 429 (1965); Can. Textile J., p. 61, May 1973.
24. E. Merian and U. Lerch, Am. Dyest. Rep. No. 19, 51, 35 (1962).

25. R. McGregor and P. W. Harris, A Chemical Engineering Approach to Dyeing Problems: The Sorption of Acid Dyes by Polyamides, Research Monograph M. 20, Allied Chemical Corporation Fibers Division, Petersburg, Va., 1969.
26. K. Thurner, Colorimetry in Textile Dyeing-Theory and Practice. Badische Anilin-u. Soda-Fabrik A. G., Ludwigshafen am Rhein, 1965.
27. H. R. Cooper and H. M. Roberts, J. Soc. Dyers Colourists, 87, 451 (1971).